INTRODUCTORY
CIRCUIT
ANALYSIS

INTRODUCTORY CIRCUIT ANALYSIS

S. Ivar Pearson
and
George J. Maler

Department of Electrical Engineering
University of Colorado
Boulder, Colorado

ROBERT E. KRIEGER PUBLISHING COMPANY
HUNTINGTON, NEW YORK
1977

Original Edition 1965
Reprint 1974, 1977 with corrections

Printed and published by
ROBERT E. KRIEGER PUBLISHING CO. INC.
645 NEW YORK AVENUE
HUNTINGTON, NEW YORK 11743

©Copyright 1965 by
JOHN WILEY & SONS, INC.
Reprinted by arrangement

Library of Congress Catalog Card Number 74-10895
ISBN No. 0-88275-175-1

All rights reserved. No reproduction in any form of this book in whole or in part (except for brief quotation in critical articles or reviews), may be made without written authorization from the publisher.

Printed in the United States of America

Library of Congress Cataloging in Publication Data

Pearson, Sven Ivar.
   Introductory circuit analysis.

   Reprint of the ed. published by Wiley, New York.
   1. Electric circuits.   I. Maler, George J., joint author.   II. Title.
[TK454.P3   1974]   621.319'2   74-10895
ISBN 0-88275-175-1

To our wives, Kayrene and Margaret Mary

1974 REPRINT STATEMENT

In this reprinting of the text, the principal errors found in the first printing have been corrected. We are grateful to the students and instructors who have found these errors. If additional errors still exist, we would appreciate being notified of them.

Preface

This text is intended for the first course in electric circuit analysis. It presumes that the student is proficient in algebra and trigonometry and has had some experience with complex numbers. The student should have had an introduction to differential and integral calculus, and should be continuing his studies in calculus. This text develops the solution of linear differential equations with constant coefficients by both the classical and Laplace transform methods on the basis that the student has not yet encountered these in his mathematics courses.

In order that the student may verify some of the deductions from the circuit model, and gain some appreciation for its shortcomings, a concurrent laboratory course is considered essential. Also, a college physics course prior to, or concurrent with, this course is considered highly desirable.

A new text in an established discipline usually comes into being because some teachers have felt strongly that a change in sequence or emphasis would improve the learning process. If such feelings lead to the creation of class notes and if such notes survive several years of class use with the resulting revisions, these notes may appear in book form. Such is the case for this text.

We had found that our students did not properly relate the time domain, the phasor domain, and the Laplace transform domain. The reason seemed to be that these subjects were often studied in isolation, the student not being required to relate these in a given situation. Also there was considerable confusion concerning the relationships, if any, between the symbol s of the Laplace transform, the symbols p or D used as differential operators, and the symbol s used as the exponent for the complex exponential driving function, Ae^{st}. Also students did **not** usually realize the close relationship between circuit theory and **elec**tromagnetic field theory. The latter condition arises because most texts on circuit theory do not stress this relationship and, unfortunately, neither do most texts on field theory.

The sequence of material in this text was selected with the hope that an improvement in the learning process would result. We have the feeling that this is so.

Chapter 1 is a short chapter whose purpose is to state rather concisely the important relationships in field theory on which circuit theory is based. We urge the student to read this chapter carefully, but not to be concerned if he feels a lack of understanding of the mathematical expressions known as Maxwell's equations. He will become more familiar with these equations in courses on electromagnetic field theory. In Chapter 2 circuit elements are defined and laws and conventions are stated. The dependence of these definitions and laws on electromagnetic field theory is shown whenever this is possible.

From a mathematical point of view, linear circuit theory requires the solution of linear differential equations with constant coefficients. There are two important methods of solution: the classical method and the Laplace transform method. The classical method has the advantage of requiring (and giving) physical insight; the Laplace transform method has advantages which arise from the facts that the differential equations are transformed into algebraic equations and that initial conditions are immediately incorporated into these equations. In order that the student may acquire both the physical insight of the phenomena and a mastery of the techniques in arriving at correct solutions, he is introduced to both methods as early as possible. He is then required to solve many problems by both methods over an extended period of time. Chapter 3 introduces the student to the two methods and gives some practice in solving simple problems by both methods. Additional practice in obtaining complete solutions by use of the two methods is provided in Chapters 9 and 11.

The early introduction of the Laplace transform permits the early presentation of the transform circuit and the definition of transform network functions. These allow the presentation of general circuit reduction methods and theorems, thus avoiding the learning of definitions and techniques which must later be modified because they were not general enough. These methods and theorems are presented in Chapter 4.

Chapter 5 is concerned with the development of the phasor domain and some of its applications. The main development is from the classical method, although the development from the transform domain is also shown. Chapter 6 deals with the loci of phasor network functions for a variable element and with the frequency characteristics of network functions. Because of the large amount of nomenclature and techniques in circuit theory, the driving functions have been limited thus far to sinusoids. In Chapter 7 these are extended to general periodic functions by use of Fourier series; however, in this chapter we limit the solutions to steady-state solutions. Chapter 8 is devoted to three-phase circuits under steady-state conditions, the previous introduction to Fourier series permitting the consideration of harmonic voltages and currents. The study of three-phase circuits causes the student to appreciate the usefulness of phasor diagrams.

In Chapter 9 complete solutions are obtained for pulses of various types by both classical and transform methods. Also the subject of impulses is presented primarily from a physical stand-point.

Analogues are not introduced until the last chapter because our experience indicates that the student needs to have a good comprehension of the language and techniques of one discipline if he is to relate these to a new discipline. This is particularly true if one expects the student to label variables, write equilibrium equations, and determine initial conditions in this new discipline.

We have tried to use standard symbols wherever possible. We have not used boldface to represent complex quantities, since this cannot be used by instructor or student. A complex quantity is designated by a caret ($\wedge$) over the quantity. Thus we use $i = I \cos(\omega t + \alpha) = \mathrm{Re}\,(\hat{I} e^{j\omega t})$, in which $\hat{I} = I e^{j\alpha}$. A detailed list of symbols is given at the beginning of the book.

Boulder, Colorado, 1965   S. I. PEARSON
                                                     G. J. MALER

Acknowledgments

We are indebted to the sophomores who struggled with the preliminary editions of this text. Their interpretation of the material, their performance, and their comments have had considerable influence on the text's final form.

We also gratefully acknowledge the encouragement and suggestions from members of the Electrical Engineering Department of the University of Colorado, particularly W. G. Worcester, L. A. Bingham, C. T. A. Johnk, P. W. Carlin, R. H. Bond, L. R. Branch, G. E. Gless, and V. C. Rideout.

We acknowledge the great contribution of Mrs. Charlotte I. Cranford, who has done the typing and who has helped with the preparation of this text in many other ways.

S. I. P.
G. J. M.

Contents

CHAPTER

1. **Electromagnetic Field Theory**   **1**
   1.1 Introduction   1
   1.2 A Discussion of Laws and Units   3
   1.3 The Laws of Electromagnetism (Maxwell's Equations)   5
   1.4 Some Deductions from Maxwell's Equations   9
   1.5 The Relationship of Circuit Theory to Field Theory   10

2. **Basic Concepts of Circuit Theory**   **13**
   2.1 Introduction   13
   2.2 Definitions and Conventions   14
   2.3 Kirchhoff's Laws   18
   2.4 Circuit Elements   21
   2.5 Initial Conditions   63
   2.6 Summary, A List of Circuit Elements, their Symbols, and their Volt-Ampere Relationship   70
   Problems   72

3. **The Solution of Linear Differential Equations with Constant Coefficients**   **86**
   3.1 Classical Method   87
   3.2 Laplace Transformation Method   102
   3.3 Summary   129
   Problems   130

4. **Network Functions, Equivalent Circuits, and General Methods**   **141**
   4.1 Transform Network Functions   141
   4.2 Equivalent Circuits and Network Reduction   144
   4.3 General Methods for Circuit Analysis (Independent Equations)   171
   4.4 Additional Items in Circuit Analysis   186
   4.5 Summary   198
   Problems   198

CHAPTER

5. Steady-State Analyses with Sinusoidal Sources — 216
- 5.1 Introduction to Sinusoidal Waves — 216
- 5.2 The Addition of Sinusoidal Functions — 217
- 5.3 Complex Quantities — 225
- 5.4 Steady-State Complex Form (Phasors) from the Time Domain — 230
- 5.5 Analysis in the Phasor Domain — 238
- 5.6 Power under Sinusoidal Steady-State Conditions (Wattmeter) — 250
- 5.7 Root-Mean-Square Values for Current and Voltage (Instruments) — 254
- 5.8 Maximum Power Transfer — 259
- 5.9 Phasor Domain from the Laplace Transform Domain — 260
- 5.10 Summary — 265
- Problems — 265

6. Graphical Methods Applied to Phasor Network Functions for a Variable Element or for Variable Frequency — 286
- 6.1 Loci of Phasor Network Functions for a Variable Element — 286
- 6.2 Frequency Characteristics of Network Functions — 301
- 6.3 Summary — 319
- Problems — 320

7. Nonsinusoidal Periodic Waves—Steady-State Response — 332
- 7.1 Introduction to Fourier Series — 333
- 7.2 Symmetry — 335
- 7.3 Determination of Coefficients by Numerical Integration — 343
- 7.4 RMS Values and Average Power of Nonsinusoidal Periodic Waves — 345
- 7.5 Frequency Spectrum of Periodic Rectangular Pulses — 348
- 7.6 Exponential Form of Fourier Series — 351
- 7.7 Summary — 352
- Problems — 353

8. Three-Phase Systems — 361
- 8.1 Balanced Loads — 362
- 8.2 Unbalanced Three-Phase Loads (Balanced Voltages) — 372
- 8.3 Power Measurements in Three-Phase Systems — 377
- 8.4 Harmonics in Three-Phase Systems — 381
- 8.5 Summary — 385
- Problems — 385

9. Pulses, Impulses, Dependent Sources — 400
- 9.1 Unit Step Function, Gate Function, and Shifted Time Function — 400
- 9.2 The Response of Circuits to Pulse Driving Functions — 403
- 9.3 Impulses — 410
- 9.4 Time Response to Single Driving Function (Pole-Zero Diagram) — 428
- 9.5 The Use of Impulse and Step Functions to Approximate a Pulse of General Shape — 438
- 9.6 The Laplace Transform and the Fourier Transform — 439
- 9.7 The Laplace Transform and Periodic Functions — 444
- 9.8 Circuit Fiction and the Frequency Spectrum of Pulses — 449
- 9.9 Dependent Sources — 449
- 9.10 Summary — 452
- Problems — 453

CHAPTER

10. Mutual Inductance and Transformers — 466
- 10.1 Review of Mutual Inductance — 466
- 10.2 One Equivalent Circuit for a Two-Winding Transformer — 468
- 10.3 A Second Equivalent Circuit for a Two-Winding Transformer — 474
- 10.4 Summary — 484
- Problems — 484

11. Analogues (Duals) — 491
- 11.1 Electrical Duals — 491
- 11.2 Electromechanical Analogues — 497
- 11.3 Electronic Analogue Computer — 514
- 11.4 Electromechanical Devices — 520
- 11.5 Summary — 521
- Problems — 522

APPENDIX

- A. Proof that $M_{12} = M_{21}$ and that $k \leqslant 1$ — 530
- B. Proof of the Uniqueness Theorem for Second-Order Linear Differential Equation with Constant Coefficients — 533
- C. Table of Transforms — 535
- D. The Basis of Operation of Certain Electrical Instruments — 537
- E. Proof of Thévenin's Theorem — 539

Index — 543

1 *Electromagnetic Field Theory*[1]

1.1 INTRODUCTION

The student's background in physics or chemistry, as well as his experiences in this age of science and engineering, have given him certain concepts of electricity and magnetism. We shall start with these concepts, expand on them, and then introduce the laws of electromagnetism which are known as Maxwell's equations. It is on these equations that electric circuit theory is based; we should have at least a nodding acquaintance with them if we are to have a good relationship with circuit theory.

To most of us the concept of electric charge is basic to the structure of the atom. Each of the building blocks of matter, such as protons, neutrons, electrons, mesons, etc., is characterized by two properties: its "rest mass" and its electric charge. This charge may be of two kinds which are arbitrarily called positive and negative, the charge of the electron being negative. Charge is quantized; that is, charge occurs in packets, the smallest one of which is equal to the charge on an electron. The effect of these atomic charges is extremely important in determining the mechanical, chemical, and electrical properties of materials.

Forces exist between charges; in fact it is because such forces were observed that the concept of charge was postulated. Probably all of us have seen the demonstration of force between charged pith balls

---

[1] It is intended that this chapter give a qualitative insight to field theory so that the student may find that a definite and significant relationship exists between field theory and circuit theory and that circuit theory is an approximation to field theory. It is not intended that this material be construed as an introductory course in field theory. Consequently, it is anticipated that only one, or possibly two, class periods be devoted to its consideration.

and noticed that like charges repel, unlike charges attract. Such forces are called electric forces. If the charges are in motion relative to each other and to the observer, additional forces are observed. These forces are called magnetic forces, magnetism thus being associated with moving electric charges or electric fields changing with time.

There is a mystery about forces which "act at a distance" such as those of electricity and magnetism. Such phenomena seem less mysterious if it is postulated that a field is associated with each charge and that forces are the result of reaction between charge and field. Such fields were first proposed by Faraday and were then adopted by Maxwell in his electromagnetic theory. We are accustomed to some terms involving fields; in particular, we speak of the earth's magnetic field and are aware that a compass will point toward the north magnetic pole.

With regard to the student's experience with electricity and magnetism, he has surely received a shock after walking on a rug and learned that charges were separated by friction. He has observed lightning which is believed to be a similar phenomenon on a grand scale, the separation of charge here being caused by relative motion of water particles and air. He knows that energy can be transmitted most efficiently in the form of electromagnetic energy; that power-generating stations are able to convert the energy of combustion of fossil fuels or the energy liberated by nuclear fission into electromagnetic energy, and that somehow this energy can be guided by transmission lines to industries and homes. This transmission usually occurs with voltages and currents varying sinusoidally with time at 60 cycles per second (cps), transformers being used to change the voltage level so that the voltage level in the homes will be relatively safe while high voltages are used for economic transmission of large amounts of power.

Electromagnetism also has a major role in communications. Telephone, telegraphy, radio, and television are all examples of the application of electromagnetism. In some of these instances, wires are used to guide the signal transmission. In others, such as radio or television, no wires are needed to guide the energy; the energy radiates in a fashion similar to that of the light radiated from a candle or electric lamp. In fact the student may know that light is simply that portion of the electromagnetic frequency spectrum to which our eyes are sensitive. Thus electromagnetism covers all the frequency spectrum from zero frequency or direct current through radio frequencies, light frequencies, x-rays, and gamma rays.

Electromagnetism has many applications in industry, in controls, in instrumentation, and in computers. In fact it is hard to find a segment of modern life in which electromagnetism is not playing an important role.

1.2 A DISCUSSION OF LAWS AND UNITS

What are the laws of a science which deals with such a wide range of phenomena, from the interaction of atomic particles to communication between distant points of the earth, or to radiation from distant galaxies? What do we mean by laws in the first place?

In a physical science we think of the laws as a group of relationships which are consistent among themselves and which lead to conclusions that are in accord with experimental evidence. Also we use the term "fundamentals" or fundamental laws as a minimum group of relations which may be considered basic and from which other relations or laws may be derived. Thus what one person calls fundamental need not be the same as what another person calls fundamental as long as there is agreement on the total group of relationships. For example, in mechanics one person may start with force relations as fundamental and then "derive" energy relations; while another person may start with energy relations and then derive the force relations. A third person may well observe that the first two are simply proving that their total groups of relations are consistent.

Since our measuring abilities improve with time, we would expect that physical laws need to be modified as time progresses. A good illustration of this modification is again in the area of mechanics in which Newton's laws have been dominant for many years. However, it has been found necessary to modify Newton's laws in accord with relativistic principles when relative velocities are not negligible in comparison with the velocity of light. Also it has been found necessary to use quantum mechanics to predict phenomena in the case of small magnitudes corresponding to the atomic and nuclear domains. These modifications do not invalidate Newton's laws for a large group of practical cases; in fact both relativistic mechanics and quantum mechanics are in accord with Newton's mechanics for these cases.

Maxwell's equations have a similar relation to electromagnetic field theory that Newton's laws have to mechanics. Newton's laws do not apply to atomic or nuclear realms because the laws do not recognize that mass (energy) occurs in discrete amounts (quanta); Maxwell's equations similarly do not apply in these realms because the equations do not recognize that charge occurs in discrete amounts.

We shall not follow the historical or experimental approach to electromagnetism. Such an approach would possibly start with Coulomb's law of force between point charges and the law of force between two current carrying conductors and then proceed to the observations of Ampere, Kirchhoff, Gauss, Faraday, and others. Maxwell[2], in

---

[2] *A Treatise on Electricity and Magnetism*, by Clerk Maxwell, Unabridged Third Edition, reprinted by Dover Publications, New York, 1954.

considering the "fundamental" laws of electricity and magnetism as they were known in his day, observed that these laws were not consistent mathematically, added a term to make them consistent and deduced that electromagnetism was a wave phenomenon having the velocity corresponding to that of light. His brilliant deductions have been confirmed and his contribution is considered such a "breakthrough" that the modified laws are known as Maxwell's equations. Thus it seems appropriate to start with Maxwell's equations as basic postulates, for in a real sense they are not derived. There are also energy relations in electromagnetism; however, we shall not try to prove that these energy relations are consistent with Maxwell's equations and the concepts of conservation of charge and energy. Such proofs properly belong in a course on electromagnetism. We shall simply accept all these relations as representing a consistent group of laws.

Maxwell's equations are expressed in terms of field quantities. The electric field is postulated to have a flux, ψ, a flux density $\bar{D}$,[3] and a field intensity $\bar{E}$. The magnetic field is postulated to have a flux, ϕ, a flux density $\bar{B}$, and a field intensity $\bar{H}$. If the medium is isotropic, $\bar{D}$ and $\bar{E}$ are in the same direction; that is, $\bar{D} = \epsilon \bar{E}$, in which ϵ is a scalar quantity called permittivity. Similarly, for an isotropic medium, $\bar{B} = \mu \bar{H}$, in which μ is permeability. If ϵ and μ are constants, independent of the magnitude or direction of the field quantities, the medium is said to be linear. If ϵ and μ are independent of location in space, the medium is said to be homogeneous. The concept of flux and field intensity may seem strange and a person may be prompted to ask whether these quantities are real. They are as real as other useful concepts. We cannot see mass, nor do we know what it is, but yet we willingly accept the concept of mass because this is helpful in predicting the behavior of material objects. The same observation may be made for the concept of charge. And so we should accept these postulates of flux and field intensity as useful concepts in predicting electromagnetic behavior.

A consistent set of units is very important. We shall use only the rationalized meter-kilogram-second (mks) system of units. Occasionally, dimensions may be given in English units; these should be changed to meters before insertion into equations. The mks system of units has been adopted internationally and has many advantages over other systems. One of these advantages is that many practical units such as ampere, volt, watt, and joule are units also in the mks system. Essentially all modern literature of significance in circuit theory is in the mks system of units. Recent literature in electromagnetic field theory is

---

[3] The bar over the quantity indicates that the quantity is a vector.

primarily in the mks rationalized system of units. The word "rationalized" implies that Maxwell's equations have no constants other than unity; the word "unrationalized" implies that a factor of 4π appears in two of Maxwell's equations. There are only a few quantities that have different magnitudes in these two systems of units. These are $\bar{D}$, $\bar{H}$, ϵ, and μ; these differ by the factor 4π. For all circuit quantities there is no difference between rationalized and unrationalized systems of units.

We sometimes consider a particular group of units to be a fundamental set, the other units being expressed in terms of these fundamental units. In mechanics it is customary to consider mass, length, and time as fundamental units and then express other quantities such as force, velocity, power, and energy in terms of these units. A dimensional check of an expression or equation is then sometimes useful in detecting an error. In electromagnetism, three fundamental units are not enough; it is necessary to add a fourth unit. This unit is frequently charge; one sometimes sees reference to the mksq system of units. However we seldom make a dimensional check in terms of these units; it is usually easier to assure ourselves that each term of a given expression has the same units, such as volts, amperes, or ohms.

1.3 THE LAWS OF ELECTROMAGNETISM (MAXWELL'S EQUATIONS)

There are only four relations which are considered to be the laws of electromagnetism and known as Maxwell's equations. The first two are concepts about magnetic and electric flux, the last two are concerned with the closed line integrals of field intensities. These laws may be expressed as follows.

(a) Magnetic flux is continuous; that is, this flux has no beginning or end. The expression for the magnetic flux, ϕ, passing through an area a may be calculated from the surface integral,

$$\phi = \int \bar{B} \cdot d\bar{a},$$

in which $d\bar{a}$ is a vector differential area and $\bar{B} \cdot d\bar{a}$ is the scalar product of two vectors meaning $Bda \cos <\genfrac{}{}{0pt}{}{\bar{B}}{d\bar{a}}$, ϕ has the units of webers and $\bar{B}$ has the units of webers per square meter. Fig. 1.1 shows the magnetic field about a long wire carrying a conduction current shown going into the paper. The direction of conduction current is taken to be the direction of motion of positive charge.

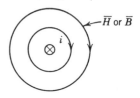

FIG. 1.1 Magnetic field around a current, i.

One way of expressing this law mathematically is to say that the net flux leaving a closed surface must be zero. Thus,

$$\oint \bar{B} \cdot d\bar{a} = 0, \tag{1.1}$$

in which now the direction of $d\bar{a}$ is taken to be outward from the volume enclosed and the circle about the integral sign means that the integration is over a closed surface.

(b) Electric flux may be continuous or may terminate on electric charge. In the latter case, the amount of flux leaving a closed surface is equal to the charge enclosed, which implies that flux leaves a positive charge and enters a negative charge. An example of electric flux being continuous would be the electric field of a dipole antenna at a distance which is large in comparison with a wavelength (see Fig. 1.2). An example of the electric field terminating on charge is the field between the two conductors of a direct current (d-c) transmission line, as shown in Fig. 1.3.

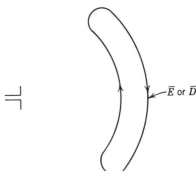

FIG. 1.2 Electric far field of antenna.

Electric flux, ψ, has the units of coulombs and electric flux density, $\bar{D}$, has the units of coulombs per square meter. We may express this law mathematically as

$$\oint \bar{D} \cdot d\bar{a} = q, \tag{1.2}$$

in which q is the charge enclosed by the closed surface.

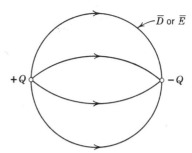

FIG. 1.3 Electric field about two d-c conductors.

This law is known as Gauss's law for electricity or simply as Gauss's law and is treated extensively in most physics texts.[4]

(c) **The closed line integral of the electric field intensity, $\oint \bar{E} \cdot d\bar{l}$, is equal to the negative time rate of change of the magnetic flux which links the closed line, the positive direction of magnetic flux being in the direction of the thumb of the right hand if the fingers of this hand are in the direction of the integration.** The designation $d\bar{l}$ is a vector differential displacement, and the circle about the integral sign means that the integration path is such that it returns to its starting point. In the sketch shown in Fig. 1.4, if ϕ is increasing with time in the direction shown, we will obtain a negative value for $\oint \bar{E} \cdot d\bar{l}$. The units of E are volts per meter, and thus the units of $\oint \bar{E} \cdot d\bar{l}$ are volts or joules per coulomb.

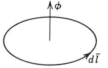

FIG. 1.4 Magnetic flux linking a closed line.

The magnetic flux linked by the closed line may be designated $\lambda' = \int_a \bar{B} \cdot d\bar{a}$, in which the area a is the area bounded by the closed line. We need to distinguish between magnetic flux, ϕ, and magnetic flux linkage, λ', since the linkage is found by integrating flux density over the area bounded by some closed line and since a particular quantity of flux may cross this area more than once. This latter condition is illustrated in Chapter 2, Fig. 2.17, page 41.

This law may then be written mathematically,

$$\oint \bar{E} \cdot d\bar{l} = \Sigma v = -\frac{d\lambda'}{dt}, \qquad (1.3)$$

---

[4] See, for example, David Halliday and Robert Resnick's, *Physics for Students of Science and Engineering*, Part II, Second Edition, Wiley, New York, 1962, Chapter 28.

in which v is the symbol for voltage. It is understood that an incremental voltage, $\Delta v = \bar{E} \cdot \Delta \bar{l}$, is plus if the first terminal is plus with respect to the second in the direction of summation. With this convention in mind as well as the fact that the positive direction for magnetic flux is in the direction of the thumb of the right hand if the fingers of this hand are in the direction of summation, **the sum of the voltages around a closed path is equal to the negative time rate of change of magnetic flux linkages.** This law is known as Faraday's law of magnetic induction and is treated extensively in texts on physics.[5] When this law is modified for electric circuits, it is called Kirchhoff's voltage law.

(d) The closed line integral of the magnetic field intensity, $\oint \bar{H} \cdot d\bar{l}$, is equal to the current enclosed, the positive sense of the current being in the direction of the thumb of the right hand if the fingers of this hand are in the direction of integration. Fig. 1.5 shows the relation between the direction of the current, i, and positive values of $\oint \bar{H} \cdot d\bar{l}$. $\bar{H}$ has the units of amperes per meter. The law may be written mathematically,

$$\oint \bar{H} \cdot d\bar{l} = i \tag{1.4}$$

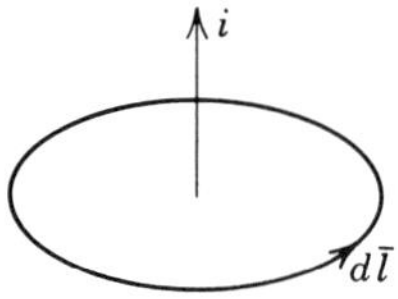

FIG. 1.5 A closed line integral around a current, i.

Current is defined as a phenomenon associated with magnetic fields. It has the symbol i and the units of amperes. There are two kinds of currents. **One kind of current is the motion of charge.** This may be conduction current as exemplified by the drift of electrons in metals or the drift of electrons or "holes" in semiconductors, or it may be convection current as exemplified by the motion of electrons in vacuum tubes or the motion of charges in plasmas. The direction of this kind of current is the direction of flow of positive charge or opposite to the flow of electrons. **The other kind of current is called displacement current and is equal to the time rate of change of electric flux.** The direction of this displacement current is the direction of increasing flux.

If the current, i, of Equation 1.4 is interpreted as consisting **only of motion of charge**, this law is referred to as Ampere's circuital law.[6] If the current, i, is interpreted as consisting of both motion of charge **and**

---

[5] See, for example, Halliday and Resnick (cited in ref. 4), Chapter 35.
[6] For discussion, see Halliday and Resnick (cited in ref. 4), Chapter 34.

displacement current, this law may be referred to as Ampere's law as modified by Maxwell. Maxwell introduced this concept of displacement current in order to make the known laws of his time consistent from a mathematical viewpoint. He then deduced the wave nature of electromagnetism.

It is important to note for Equation 1.4 that the current may be evaluated over any surface bounded by the closed path of integration. This implies that **current must be continuous**. Another way of stating that current is continuous is to say that the **net current leaving (or entering) any closed surface must be zero**. When this concept is applied to a closed surface surrounding a junction of several elements in an electric circuit, it is called Kirchhoff's current law, and written $\Sigma i = 0$.

1.4 SOME DEDUCTIONS FROM MAXWELL'S EQUATIONS

We shall not attempt to obtain a solution to any particular physical problem by applying Maxwell's equations and suitable boundary conditions. Such solutions are difficult to obtain and are better left to a succeeding course in electromagnetic field theory. Note that each vector quantity may have three components and each component may be a function of three space coordinates as well as time. Thus solutions to Maxwell's equations are not obtained easily. We shall give without proof the nature of the solutions obtained under certain simplifying assumptions.

First, assume that steady-state sinusoidal variations with time exist and also that there is no charge in the medium. These assumptions are necessary to reduce the mathematical difficulties to a reasonable level. It means that the solutions will not apply to ionized space having a net charge density, or to the space inside vacuum tubes. However, the solutions will apply to conducting media such as metals. In metals, from the large-scale view of Maxwell's equations, there is no net charge density. However there may be a flow of charge with a corresponding electric field intensity. The relation between current density, $\bar{J}$, and electric field intensity, $\bar{E}$, is given for isotropic media to be $\bar{J} = \sigma \bar{E}$, in which σ is the conductivity of the material with the units of mhos per meter and $\bar{J}$ has the units of amperes per square meter. This relation, $\bar{J} = \sigma \bar{E}$, is the point form of Ohm's law.

In addition to the above assumptions of sinusoidal time variations and zero charge density, it is desirable to assume that the medium is boundless, linear, isotropic, and homogeneous. With these assumptions the following deductions of interest to us are obtained.

(a) The ratio of conduction current density to displacement current density depends on the properties of the medium and the frequency of the sinusoidal variations. This ratio is

$$\frac{\text{conduction current density}}{\text{displacement current density}} = \frac{\sigma}{\omega\epsilon},$$

in which $\omega = 2\pi f$ with f being the frequency in cycles per second. If $\sigma/\omega\epsilon$ is very large we call the medium a conducting medium; if the ratio is very small, we call the medium an insulating or dielectric medium. For metals such as copper with $\sigma = 5.8(10^7)$ mhos per meter and $\epsilon = 8.85(10^{-12})$ farads per meter, the displacement current will not become one per cent of the conduction current until a frequency of 10^{16} cycles per second is reached. On the other hand, for insulating materials such as air, mica, paper, and oil, we are generally safe in ignoring the conduction current.

(b) The field functions are traveling waves, traveling at a finite velocity. For a good dielectric the velocity is $1/\sqrt{\mu\epsilon}$ which is equal to $3(10^8)$ meters per second for free space. For a good conductor the velocity is appreciably less, being equal to $\sqrt{2\omega/\mu\sigma}$. For copper at 10^6 cycles per second this velocity is 416 meters per second. This velocity is **not** the velocity of electrons constituting the conduction current. The average drift velocity of such electrons is a function of the current density and of the number of free electrons in the medium but is usually much less than 1 meter per second. In a conductor the wave attenuates or decreases in strength as it progresses because some energy is transformed into heat.

In practical problems the media are not boundless. We may have a pair of wires connecting a power plant to a lamp or connecting two telephones together. The solution from the field approach tends to tell us that the energy travels through the dielectric which separates the wires and that the wires simply serve as guides, the actual velocity being somewhat less than $1/\sqrt{\mu\epsilon}$. Enough power is fed from the field into the wires to supply the power dissipated as heat in these conductors.

(c) The energy stored in the electric field may be calculated from the volume integral, $\int \frac{1}{2}\epsilon E^2 \, d\tau$, in which E is the magnitude of $\bar{E}$ and $d\tau$ is a differential volume. The energy stored in the magnetic field has a similar expression, $\int \frac{1}{2}\mu H^2 \, d\tau$. The expression for the power dissipated as heat in a conducting medium is $\int (J^2/\sigma) \, d\tau$. Energy has the units of joules; power has the units of watts.

1.5 THE RELATIONSHIP OF CIRCUIT THEORY TO FIELD THEORY

The electric and magnetic field quantities, which represent our best model of reality for such phenomena, are vector quantities which may vary with both space and time and are therefore difficult to handle mathematically. However, for the majority of electrical engineering applications, we are justified in replacing the field model with a "fictional" circuit model in which the variables, voltage and current, are scalar quantities which vary only with time. Thus circuit theory is easier and may be studied successfully without the mathematical background needed for field theory. However, it is important that we realize the limitations of this circuit model. We use the words "circuit fiction" to remind ourselves that circuit theory is an approximation to field theory.

Certain laws are borrowed directly from field theory. Faraday's law of magnetic induction is modified to become Kirchhoff's voltage law and is also used to obtain a definition of the circuit element, inductance. Ampere's circuital law as modified by Maxwell is interpreted to mean that current must be continuous, a statement of Kirchhoff's current law. Also the energy relations in field theory are useful in obtaining expressions for some of the circuit elements. These circuit elements, such as resistance, inductance, and capacitance are lumped elements which are used to represent such phenomena as the dissipation of power into heat in some region of space, or the magnetic or electric fields throughout some region of space.

Neither field theory nor circuit theory is concerned directly with certain aspects of energy sources, such as electrochemical or photoelectric processes or high frequency oscillators. Such subjects are usually considered in courses concerned with energy conversion or devices.

One of the real limitations of circuit theory arises out of the wave nature of electromagnetism. In order that a particular arrangement of an energy source, conducting wires, and insulating plates may be represented adequately by a circuit consisting of elements in series, the dimensions of our physical arrangement must be very small in comparison with a wavelength. Since wavelength is equal to velocity (meters per second) divided by frequency (cycles per second), a particular circuit will give results corresponding to physical reality only for frequencies below some value. Suppose we consider a physical system whose largest dimension is 1 meter and suppose we consider 1 per cent as being "very small." Then the frequency above which a simple circuit would not apply would correspond to a wavelength of 100 meters. If the velocity of wave propagation may be treated as equal to that of a plane wave in free space, $3(10^8)$ meters per second, this upper frequency limit is $3(10^6)$ cycles per second. Following from this kind of argument a transmission line operating at 60 cycles per second should not be longer than 50,000 meters, or approximately 30 miles, if it is to have a simple equivalent circuit; if the line is operating at $60(10^6)$ cycles per second it should not be longer than 0.05 meters.

If the dimensions are large enough so that one is not justified in replacing the system by a single simple circuit, is it possible to subdivide the system into many portions, each of which may be represented by a circuit? Yes, this is possible although a more sophisticated engineering judgment is required to decide on appropriate circuits. Also the analysis will probably become so cumbersome that a solution may be more readily obtained from field theory or from experimental methods. One important exception exists in the case of parallel wire transmission lines. For these lines it is possible to consider each differential length of line as having inductance, capacitance, and resistance per unit length and arrive at wave solutions which are equivalent to those obtained from field theory. This exception is outside the scope of this text.

It may occur to us that there are single pulses of voltage or current, such as

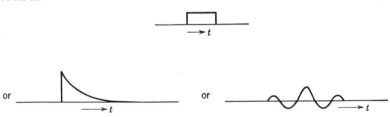

It has been shown by the French mathematician Fourier that such time functions can be expressed in terms of their frequency content or frequency spectrum. Thus it becomes possible to analyze such functions in terms of frequency. As we pursue our study of electricity and magnetism we find that for a particular case nature often limits the frequency range which is significant. The Federal Communication Commission and other agencies also place restrictions on the frequency limits which we may use, a restriction which serves as a blessing to the analyst. Let us try to keep in mind that for a particular phenomenon we are dealing with a limited band of frequencies.

FURTHER READING

The laws of electricity and magnetism are discussed extensively in all college physics texts. A good example of such a text is David Halliday and Robert Resnick's *Physics for Students of Science and Engineering*, Combined Edition, Part II, Wiley, New York, 1962. An excellent presentation of these concepts is also given in the first three chapters of P. R. Clement and W. C. Johnson's *Electrical Engineering Science*, McGraw-Hill, New York, 1960.

2 Basic Concepts of Circuit Theory

2.1 INTRODUCTION

It was the purpose of Chapter 1 to introduce us to the general concepts of electromagnetic field theory. These general concepts or postulates are to be accepted on faith. We shall find that deductions from them lead to results in accord with reality and thus permit us to predict behavior of new physical systems or devices. However, because of the mathematical difficulties in applying electromagnetic field theory directly to physical problems we shall limit ourselves to a branch of electromagnetic field theory known as electric circuit theory.

There are really two parts to solving a physical problem by means of circuit theory. The first part is to use electromagnetic theory and engineering judgment in representing an actual system with a sketch of interconnected idealized circuit elements. The second part is to use the volt-ampere characteristics of the elements and Kirchhoff's Laws to obtain solutions for voltages, currents, and powers. If both parts are well done these solutions will be good approximations to the corresponding voltages, currents, and powers which can be measured in the physical system.

The student cannot perform the first part of this process at this time; his ability to do so will develop as he gains understanding of circuit theory and field theory and as he gains experience with actual physical systems in the laboratory and in practice. This text is therefore devoted primarily to the second part—the techniques for calculating the performance of given electrical circuits. However, we shall try to point out the assumptions that are made in defining circuit elements and we shall stress the fact that, for a given system, the complexity of the representative circuit increases with the frequency.

2.2 DEFINITIONS AND CONVENTIONS

An electric circuit or network is a mathematical model used to represent some physical system, and consists of an interconnection of circuit elements, each of which represents some physical phenomenon in the physical system.

A circuit element has two terminals and may be sketched as shown in Fig. 2.1. The nature of the element is defined by the relationship between two quantities. These quantities are the voltage difference between the terminals and the current which is directed into one terminal and out of the other terminal.

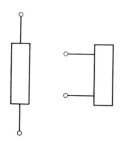

FIG. 2.1 Circuit elements.

Voltage and current are scalar quantities which vary only with time. We could use the functional notation $v(t)$ and $i(t)$; however, it is customary to drop the functional notation, using the small letters v and i to represent functions of time and capital letters to represent fixed values, thus for $v = V \cos \omega t$, V represents the maximum value or amplitude of the sinusoidal variation.

The functional notation is used occasionally. For instance, to indicate the voltage at some specific time, t_1, we may write $v(t_1)$. In particular we shall have occasion to indicate voltage or current immediately before or after zero time. These will be indicated by $v(0-)$, $v(0+)$, $i(0-)$, and $i(0+)$ respectively.

Current is a phenomenon associated with magnetic fields and is either charge in motion or the time rate of change of electric flux. The direction of current is, by convention, the direction of motion of positive charge (opposite to the direction of motion of negative charge) or the direction of positive time rate of change of electric flux. The unit of current is the ampere (coulomb per second), the standard ampere, by international agreement, being that current which develops a certain force between two coils in a specified geometry. From Maxwell's equations, current is continuous; this concept is maintained in our definitions of circuit elements so that at any instant of time the current entering one terminal of an element is equal to the current leaving the other terminal.

14

It is important to indicate reference or assumed positive direction for current; it is also important to indicate reference or assumed positive voltage difference. The reference or assumed positive direction for current is indicated with an arrow as shown in Fig. 2.2. This arrow does not mean that the current is only in this sense; in general the current will be in one direction part of the time, in the other direction part of the time, as shown on the sketch. At those times at which the current is +, such as t_1, the current is in the direction indicated by the arrow; at those times at which the current is −, such as t_2, the current is in the opposite sense.

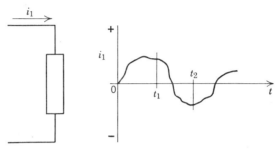

FIG. 2.2 Current through an element.

The voltage difference between two points **in space** could be defined as

$$v_a - v_b = -\int_b^a \bar{E} \cdot d\bar{l}.$$

This is equal to the **power per ampere** or **energy per coulomb** along the particular path between these points. However, if the magnetic field in the neighborhood of these points is changing with time, the value of the integral, $\int_b^a \bar{E} \cdot d\bar{l}$, will depend on the path taken. Thus, the voltage difference or energy per coulomb will usually be different for every different path, and it is not useful to speak of the actual voltage difference between two points unless the path is specified. For circuit theory we so define our elements that the voltage difference between any two terminals of our circuit is independent of the path taken between them. The polarity convention is such that if energy is required to move a positive charge from terminal a to terminal b, we then consider a to be plus with respect to b. This is demonstrated in Fig. 2.3a showing energy flow or power into the element for the current direction and voltage polarity shown. If energy flows from the element the current direction is from the negative to the positive terminal of the element as shown in Fig. 2.3b.

For voltage differences, two conventions will be used. One convention is shown in Fig. 2.4a in which polarity marks are shown. The

Basic Concepts of Circuit Theory 15

voltage v_1 is positive when terminal e is + with respect to terminal f, and negative when the reverse is true. In Fig. 2.4b, no polarity marks are shown and here we could use double subscripts to indicate the voltage difference. The convention used in this text is that v_{ef} is + when e is + with respect to f. We may think of this as $v_{ef} = v_e - v_f$. This is known as the "voltage drop" convention. It should be apparent that $v_1 = v_{ef} = -v_{fe}$.

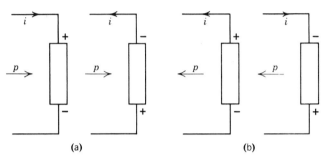

FIG. 2.3 (a) Power (energy flow) into element. (b) Power (energy flow) out of element.

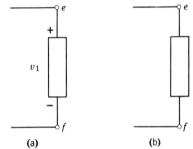

FIG. 2.4 Voltage across an element.   (a)   (b)

A circuit node is the common connection point of two or more elements. In Fig. 2.5a is shown a circuit with the four elements, A, B, C, and D and with the three nodes a, b, and c. Fig. 2.5b shows the same circuit, some additional lines, b–b', c–c', and c'–c'', being drawn to make a more open sketch. There are still only three nodes; for example, the terminals b and b' constitute only one node, the understanding being that there can be no voltage difference between terminals b and b' regardless of the amount of current in the line b–b'. This defines the line b–b' as a "short circuit." The lines c–c' and c'–c'' are also "short circuits." These idealized short circuits should be distinguished from actual short circuits which always have some small voltage difference associated with the current flow.

Elements which have a common current are said to be in series. In Fig. 2.5 elements A and B are in series since the current in A must be always identically the same as the current in B.

16

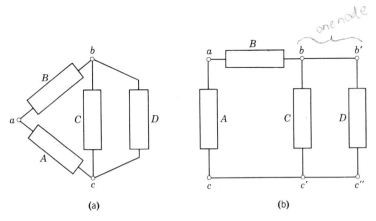

FIG. 2.5 A circuit of four elements.

Elements which have a common voltage difference are said to be in parallel. In Fig. 2.5 elements C and D are in parallel since each has identically the same voltage difference, v_{bc}.

A circuit branch consists of one or more elements in series. In Fig. 2.5 there are three branches, one with elements A and B in series, one with element C, and one with element D.

An "open circuit" exists in a path between two terminals if there is no current in this path regardless of the voltage difference between the terminals. As an example, if we omit the connection b–b' in Fig. 2.5b, sketching the circuit as in Fig. 2.6a, we would say an open circuit existed between b and b'.

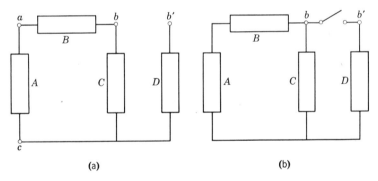

FIG. 2.6 An "open circuit" and a switch in a circuit. (*a*) "Open circuit" between b and b'. (*b*) Switch between b and b'.

An ideal switch is defined as an open circuit when open and a "short circuit" when closed and is presumed to move from one position to the other in zero time. An actual switch will take a finite time to move, will usually arc during the time the contacts are separating, and will have a relatively small voltage across it when closed. Nevertheless, in many cases we are justified in replacing an actual switch by an ideal switch or by an ideal switch in combination with some circuit elements.

2.3 KIRCHHOFF'S LAWS

After a certain circuit has been selected to represent a physical situation, two important concepts from field theory are used to determine the unknown voltages and currents. These two concepts are known as Kirchhoff's laws.

2.3.1 Kirchhoff's Current Law

One of Maxwell's equations states that the closed line integral of magnetic field intensity, $\oint \bar{H} \cdot d\bar{l}$, is equal to the current enclosed. Since any surface bounded by the closed line may be used in evaluating the current and since the current must be the same for any of these surfaces it follows that **current is continuous**, or that **the net current leaving, or entering, a closed surface is zero**. These are statements of Kirchhoff's current law. Mathematically we write $\Sigma i = 0$.

We usually apply Kirchhoff's current law to each node of a circuit. This will be done in one of two ways, each of which is demonstrated in Fig. 2.7. If we are seeking solutions for unknown voltages we would label currents as demonstrated in Figs. 2.7a and 2.7b. If we are seeking solutions for unknown currents, we would label independent currents as in Figs. 2.7c and 2.7d. The branch currents, $i_1 + i_2$ and $i_4 - i_3$ are not independent currents since they are obtained by algebraic addition of other currents.

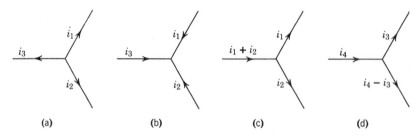

FIG. 2.7 Kirchhoff's current law applied at a node. (*a* and *b*) Labeling of all currents. $\Sigma i = 0$, $i_1 + i_2 + i_3 = 0$. (*c* and *d*) Use of Kirchhoff's current law to label independent currents. $\Sigma i = 0$.

Kirchhoff's current law may be applied to a closed surface enclosing several nodes. This is demonstrated in Fig. 2.8. In Fig. 2.8b, Kirchhoff's current law is applied at each node to demonstrate that if this is done, Kirchhoff's current law applies to a region enclosing several nodes.

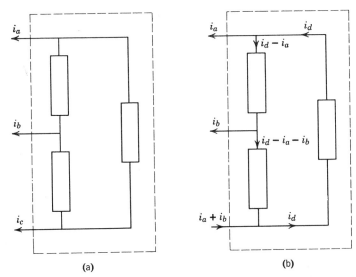

FIG. 2.8 Kirchhoff's current law applied to closed surface about several nodes. (a) Labeling of all currents leaving closed surface, $i_a + i_b + i_c = 0$. (b) Labeling in terms of independent currents.

2.3.2 Kirchhoff's Voltage Law

Faraday's law of magnetic induction is that the sum of the voltages around a closed path is equal to the negative time rate of change of the magnetic flux linking this path. Mathematically we write $\Sigma v = -d\lambda'/dt$. The implication of this law is that for physical systems it is not possible in general to specify the voltage between two points since different magnitudes of voltage are obtained for different paths between the two points. In circuit theory we shall so define our circuit elements that the voltage difference between two points is independent of the path between these points; or in other words: the sum of the voltages around a closed path is equal to zero. This is called Kirchhoff's voltage law and is written, $\Sigma v = 0$.

2.3.3 The Application of Kirchhoff's Laws to Circuits

Kirchhoff's laws are used for the solution of every electric circuit. We shall illustrate this for the circuit of Fig. 2.9a, which consists of six elements and which may be considered an expansion of Fig. 2.8b since the right-half portion of Fig. 2.9a has the same current labeling as used in Fig. 2.8b. Let us assume that the nature of each of the circuit elements is known but that all voltages and currents are unknown. (This is not generally the case since for independent sources, either the voltage or the current will be known; however, the effect of such sources is simply to reduce the total number of unknown quantities.)

Basic Concepts of Circuit Theory 19

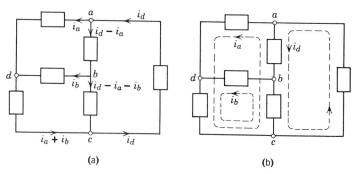

FIG. 2.9 Circuit with six elements (an expansion of Fig. 2.8b). (a) Independent branch currents. (b) Independent loop currents.

This circuit with six elements has six unknown currents and six unknown voltages, a total of twelve unknowns! However, if we knew all the currents, the voltages could be determined readily because each network element is known; or if we knew all the voltages, the currents could be determined readily. Furthermore, the six unknown currents are not independent; neither are the six unknown voltages.

Inspection of Fig. 2.9a shows application of Kirchhoff's current law to the circuit has permitted all branch currents to be labeled in terms of three independent branch currents: i_a, i_b, and i_d. We then need three independent equations which involve these three currents; these equations are obtained by applying Kirchhoff's voltage law to the circuit. There are **seven** possible closed paths around which we could write Kirchhoff's voltage law! How can we be sure that the three equations we write are independent? **One** approach is to consider the independent currents as having continuous paths or loops as illustrated in Fig. 2.9b, and then to write the independent voltage equations around these loops. This ensures that the equations will be independent since each equation has a term that is not in any other equation. This method, which starts with the labeling of independent branch currents, is known as the branch current method or loop current method.

Suppose we decided to solve for the unknown voltages for Fig. 2.9. The number of independent unknown voltages is one less than the number of nodes. This follows from the fact that if there were only two nodes, there would be only one unknown voltage; if one additional node is added, one additional independent unknown voltage is required. Thus, if the voltage across every element is unknown, the number of independent unknown voltages is the number of nodes less one. For the circuit of Fig. 2.9 there are four nodes; a, b, c, and d; and thus there are three independent unknown voltages. How can we choose these voltages so as to ensure that the voltages are independent? If we choose the voltages v_{ad}, v_{ab}, v_{bc}, these are independent. If we choose the voltages v_{ad}, v_{db}, v_{ab}, these are not independent because one of these is determined by the other two; that is, $v_{ab} = v_{ad} + v_{db}$, and thus our choice gave only two independent quantities. **One** method of selecting the proper number of independent unknown voltages is to select one node as a reference node; the independent unknown voltages are then the voltages between each remaining node and the reference node. If we select node *d* as the reference node, the independent unknown voltages will be v_{ad}, v_{bd}, and v_{cd}. Now how do we write three independent equations involving these three unknowns? These equations are obtained by applying Kirchhoff's current law to surfaces which separate the circuit into two parts, and somehow assuring ourselves that the equations are independent. **One** method is to apply Kirchhoff's current law about three of the four nodes; we shall always exclude the reference node, and thus in this case apply Kirchhoff's current law to nodes *a*, *b*, and *c*.

2.4 CIRCUIT ELEMENTS

Only six circuit elements are needed for much of linear circuit analysis. These are an independent voltage source, an independent current source, resistance *R*, self-inductance *L*, mutual inductance *M*, and capacitance *C*.

2.4.1 Independent Voltage Sources

An independent voltage source is a voltage source which is independent of other circuit quantities and is a function of time only. Specifically, it is not a function of the magnitude or direction of the current passing through it. It will be pictured as shown in Fig. 2.10a. The source voltage *v* shown is a function of time. If the voltage is a constant (a direct voltage) it is customary to show it as in Fig. 2.10b, and indicate that it is a constant by using the capital letter.

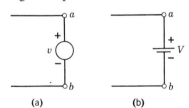

FIG. 2.10 Independent voltage sources. (a) (b)

A voltage source is not necessarily a power source. The current through a voltage source is determined by other elements as well as by the voltage source. At those instants of time at which the current direction is from − to + through the voltage source the energy flows from the voltage source; when the converse is true, energy flows into the voltage source.

A practical voltage source differs from the idealized one that we have defined in that the voltage changes with the current through the source. However, there are some actual sources which have a relatively constant voltage over the normal range of current requirements. As an example, the voltage supplied by a power utilities company to a home is nearly independent of the number of lamps or appliances which are switched "on." Any practical source can be approximated by some combination of a voltage source and other circuit elements.

We do not allow a short circuit to be placed across our idealized voltage source since this would violate our definitions of these two terms. For the same reason we do not place unequal ideal voltage sources in parallel.

2.4.2 Independent Current Sources

An independent current source is a current source which is independent of other circuit quantities and is a function of time only. Specifically, it is not a function of the voltage across its terminals. It will be pictured as shown in Fig. 2.11 connected to some other circuit element.

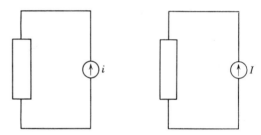

FIG. 2.11 Independent current sources.

A current source is not necessarily a power source since the polarity of the voltage across it is determined by other elements as well as by the current source.

The student has had some experience with electrical sources, particularly the 60 cps power supplied to the home, storage batteries, and dry cells. Since these particular sources are approximately constant voltage sources, the average student may be inclined to think that current sources are not as essential to circuit theory as voltage sources. He should try to overcome his prejudices; there are some practical sources which have nearly a constant current output over the normal range of power requirements. Also any practical electrical source can be approximated by some combination of a current source and other circuit elements.

We do not allow our idealized current source to be open circuited since this would violate our definitions of these two terms. For the same reason we do not place unequal ideal current sources in series.

The current through a current source and the voltage across a voltage source are referred to as "driving" functions or "forcing" functions, the other currents and voltages are referred to as "response" functions.

2.4.3 Resistance, R

As conduction current exists in a medium, some energy is dissipated in the form of heat. For a current density with a magnitude of J amperes per square meter the power absorbed per cubic meter is J^2/σ, σ being the conductivity of the medium in mhos per meter.

If the total current, i, is constrained to move in a continuous path or circuit the total power absorbed is the volume integral taken throughout the region in which the current occurs, or

$$p = \int \frac{J^2}{\sigma} d\tau = Ri^2, \tag{2.1}$$

or

$$R = \frac{p}{i^2}. \tag{2.2}$$

R is defined as the resistance of the circuit and has the units of ohms or watts/ampere². In this definition we have assumed that displacement current is negligible. This is permissible only at such low frequencies that the largest dimension of the circuit is very small in comparison with a wavelength. However, it is often convenient to define the resistance of a **portion** of a circuit for which the above assumption holds. We shall show this element as in Fig. 2.12. The power into this element is $v_{ab}i = Ri^2$, or

$$v_{ab} = Ri. \tag{2.3}$$

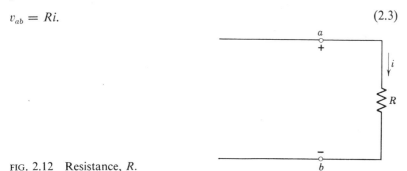

FIG. 2.12 Resistance, R.

The polarity of v_{ab} is plus to minus in the direction of current flow through R and if the current i should be negative, that is, if it is in the direction opposite to that indicated, the voltage v_{ab} is also negative. The mathematical expression of Equation 2.3 correctly expresses this relationship, and is known as Ohm's Law.

If a direct current I exists in a wire of constant cross-section a and length l, the current density, J, will be constant throughout the conductor volume and equal to I/a. Equation 2.1 becomes

$$\int \frac{J^2}{\sigma} d\tau = \frac{la}{\sigma}\left(\frac{I}{a}\right)^2 = RI^2,$$

whence for this case,

$$R = \frac{l}{\sigma a}. \tag{2.4}$$

The conductivity is a function of temperature. As temperature increases, the conductivity for most conductors decreases, and for most semiconductors increases. Certain materials have infinite conductivities at temperatures near absolute zero. This phenomenon, known as superconductivity, permits refrigerated circuits to be constructed having zero resistance.

For a sinusoidal current in metallic conductors at a fixed temperature the resistance R is a constant and the relationship between voltage and current is a straight line as shown in Fig. 2.13.

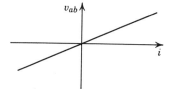

FIG. 2.13 Linear relationship between voltage and current for Fig. 2.12.

As the frequency is increased from zero the resistance of a circuit will increase because the current density will become nonuniform, tending to be higher at those locations where the magnetic field intensity is higher. This phenomenon is generally known as "skin effect;" some of its useful applications are listed in the literature under the term "induction heating."

Some materials have conductivities which vary with the magnitude of the current density; thus the voltage-current curve will not be a straight line, or will be nonlinear. Also certain devices which are essentially resistive in character, have a voltage current curve which is a function of the direction of current flow. An example of this is a so-called "rectifier."

We shall in general treat resistance R as a constant, not varying with temperature, frequency, or direction, or magnitude of current flow. There will be a few exceptions to this and we shall try to be careful to point these out as exceptions.

The Greek letter Ω (omega) is used to represent ohms. The prefix k is used to represent kilo or thousand and the prefix meg is used to represent million. Thus 500,000 ohms may be written 500 $k\Omega$ or 0.5 meg Ω.

A resistor is a physical device which is constructed to have primarily the property of resistance. It is often made of metal wire, or of carbon, or of a thin deposited film of metal. In all cases the design must be such that the temperature rise caused by the absorbed heat will not damage the resistor.

We need to distinguish between resistance and resistor since resistance is an idealized circuit element which can only dissipate energy and is not capable of storing energy, while a resistor is a physical device which is intended to **act** as a resistance below some frequency but which will act quite differently when this frequency is exceeded.

2.4.3.1 *The Solution of Circuits with Resistances and Independent Sources.* With our definitions of independent voltage and current sources and the element resistance it is possible to solve simple circuits by applying Kirchhoff's laws. We need to assume that $d\lambda'/dt$ is zero or negligible. It will be zero if all voltages and currents are constants, not varying with time; it may be negligible in other cases.

There are two general methods for solving circuits; these methods have already been discussed under Section 2.3.3.

The **branch current method or loop method** consists of determining the number of independent unknown currents by applying Kirchhoff's current law at all nodes, labeling independent unknown branch currents as needed. Then Kirchhoff's voltage law is applied around the paths or loops of the independent unknown currents. After the independent unknown currents have been determined, all other currents and thus all voltages and powers can readily be determined. (It is possible to approach the solution for unknown currents from the field of topology. We shall do this in Chapter 4, the method being called the **link-current method**. In this method, the selection of unknown currents is obtained differently than we have outlined, but the independent voltage equations are obtained in exactly the way we are using for the branch current method.)

The **node voltage method** consists of determining the number of independent unknown voltages by the application of Kirchhoff's voltage law. Then Kirchhoff's current law is used to write a number of independent equations equal to the number of independent unknowns. The best way to ensure independence as well as symmetry of the equations is to select a reference node and then to apply Kirchhoff's current law about those nodes which are the first subscripts of the independent unknown voltages. After the independent voltages have been obtained, all other voltages, and all currents and powers may be determined.

These methods can best be illustrated by the use of examples.

Example 2.1

GIVEN: The network of Fig. E-1.1, the switch being closed at $t = 0$. V, R_1, R_2, and R_3 are known. Assume that terms of $d\lambda'/dt$ are negligible so that Kirchhoff's voltage law may be used, $\Sigma v = 0$.

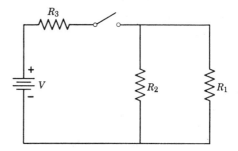

FIG. E-1.1

TO FIND: (a) The literal expressions for the voltages and currents by the node voltage method for $t \gg 0$.

(b) The literal expressions for the voltages and currents by the branch current method for $t \gg 0$.

(c) The numerical values for all voltages and currents for $V = 10$ volts $R_1 = 9 \, \Omega$, $R_2 = 1 \, \Omega$, $R_3 = 1.1 \, \Omega$.

SOLUTION: (a) In the **node voltage method**, the first thing to do is to resketch the circuit as in Fig. E-1.2 with the switch closed and to label all nodes. The nodes are arbitrarily labeled as shown, and node o is selected as the reference node. The next thing to do is to determine the independent unknown voltages. In this case v_{bo} is known, since $v_{bo} = V$, and v_{ao} is the only independent unknown voltage. v_{ba} is not independent since $v_{ba} = v_{bo} + v_{oa} = v_{bo} - v_{ao} = V - v_{ao}$. We now need to use Kirchhoff's current law about a node so that we may solve for the independent unknown voltage. There is only one independent unknown voltage and thus only one independent equation can be written.

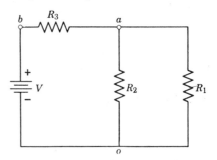

FIG. E-1.2

This equation could be written about either node a or the reference node o. In order to ensure that the equations will always be independent for any number of unknown voltages we shall never write

Kirchhoff's current law about the reference node but we shall write about those nodes which are the first subscripts of our independent unknown voltages. In this case we write Kirchhoff's current law about node a. It is convenient to label all the branch currents away from the node as shown in Fig. E-1.3. Kirchhoff's current law about node a is

$$i_1 + i_2 + i_3 = 0. \tag{1}$$

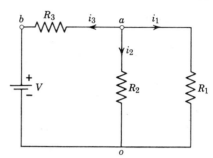

FIG. E-1.3

In order for this equation to permit us to solve for v_{ao}, each current must be expressed in terms of v_{ao} and known quantities. This can be done by use of Ohm's law and Kirchhoff's voltage law.

$$i_1 = \frac{v_{ao}}{R_1}, \tag{2}$$

and

$$i_2 = \frac{v_{ao}}{R_2} \tag{3}$$

from Ohm's law.

In order to determine i_3, we note that $v_{ao} = v_{ab} + v_{bo}$, or

$$v_{ao} = R_3 i_3 + V, \quad \text{or} \quad i_3 = \frac{v_{ao} - V}{R_3}. \tag{4}$$

Substitution of Equations 2, 3, and 4 into Equation 1 results in,

$$\frac{v_{ao}}{R_1} + \frac{v_{ao}}{R_2} + \frac{v_{ao} - V}{R_3} = 0, \tag{5}$$

With some practice, Equation 5 could be written directly from Fig. E-1.2. Equation 5 may be solved for v_{ao}.

$$v_{ao} = \frac{V R_1 R_2}{R_1 R_2 + R_1 R_3 + R_2 R_3}. \tag{6}$$

The remaining voltages v_{ab}, and the currents i_1, i_2, and i_3 may now readily be written.

$$v_{ab} = v_{ao} + v_{ob} = v_{ao} - V = -\frac{V(R_1R_3 + R_2R_3)}{R_1R_2 + R_1R_3 + R_2R_3}. \tag{7}$$

$$i_1 = \frac{v_{ao}}{R_1} = \frac{VR_2}{R_1R_2 + R_1R_3 + R_2R_3} \tag{8}$$

$$i_2 = \frac{v_{ao}}{R_2} = \frac{VR_1}{R_1R_2 + R_1R_3 + R_2R_3} \tag{9}$$

$$i_3 = \frac{v_{ab}}{R_3} = -\frac{V(R_1 + R_2)}{R_1R_2 + R_1R_3 + R_2R_3}. \tag{10}$$

The minus sign for v_{ab} means that b is $+$ with respect to a. The minus sign for i_3 means that the current is actually in the direction opposite to that indicated in Fig. E-1.3.

(b) In the **branch current method or loop method** the first thing to do is to label each branch current in terms of independent unknown currents and known currents. This is done in Fig. E-1.4. The direction assumed positive for the currents is arbitrary; also which currents to consider independent is arbitrary. The choices shown in Fig. E-1.4 are made primarily to agree with currents of part (a).

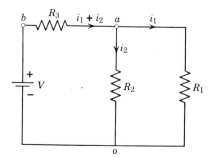

FIG. E-1.4

In this case there are two independent currents, i_1 and i_2, and therefore two independent equations must be written. **One way to ensure that the Kirchhoff's voltage law equations will always be independent is to write these around the paths or "loops" of the independent currents.** We illustrate this in Fig. E-1.5, showing that the branch currents i_1 and i_2 may be considered to be in loops as illustrated. The actual current in R_3 is the sum of the two loop currents, or $i_1 + i_2$ and thus the branch labeling of Fig. E-1.4 may be interpreted also as the loop labeling of Fig. E-1.5. The important concept, however, is that independent voltage equations are assured if these are written around the paths of the loop currents shown in Fig. E-1.5. The student is cautioned against using loop labeling to determine the independent unknown currents. Many mistakes are possible. One way to avoid these mistakes is to label branch currents in terms of unknown independent branch currents and known branch currents. Following the closed path of i_1, we obtain

$$v_{ob} + v_{ba} + v_{ao} = 0,$$

or

$$-V + R_3(i_1 + i_2) + R_1 i_1 = 0. \tag{11}$$

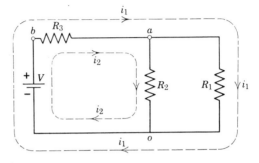

FIG. E-1.5

Following the closed path of i_2, we obtain

$$v_{ob} + v_{ba} + v_{ao} = 0,$$

or

$$-V + R_3(i_1 + i_2) + R_2 i_2 = 0. \tag{12}$$

Equations 11 and 12 may be rearranged into standard form.

$(R_1 + R_3)i_1 + R_3 i_2 = V,$

and

$R_3 i_1 + (R_2 + R_3)i_2 = V.$

These equations are independent and may be solved simultaneously by Cramer's rule to obtain,

$$i_1 = \frac{\begin{vmatrix} V & R_3 \\ V & R_2 + R_3 \end{vmatrix}}{\begin{vmatrix} R_1 + R_3 & R_3 \\ R_3 & R_2 + R_3 \end{vmatrix}} = \frac{V(R_2 + R_3 - R_3)}{(R_1 + R_3)(R_2 + R_3) - R_3^2}$$

$$= \frac{VR_2}{R_1 R_2 + R_1 R_3 + R_2 R_3} \tag{13}$$

and

$$i_2 = \frac{\begin{vmatrix} R_1 + R_3 & V \\ R_3 & V \end{vmatrix}}{\begin{vmatrix} R_1 + R_3 & R_3 \\ R_3 & R_2 + R_3 \end{vmatrix}} = \frac{VR_1}{R_1 R_2 + R_1 R_3 + R_2 R_3} \tag{14}$$

It is important to observe that the matrix in the denominator is symmetric. **For the systematic approaches we are using,** and for all the circuits we shall encounter, except those involving dependent sources, the solution by determinants will yield a symmetric matrix.

The current, $(i_1 + i_2)$, in R_3 is

$$i_1 + i_2 = \frac{V(R_1 + R_2)}{R_1 R_2 + R_1 R_3 + R_2 R_3}. \tag{15}$$

Also

$$v_{ba} = R_3(i_1 + i_2) = \frac{VR_3(R_1 + R_2)}{R_1 R_2 + R_1 R_3 + R_2 R_3}, \tag{16}$$

and

$$v_{ao} = R_1 i_1 = \frac{VR_1 R_2}{R_1 R_2 + R_1 R_3 + R_2 R_3}. \tag{17}$$

These results are the same as those obtained by the node voltage method. We need to be familiar with both methods; for each circuit we would probably choose that method which had the smaller number of independent unknowns.

(c) For $V = 10$ volts, $R_1 = 9\,\Omega$, $R_2 = 1\,\Omega$, and $R_3 = 1.1\,\Omega$, Equations 13 and 14 result in

$$i_1 = \frac{10(1)}{9 + 9.9 + 1.1} = 0.5 \text{ amperes},$$

and

$$i_2 = \frac{10(9)}{20} = 4.5 \text{ amperes}.$$

Then $i_1 + i_2 = 5.0$ amperes,

$$v_{ba} = R_3(i_1 + i_2) = 1.1(5.0) = 5.5 \text{ volts},$$

and

$$v_{ao} = R_1 i_1 = 9(0.5) = 4.5 \text{ volts}.$$

In this example the currents through and the voltages across the resistances changed **immediately** from zero values to some finite value. This is contrary to the field concept that the velocity of propagation is finite. Nevertheless in a number of cases the change is so rapid that we are justified in considering it instantaneous. Just before the switch was closed the current in R_3 was zero because R_3 is in series with the switch. The currents in R_1 and R_2 must also be zero before the switch is closed because Kirchhoff's voltage law may be applied to this part of the network (shown in Fig. E-1.6) to give $i_o(R_1 + R_2) = 0$ or $i_o = 0$.

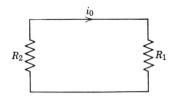

FIG. E-1.6

Example 2.2

GIVEN: The circuit of Fig. E-2.1. The current sources I_1 and I_2 are direct current.

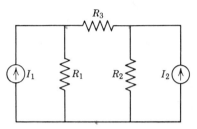

FIG. E-2.1

TO FIND: The solution for the response functions, the unknown voltages and currents, by both the node voltage method and the branch current method using $I_1 = 10$ amp, $I_2 = 5$ amp, $R_1 = 1\ \Omega$, $R_2 = 4\ \Omega$, and $R_3 = 5\ \Omega$.

SOLUTION: For the node voltage method we label nodes as shown in Fig. E-2.2, selecting o as the reference node. There are two independent unknown voltages, v_{ao} and v_{bo}, and we need to write Kirchhoff's current law about nodes a and b. In Fig. E-2.2 the currents in the resistances

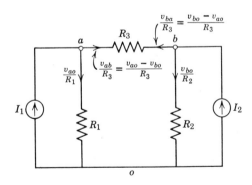

FIG. E-2.2

are indicated positive in the direction away from each node. Kirchhoff's current law about nodes a and b gives,

$$\frac{v_{ao}}{R_1} + \frac{v_{ao} - v_{bo}}{R_3} - I_1 = 0 \tag{1}$$

Basic Concepts of Circuit Theory 31

and

$$\frac{v_{bo}}{R_2} + \frac{v_{bo} - v_{ao}}{R_3} - I_2 = 0 \tag{2}$$

Rearrangement of Equations 1 and 2 to put them in standard form results in

$$\left(\frac{1}{R_1} + \frac{1}{R_3}\right)v_{ao} - \left(\frac{1}{R_3}\right)v_{bo} = I_1 \tag{3}$$

and

$$-\left(\frac{1}{R_3}\right)v_{ao} + \left(\frac{1}{R_2} + \frac{1}{R_3}\right)v_{bo} = I_2 \tag{4}$$

The known quantities may be replaced by their numerical values to give,

$$1.2v_{ao} - 0.2v_{bo} = 10, \tag{5}$$

and

$$-0.2v_{ao} + 0.45v_{bo} = 5. \tag{6}$$

These equations are solved simultaneously to obtain

$$v_{ao} = \frac{\begin{vmatrix} 10 & -0.2 \\ 5 & 0.45 \end{vmatrix}}{\begin{vmatrix} 1.2 & -0.2 \\ -0.2 & 0.45 \end{vmatrix}} = \frac{5.5}{0.5} = 11 \text{ volts}$$

and

$$v_{bo} = \frac{\begin{vmatrix} 1.2 & 10 \\ -0.2 & 5 \end{vmatrix}}{0.5} = \frac{8}{0.5} = 16 \text{ volts}$$

Then the currents $\dfrac{v_{ao}}{R_1} = \dfrac{11}{1} = 11$ amp,

$$\frac{v_{bo}}{R_2} = \frac{16}{4} = 4 \text{ amp},$$

and

$$\frac{v_{ao} - v_{bo}}{R_3} = \frac{11 - 16}{5} = -1 \text{ amp}.$$

The minus sign means that the current is in the direction opposite to that indicated in Fig. E-2.2.

For the branch current method we label independent unknown currents. In Fig. E-2.3 we arbitrarily choose the current in R_1 as the independent current, i. We also arbitrarily choose i as positive in the direction shown. The use of Kirchhoff's current law about each node permits the labeling of all other branch currents.

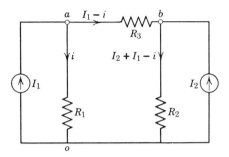

FIG. E-2.3

Figure E-2.4 shows the loops of the currents I_1, I_2, and i. We do not write a voltage equation around the loops of I_1 and I_2 since these are known currents and since such equations would involve unknown voltages, v_{ao} or v_{bo}. We do write Kirchhoff's voltage equation around the path of the independent unknown i. Thus, $v_{ao} + v_{ob} + v_{ba} = 0$,

or

$$iR_1 + (i - I_1 - I_2)R_2 + (i - I_1)R_3 = 0,$$

or

$$(R_1 + R_2 + R_3)i = (R_2 + R_3)I_1 + R_2 I_2. \tag{7}$$

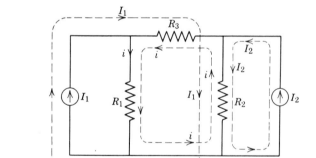

FIG. E-2.4

Equation 7 may be solved for i by substituting numerical values for the known quantities. Thus, $(1 + 4 + 5)i = (4 + 5)10 + 4(5)$,

or

$i = 11$ amp.

Then

$I_1 - i = 10 - 11 = -1$ amp,

$I_2 + I_1 - i = 5 + 10 - 11 = 4$ amp,

$v_{ao} = R_1 i = 1(11) = 11$ volts

Basic Concepts of Circuit Theory 33

and

$$v_{bo} = R_2(I_2 + I_1 - i) = 4(4) = 16 \text{ volts.}$$

These values agree, as they must, with the results obtained by the node voltage method.

Example 2.3

GIVEN: The network of Fig. E-3.1 with the elements indicated.

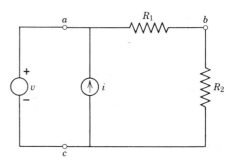

FIG. E-3.1

TO FIND: (a) The literal expressions for the voltage, current, and power for each element in terms of the source voltage v, the source current i, and resistances R_1 and R_2. Use any method.

(b) The numerical expressions and sketches for the current and power for the voltage source if $v = 10 \cos 2t$ volts, $i = 1$ ampere, $R_1 = 1$ ohm, and $R_2 = 4$ ohms.

SOLUTION: (a) In selecting a method for solving the problem we look closely at Fig. E-3.1. Since one current is known there is only one independent unknown current if we were to use the branch current method. If we were to use the node voltage method, and select c as the reference node we note that the voltage v_{ac} is known and that there is apparently only one independent voltage v_{bc}. Since there is no apparent difference, let us use the branch current method. Fig. E-3.2 shows our choice of unknown current and its direction. Kirchhoff's voltage law around the path of i_1 yields,

$$\sum v = 0 \quad \text{or} \quad v_{ca} + v_{ab} + v_{bc} = 0,$$

or

$$-v + R_1 i_1 + R_2 i_1 = 0. \tag{1}$$

Then

$$i_1 = \frac{v}{R_1 + R_2} \tag{2}$$

*Current goes
– to + through
voltage diff.,
then element
gives up energy*

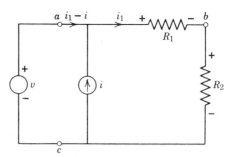

FIG. E-3.2

The voltage source has the voltage v and the current,

$$i_1 - i = \frac{v}{R_1 + R_2} - i,$$

and **supplies to** the system the power,

$$p_v = v(i_1 - i) = \frac{v^2}{R_1 + R_2} - vi. \tag{3}$$

The current source has the current i and the voltage v and **supplies to** the system the power,

$$p_i = vi. \tag{4}$$

The resistance R_1 has the current

$$i_1 = \frac{v}{R_1 + R_2},$$

the voltage

$$v_{ab} = R_1 i_1 = \frac{vR_1}{R_1 + R_2},$$

and **absorbs from** the system the power, *(always)*

$$p_{R_1} = v_{ab} i_1 = \frac{v^2 R_1}{(R_1 + R_2)^2}. \tag{5}$$

The resistance R_2 has the current

$$i_1 = \frac{v}{R_1 + R_2},$$

the voltage

$$v_{bc} = R_2 i_1 = \frac{vR_2}{R_1 + R_2},$$

and absorbs from the system the power,

$$p_{R_2} = v_{bc} i_1 = \frac{v^2 R_2}{(R_1 + R_2)^2}. \tag{6}$$

Basic Concepts of Circuit Theory 35

We note that the sum of the powers supplied to the system is equal to the sum of the powers absorbed from the system.

(b) For the constants given,

$$i_1 = \frac{v}{R_1 + R_2} = \frac{10 \cos 2t}{1 + 4} = 2 \cos 2t \text{ amperes,} \tag{7}$$

$$i_1 - i = 2 \cos 2t - 1 \text{ amperes,} \tag{8}$$

and

$$p_v = v(i_1 - i) = 10 \cos 2t(2 \cos 2t - 1)$$
$$= 20 \cos^2 2t - 10 \cos 2t \text{ watts.} \tag{9}$$

Fig. E-3.3 shows v, $i_1 - i$, and p_v sketched as functions of time for one period of the sinusoidal function.

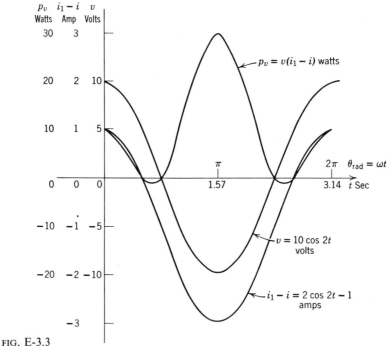

FIG. E-3.3

Observe that the power **supplied to** the system by the voltage source is not always positive; there are time intervals during which power is absorbed by the voltage source.

2.4.4 Self-Inductance, L

The definition of inductance arises from Faraday's law of magnetic induction,

$$\oint \bar{E} \cdot d\bar{l} = \Sigma v = -\frac{d\lambda'}{dt}.$$

We shall first consider a simple circuit of an independent voltage source in series with a loop of metal wire (see Fig. 2.14). The ideal voltage source is considered to be of a sinusoidal form, such as $V \cos \omega t$ volts, and steady state is assumed to exist; that is any temporary disturbances which occurred at the time this circuit was energized are presumed to have disappeared.

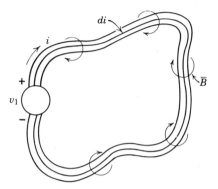

FIG. 2.14 Simple circuit with voltage source and loop of wire.

From Chapter 1 we recall that the velocity of propagation is finite, approximately equal to $1/\sqrt{\mu\epsilon}$. Also we note that frequency, velocity, and wave length are related by

$$\text{wavelength} = \frac{\text{velocity}}{\text{frequency}} \cong \frac{1}{f\sqrt{\mu\epsilon}}.$$

The implications are that the current will vary with distance from the source; in a distance of one half wave length it will vary from a maximum in one direction to a maximum in the other direction. If the largest dimension of our system is very small in comparison with a wave length then it becomes possible to neglect this variation in current and to assume that the same current exists in all parts of our circuit. **This is an important assumption and limitation for circuit theory, since it specifically places an upper limit to the frequency range for which our circuit fiction holds.** In this case it implies also that the displacement current between elements of the wire is negligible in comparison with the conduction current, i.

The conduction current i may be considered to set up a magnetic field which is in such a direction that a certain amount of flux, $\int \bar{B} \cdot d\bar{a}$, will link each filament of current. The implications of Faraday's law of magnetic induction,

$$\oint \bar{E} \cdot d\bar{l} = \sum v = -\frac{d\lambda'}{dt},$$

are that we should follow each differential current filament, di, around

Basic Concepts of Circuit Theory 37

its closed path, adding up all the voltages that are encountered in the plus to minus sense and then setting these equal to the negative time rate of change of the flux which links this path. In doing this we obtain

$$-v_1 + \oint \frac{J}{\sigma} dl = -\frac{d\lambda'}{dt}\bigg|_{\text{for filament } di} \tag{2.5}$$

Thus for each di filament, $\oint J/\sigma \, dl$ and λ' will be different. If each voltage expression be multiplied by the differential current di, $(J \, da)$, for which it is written and these expressions integrated over the cross-section of the conductor a power expression results which is written:

$$-v_1 \int_a J \, da + \int_a \oint \frac{J^2}{\sigma} dl \, da = -\int_a \frac{d\lambda'}{dt} J \, da,$$

or

$$-v_1 i + \int_\tau \frac{J^2}{\sigma} d\tau = -\int \frac{d\lambda'}{dt} di. \tag{2.6}$$

In performing the integration the differential length dl must have the same direction as J at any point and the differential area da must be normal to J. We recognize the second term of the equation as Ri^2. We define $d\lambda/dt$ as the average value of $d\lambda'/dt$ for the di filaments, or

$$\frac{d\lambda}{dt} = \frac{\int_i \frac{d\lambda'}{dt} di}{i}. \tag{2.7}$$

Equation 2.6 becomes

$$-v_1 i + Ri^2 = -i\frac{d\lambda}{dt},$$

or

$$v_1 i = Ri^2 + i\frac{d\lambda}{dt}. \tag{2.8}$$

The term $v_1 i$ represents the power being supplied by the voltage source v_1, the term Ri^2 represents the power being dissipated as heat, and the term $i(d\lambda/dt)$ represents power being supplied to the magnetic field. The flux linkage, λ, is an average of the values of flux linkages, λ', for the di filaments. For a given frequency in linear media, λ is directly proportional to the current. The constant of proportionality is called self-inductance, L,[1] with the units of webers per ampere or henries.

$$L = \frac{\lambda}{i} \text{ henries.} \tag{2.9}$$

---

[1] It should be pointed out that the integrations indicated in Equations 2.6 or 2.11 are not easy for physical circuits. Calculations of inductance are therefore postponed until the student has had a good introduction to electromagnetic field theory.

An alternate definition, which is identical in theory, is obtained by using the volume integral, $\int \frac{1}{2}\mu H^2 \, d\tau$, as the energy stored in the magnetic field and equating this to the integral of the power being supplied to the magnetic field. Thus,

$$W_m = \tfrac{1}{2}\int \mu H^2 \, d\tau = \int_{-\infty}^{t} p \, dt = \int_{-\infty}^{t} i\frac{d\lambda}{dt} \, dt = \int_{0}^{i} Li \, di = \tfrac{1}{2}Li^2, \quad (2.10)$$

or

$$L = \frac{\int \mu H^2 \, d\tau}{i^2}. \quad (2.11)$$

Equation 2.8 may be written

$$\sum v = -v_1 + Ri = -\frac{d\lambda}{dt} = -L\frac{di}{dt}. \quad (2.12)$$

The sum of all the voltages found in traversing the circuit is not zero but equal to $-L(di/dt)$. This means that wherever there is a magnetic field changing with time the voltage difference or potential difference between two points, $\int \bar{E} \cdot d\bar{l}$, is a function of the path taken between these points. If our circuit fiction is to be useful it is highly desirable to have the voltage difference between points in our circuit independent of path. This can be done by **arbitrarily** introducing the voltage $-(d\lambda/dt)$ or $-L(di/dt)$ into our circuit. Thus our equivalent circuit for Fig. 2.14 becomes that of Fig. 2.15. The equation for this circuit is in accord with Equation 2.12 but now we write Faraday's law of magnetic induction as Kirchhoff's voltage law.

$$\sum v = 0,$$
$$v_{ca} + v_{ab} + v_{bc} = 0,$$

or

$$-v_1 + Ri + L\frac{di}{dt} = 0. \quad (2.13)$$

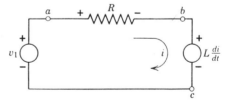

FIG. 2.15 An equivalent circuit for Fig. 2.14.

The introduction of $L(di/dt)$ as a circuit voltage is a fiction which permits us to sketch the circuit shown and to define circuit potential as a scalar quantity, so that the voltage difference between points is independent of path.

It is customary to replace the voltage source $L(di/dt)$ with the element inductance, having the symbol L, so Fig. 2.15 is redrawn as shown in Fig. 2-16a. The polarity of $v_{bc} = L(di/dt)$ is plus to minus in the assumed positive direction of current flow when the derivative of this current with respect to time is positive. Note that the current i may be solved for as

$$i = \frac{1}{L}\int_{-\infty}^{t} v_{bc}\, dt,$$

in which the lower limit, $-\infty$, is used to indicate that all voltages across the element which have ever appeared in this circuit must be considered.

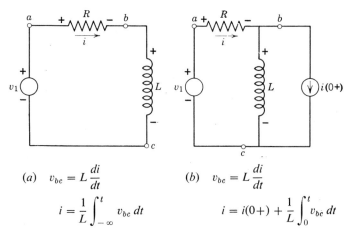

(a) $v_{bc} = L\dfrac{di}{dt}$

$i = \dfrac{1}{L}\int_{-\infty}^{t} v_{bc}\, dt$

(b) $v_{bc} = L\dfrac{di}{dt}$

$i = i(0+) + \dfrac{1}{L}\int_{0}^{t} v_{bc}\, dt$

FIG. 2.16 Equivalent circuits for Fig. 2.14.

It is generally convenient to use time equal to zero as the time of some switching operation and to use $i(0+)$ to represent the current immediately after the switching operation. Then

$$i = \frac{1}{L}\int_{-\infty}^{0+} v_{bc}\, dt + \frac{1}{L}\int_{0+}^{t} v_{bc}\, dt = i(0+) + \frac{1}{L}\int_{0+}^{t} v_{bc}\, dt, \qquad (2.14)$$

and a second equivalent circuit, shown in Fig. 2.16b, may be used. Note that this second circuit has the advantage that the initial condition, $i(0+)$, may be shown directly on the circuit diagram and appears as an independent current source of fixed magnitude. In this circuit the current through the element L itself is zero at time $0+$. Equation 2.14 is a statement of Kirchhoff's current law as applied to Fig. 2.16b at junctions b or c.

The reason that we distinguish between $i(0-)$ and $i(0+)$ is that, for idealized circuit conditions, there may be cases in which these are not equal. $i(0-)$ is defined as the limit $i(0-\epsilon)$ and $i(0+)$ is similarly
$\epsilon \to 0$

40

defined as the limit $i(0 + \epsilon)$, ϵ being a finite positive value of time. We could write Equation 2.14 as

$$i = \frac{1}{L}\int_{-\infty}^{t} v_{bc}\, dt = \frac{1}{L}\int_{-\infty}^{0-\epsilon} v_{bc}\, dt + \frac{1}{L}\int_{0-\epsilon}^{0+\epsilon} v_{bc}\, dt + \frac{1}{L}\int_{0+\epsilon}^{t} v_{bc}\, dt$$

The limit of this equation as $\epsilon \to 0$ is

$$i = \frac{1}{L}\int_{-\infty}^{t} v_{bc}\, dt = \frac{1}{L}\int_{-\infty}^{0-} v_{bc}\, dt + \frac{1}{L}\int_{0-}^{0+} v_{bc}\, dt + \frac{1}{L}\int_{0+}^{t} v_{bc}\, dt$$

$$= i(0-) + \frac{1}{L}\int_{0-}^{0+} v_{bc}\, dt + \frac{1}{L}\int_{0+}^{t} v_{bc}\, dt \qquad (2.15)$$

Then

$$i(0+) = i(0-) + \frac{1}{L}\int_{0-}^{0+} v_{bc}\, dt, \qquad (2.16)$$

and $i(0+)$ will be equal to $i(0-)$ if $\frac{1}{L}\int_{0-}^{0+} v_{bc}\, dt = 0$. From the standpoint of time there is no differences between $t = 0-$, $t = 0$, and $t = 0+$ and thus the integral will be zero unless v_{bc} is infinitely large. In the latter case it is possible for the integral $\int_{0-}^{0+} v_{bc}\, dt$ to be finite. This is called a voltage impulse since it is analogous to a force impulse. We shall postpone a discussion of impulses till Chapter 9 and shall temporarily avoid circuits in which these may occur.

In the case of a single loop of wire it is fairly easy to picture the flux which links this loop but in the case of more complex circuits, flux linkages are also more complicated. Consider the circuit of Fig. 2.17 in which the loop has several turns. The area bounded by the conductor is a fairly complex surface and the differential flux, $d\phi_1$, links this surface three times whereas $d\phi_2$ links it only once.

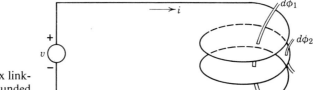

FIG. 2.17 Flux linking the area bounded by a closed loop.

Inductors are physical devices which are constructed to have primarily the property of inductance. They usually consist of insulated copper wire wound in a coil. This coil sometimes has a core of ferromagnetic material whose primary purpose is to increase the magnitude of the inductance. This ferromagnetic material may be a nonconducting ferrite or may be laminated or powdered iron. The iron is laminated or powdered to reduce the "eddy" currents which would otherwise exist in the iron. The magnetic flux in the iron is not a linear function of current in the coil because of hysteresis and saturation; however, we shall in this text assume that there is a linear relation and that therefore inductance is a constant.

An inductor may be represented at low frequencies by an inductance in series with a resistance; at higher frequencies the displacement currents become appreciable and therefore this simple representation no longer is adequate.

The prefix m is used to represent milli or 10^{-3}; the prefix μ is used to represent 10^{-6}. Thus 0.005 henries = 5 mh = 5000 μh.

2.4.4.1 *Circuits with Inductance, Resistance, and Sources.* Circuits which consist only of resistances and sources are relatively easy to solve since the currents through and the voltage difference across the resistors can change immediately with time. In the case of inductance the voltage-current relationship, $v = L(di/dt)$, implies that the current can not change instantaneously with finite voltages. This relationship also tells us that solutions are relatively easy if either the current through or the voltage across the inductance element are known. For the many cases in which this is not true, the equations to be solved are differential equations.

In the following three examples, the first two illustrate cases in which the current through the inductance is known and results can be readily calculated; the last example demonstrates the writing of a simple differential equation.

Example 2.4

GIVEN: The circuit of Fig. E-4.1 with $i = I \cos \omega t$ amperes.

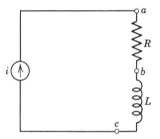

FIG. E-4.1

TO FIND: (a) The expressions for the voltages v_{ab}, v_{bc}, and v_{ac} in literal form.

(b) Evaluate these voltages if $I = 10$ amp, $R = 4$ ohms, $L = 2$ henries, $\omega = 2$ radians per second. Sketch the voltages as a function of time.

SOLUTION: (a) We do not consider the node voltage method suitable for this problem because there are two independent unknown voltages. Since the current through each of the elements is known, the voltages are readily calculated from the volt-ampere relationship of the elements.

$$v_{ab} = Ri = RI \cos \omega t, \tag{1}$$

$$v_{bc} = L\frac{di}{dt} = L\frac{d}{dt}(I \cos \omega t) = -\omega LI \sin \omega t$$

$$= \omega LI \cos(\omega t + 90°), \tag{2}$$

and

$$v_{ac} = v_{ab} + v_{bc} = RI \cos \omega t - \omega LI \sin \omega t. \tag{3}$$

Equation 3 may be written:

$$v_{ac} = \sqrt{R^2 + \omega^2 L^2}\, I \cos\left(\omega t + \tan^{-1}\frac{\omega L}{R}\right). \tag{4}$$

The student should show that Equations 3 and 4 are equal by expanding Equation 4, and should note that all voltages vary sinusoidally with the same frequency but differ in phase angle. The phase angle may be defined as the angle ϕ if the sinusoidal function is written in the form, $A \cos(\omega t + \phi)$, A being a positive number.

(b) With the values given, the voltages are written:

$$v_{ab} = RI \cos \omega t = 40 \cos 2t \text{ volts}, \tag{5}$$

$$v_{bc} = \omega LI \cos(\omega t + 90°) = 40 \cos(2t + 90°) \text{ volts}, \tag{6}$$

and

$$v_{ac} = \sqrt{R^2 + \omega^2 L^2}\, I \cos\left(\omega t + \tan^{-1}\frac{\omega L}{R}\right),$$

$$= 56.6 \cos(2t + 45°) \text{ volts}. \tag{7}$$

These are sketched in Fig. E-4.2.

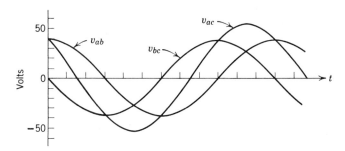

FIG. E-4.2

Example 2.5

GIVEN: The circuit shown in Fig. E-5.1 with $R = 5$ ohms, $L = 2$ henries, and the current specified by the graph shown in Fig. E-5.2.

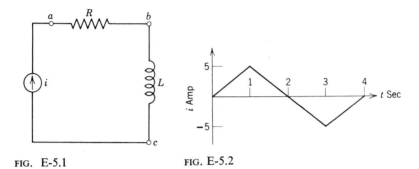

FIG. E-5.1   FIG. E-5.2

TO FIND: (a) Sketches of v_{ab} and v_{bc} versus time.
(b) Analytical expressions for v_{ab} and v_{bc} as functions of time.
(c) Evaluation of the magnitude and direction of the power for each element at $t = 2.5$ seconds.
(d) Evaluation of the energy stored in the field of the inductance at $t = 2.5$ seconds.

SOLUTION: (a) The sketch of v_{ab} versus t is obtained by noting that v_{ab} is directly proportional to i. The sketch of v_{bc} vs. t is obtained by noting that v_{bc} is directly proportional to the derivative of i with respect to t or proportional to the slope of the function i vs. t. These sketches are shown in Fig. E-5.3. The voltage scales and analytical expressions are written in later, from the solution to part (b).

(b) In the interval $0 < t < 1$, $i = 5t$ amperes,

$$v_{ab} = Ri = 5(5t) = 25t \text{ volts,}$$

and

$$v_{bc} = L\frac{di}{dt} = 2\frac{d}{dt}(5t) = 2(5) = 10 \text{ volts.}$$

In the interval $1 < t < 3$, $i = -5t + 10$ amperes,

$$v_{ab} = Ri = 5(-5t + 10) = -25t + 50 \text{ volts,}$$

and

$$v_{bc} = L\frac{di}{dt} = 2\frac{d}{dt}(-5t + 10) = -10 \text{ volts.}$$

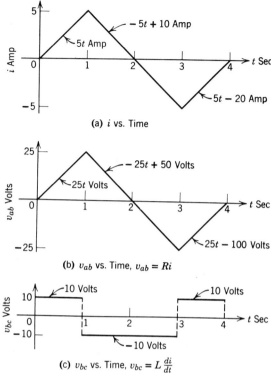

FIG. E-5.3 Waveforms of voltage and current for Fig. E-5.1.

In the interval $3 < t < 4$, $i = 5t - 20$ amperes,
$v_{ab} = Ri = 5(5t - 20) = 25t - 100$ volts,
and
$$v_{bc} = L\frac{di}{dt} = 2\frac{d}{dt}(5t - 20) = 10 \text{ volts.}$$

(c) At $t = 2.5$ seconds,
$i = -5t + 10 \big|_{t=2.5} = -12.5 + 10 = -2.5$ amp,
$v_{ab} = Ri = 5(-2.5) = -12.5$ volts,
and
$v_{bc} = -10$ volts.

The power into the resistance at $t = 2.5$ seconds is
$p_R = v_{ab}i = -12.5(-2.5) = 31.25$ watts.

The power into the field of the inductance from the system is
$p_L = v_{bc}i = -10(-2.5) = 25$ watts.

The power supplied to the system by the current source is

$$p_i = v_{ac}i = (v_{ab} + v_{bc})i = (-12.5 - 10)(-2.5) = +56.25 \text{ watts}.$$

(d) The energy stored in the field of the inductance at $t = 2.5$ seconds is

$$w_m = \tfrac{1}{2}Li^2 = \tfrac{1}{2}(2)(-2.5)^2 = 6.25 \text{ joules}.$$

Example 2.6

GIVEN: The circuit of Fig. E-6.1 with the switch closed at $t = 0$.

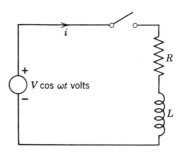

FIG. E-6.1.

TO FIND: The equation which needs to be solved to determine the current i for $t > 0$.

SOLUTION: With the switch closed the circuit is shown in Fig. E-6.2. Kirchhoff's voltage law for the circuit is

$$-V\cos \omega t + Ri + L\frac{di}{dt} = 0,$$

or

$$L\frac{di}{dt} + Ri = V\cos \omega t. \tag{1}$$

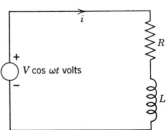

FIG. E-6.2.

This equation is a differential equation of first order with constant coefficients. The current at $t = 0+$ must equal zero since $i(0-) = 0$ and since no voltage impulse is present. The technique for solving this equation will be shown in Chapter 3.

46

2.4.5 Capacitance C

In the discussion of self-inductance, L, for a loop of wire the displacement current between elements of the wire was neglected in comparison with the conduction current. Fortunately this is a good approximation when the largest dimension of the circuit is small in comparison with the wavelength. Now, let us place into this circuit two metal plates separated by a thin insulating material, the **largest dimension of the plates being small in comparison with a wavelength.** This is illustrated in Fig. 2.18.

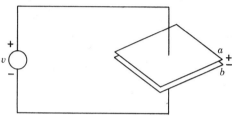

FIG. 2.18 Circuit with insulating material between plates a and b.

Under this circumstance the voltage v_{ab}, as determined by integrating $\bar{E} \cdot d\bar{l}$ between the metal plates and in the insulating material, is approximately constant and there are equal, but opposite, charges on the two conducting plates. **Capacitance is defined as the positive ratio of the charge on one conducting plate to the voltage difference between the two plates.** It is pictured as a circuit element either by ⊣⊢ or by ⊣⊦.

The volt-ampere relationship is

$$C = \frac{q}{v_{ab}} = \frac{\int_{-\infty}^{t} i \, dt}{v_{ab}} \tag{2.17}$$

or
$$v_{ab} = \frac{1}{C} \int_{-\infty}^{t} i \, dt \tag{2.18}$$

or
$$i = C \frac{dv_{ab}}{dt}, \tag{2.19}$$

the current i being a displacement current between the plates and a conduction current in the leads to the plates.

Figure 2.19a shows the capacitance element with reference current, charge, and field intensity $\bar{E}$ shown. Because of our definition for voltage difference and because of the convention that electric flux leaves a positive charge, $dv = -\bar{E} \cdot d\bar{l}$. Then the voltage difference, v_{ab}, is

$$v_{ab} = v_a - v_b = \int_{v_b}^{v_a} dv = -\int_{b}^{a} \bar{E} \cdot d\bar{l}.$$

If the two metal plates have an area a, a uniform spacing d, and if the spacing d is very small in comparison with the area dimensions of

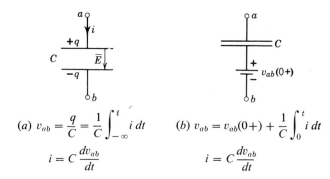

FIG. 2.19 Circuits and volt-ampere relationship for capacitance, C.

the plates, the expression for capacitance may be readily determined because E may be considered uniform. The voltage v_{ab} is

$$v_{ab} = -\int_d^0 \bar{E} \cdot d\bar{l} = Ed \tag{2.20}$$

The electric flux, ψ, between the two plates is equal to the charge on one plate (see Section 1.3, part b). Thus

$$q = \psi = Da = \epsilon Ea, \tag{2.21}$$

in which ϵ is the permittivity of the dielectric.

From Equations 2.17, 2.20, and 2.21 the capacitance of two parallel plates is,

$$C = \frac{q}{v_{ab}} = \frac{\epsilon E a}{Ed} = \frac{\epsilon a}{d}. \tag{2.22}$$

In addition to the parallel-plate case, a few other geometries permit easy calculation of capacitance. However, the majority of geometries require either advanced mathematical techniques or measurements. For any given geometry there will be some frequency above which the previous type of calculations no longer hold and the system can no longer be represented by a single element such as capacitance. This occurs when the dimensions involved are no longer small in comparison with the wavelength. Another way of stating the same thing is to say that the inductive effects become large enough so that the integral, $\int \bar{E} \cdot d\bar{l}$, between the two conductors becomes a function of the path between the plates.

Since switching operations are usually said to occur at $t = 0$, we may write the equation for v_{ab} (Equation 2.18) as follows.

$$v_{ab} = \frac{1}{C}\int_{-\infty}^{t} i\, dt = v_{ab}(0-) + \frac{1}{C}\int_{0-}^{0+} i\, dt + \frac{1}{C}\int_{0+}^{t} i\, dt$$

$$= v_{ab}(0+) + \frac{1}{C}\int_{0+}^{t} i\, dt \tag{2.23}$$

This equation leads to the equivalent circuit shown in Fig. 2.19b which has the initial voltage across the capacitance as an independent voltage source in series with a capacitance C which is uncharged at time 0+. We distinguish between $v_{ab}(0-)$ and $v_{ab}(0+)$ because our idealized circuit conditions lead to cases in which $\int_{0-}^{0+} i\, dt \neq 0$, or to current impulses. As we did for voltage impulses, we postpone discussion of these current impulses until Chapter 9 and shall temporarily avoid situations in which these occur.

Capacitance may also be defined in terms of the energy stored in the electric field. From Equation 2.19 for two conductors separated by a dielectric,

$$i = C \frac{dv_{ab}}{dt}.$$

Then $\quad p = v_{ab} i = C v_{ab} \dfrac{dv_{ab}}{dt},$

and the energy stored in the electric field is

$$w_e = \int_{-\infty}^{t} p\, dt = \int_{0}^{v_{ab}} C v_{ab}\, dv_{ab} = \tfrac{1}{2} C v_{ab}^2. \tag{2.24}$$

This expression may be equated to $\tfrac{1}{2} \int \epsilon E^2\, d\tau$ to give

$$C = \frac{\int \epsilon E^2\, d\tau}{v_{ab}^2}.$$

Capacitors are devices which are constructed to have primarily the property of capacitance. They usually consist of two strips of metal foil separated from each other by a thin sheet of insulation and wrapped in a coil to minimize space requirements.

The symbol μ is used to represent 10^{-6} and the symbol $\mu\mu$ or p (for pico) is used to represent 10^{-12}. Thus $0.005(10^{-6})$ farads $= 0.005\ \mu f = 5000\ \mu\mu f = 5000$ pf.

At low values of frequency a capacitor is often represented as a pure capacitance, unless it is important to consider power losses in which case a capacitor may be represented by the circuit of Fig. 2.20, in which R_1 is used to account for the power loss in the conductive parts and R_2 is used to account for the power loss in the insulation. At higher frequencies the inductive effects become appreciable and this simple representation no longer holds.

.4.5.1 *Circuits with Capacitance, Inductance, Resistance and Sources.* The writing of equations for circuits containing capacitance can best be demonstrated with examples.

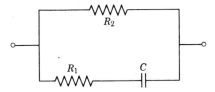

FIG. 2.20 Equivalent circuit for capacitor.

Example 2.7

GIVEN: The current i (same as in Example 2.5) flowing through the capacitance, $C = 0.5$ farads, as shown in Fig. E-7.1. $v_{ab}(0+) = 0$.

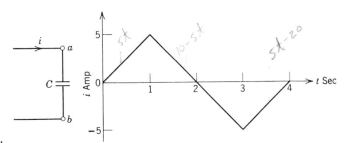

FIG. E-7.1.

TO FIND: (a) The voltage v_{ab} as a function of time.
(b) A sketch of voltage v_{ab} versus time.
(c) The energy stored in the field of the capacitance at $t = 2.5$ seconds.

SOLUTION: (a) For the interval $0 < t < 1$

$i = 5t$,

$$v_{ab} = \frac{1}{C} \int_0^t i\, dt + v_{ab}(0+),$$

$$= \frac{1}{0.5} \int_0^t 5t\, dt + 0 = \frac{5t^2}{0.5(2)} \bigg|_0^t = 5t^2 \text{ volts.}$$

At $t = 1$ second, $v_{ab}(1) = 5$ volts. For the interval $1 < t < 3$,

$i = -5t + 10$ amperes,

$$v_{ab} = \frac{1}{C} \int_1^t i\, dt + v_{ab}(1) = \frac{1}{0.5} \int_1^t (-5t + 10)\, dt + 5$$

$$= -5t^2 \bigg|_1^t + 20t \bigg|_1^t + 5 = -5t^2 + 20t - 10 \text{ volts.}$$

At $t = 3$ seconds, $v_{ab}(3) = -5(3)^2 + 20(3) - 10 = 5$ volts. For the interval $3 < t < 4$,

$$i = 5t - 20 \text{ amp.}$$

$$v_{ab} = \frac{1}{C} \int_3^t i\, dt + v_{ab}(3)$$

$$= \frac{1}{0.5} \int_3^t (5t - 20)\, dt + 5$$

$$= 5t^2 \big|_3^t - 40t \big|_3^t + 5 = 5t^2 - 40t + 80 \text{ volts.}$$

(b) The sketch of i and v_{ab} are shown in Fig. E-7.2. The sketch v_{ab} may be obtained from the analysis or may be obtained by considering v_{ab} as a constant $\left(\dfrac{1}{C}\right)$ times the area of the i vs. t curve.

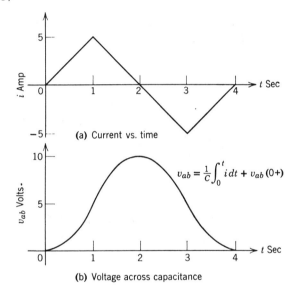

FIG. E-7.2.

(a) Current vs. time

(b) Voltage across capacitance

(c) At $t = 2.5$, $v_{ab} = -5t^2 + 20t - 10 \big|_{t=2.5} = 8.75$ volts. The energy stored in the field of the capacitance at $t = 2.5$ seconds is

$$w = \tfrac{1}{2}Cv^2 = \tfrac{1}{2}(0.5)(8.75)^2 = 19.1 \text{ joules.}$$

Example 2.8

GIVEN: The circuit of Fig. E-8.1 with $V \cos \omega t$, R, L, C, $i(0+)$ and $v_{ab}(0+)$ known.

TO FIND: The equation whose solution will give i for $t > 0$.

Basic Concepts of Circuit Theory 51

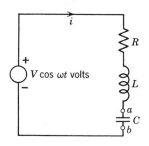

FIG. E-8.1.

SOLUTION: Kirchhoff's voltage law applied to the circuit gives

$$-V\cos \omega t + Ri + L\frac{di}{dt} + \frac{1}{C}\int_0^t i\, dt + v_{ab}(0+) = 0.$$

or $\quad L\dfrac{di}{dt} + Ri + \dfrac{1}{C}\displaystyle\int_0^t i\, dt = V\cos \omega t - v_{ab}(0+)$ \hfill (1)

Equation 1 is an integro-differential equation which may be differentiated to give a differential equation of second order.

$$L\frac{d^2 i}{dt^2} + R\frac{di}{dt} + \frac{i}{C} = -\omega V \sin \omega t \tag{2}$$

The technique for solving this equation will be shown in Chapter 3.

Note that the branch current method used has only one unknown current. If the node voltage method had been used there would have been two unknown node voltages.

Example 2.9

GIVEN: The circuit of Fig. E-9.1 with the switch closed at $t = 0$.

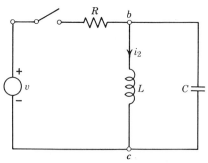

FIG. E-9.1.

TO FIND: (a) By the node-voltage method, the equation whose solution will give v_{bc}.

(b) By the branch current method, the equation whose solution will give i_2.

SOLUTION: (*a*) Fig. E-9.2 shows the currents about the node *b* labeled in terms of the voltage v_{bc}. Kirchhoff's current law applied to node *b* results in

$$C\frac{dv_{bc}}{dt} + \frac{v_{bc} - v}{R} + \frac{1}{L}\int_0^t v_{bc}\, dt + i_2(0+) = 0,$$

or

$$C\frac{dv_{bc}}{dt} + \frac{v_{bc}}{R} + \frac{1}{L}\int_0^t v_{bc}\, dt = \frac{v}{R} - i_2(0+). \tag{1}$$

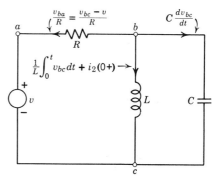

FIG. E-9.2.

Equation 1 is an integro-differential equation which may be differentiated to obtain

$$C\frac{d^2v_{bc}}{dt^2} + \frac{1}{R}\frac{dv_{bc}}{dt} + \frac{v_{bc}}{L} = \frac{1}{R}\frac{dv}{dt}. \tag{2}$$

Equation 2 is a differential equation whose solution will give v_{bc}. The technique for doing this is given in Chapter 3.

(*b*) Fig. E-9.3 shows the circuit resketched with branch currents labeled.

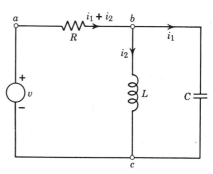

FIG. E-9.3.

Kirchhoff's voltage law applied around the paths of the currents gives

$$v = R(i_1 + i_2) + L\frac{di_2}{dt}, \tag{3}$$

Basic Concepts of Circuit Theory 53

and
$$v = R(i_1 + i_2) + \frac{1}{C}\int_0^t i_1\, dt + v_{bc}(0+). \tag{4}$$

These are the two independent equations which may be solved simultaneously to obtain a solution for i_2. In this case we note that Equation 3 may be solved for i_1.

$$i_1 = \frac{v - Ri_2 - L\dfrac{di_2}{dt}}{R}. \tag{5}$$

Equation 5 may be substituted into Equation 4, but it is easier to first differentiate Equation 4 to obtain

$$\frac{dv}{dt} = R\frac{di_1}{dt} + R\frac{di_2}{dt} + \frac{i_1}{C}. \tag{6}$$

Now, substitution of Equation 5 into Equation 6 gives the equation,

$$\frac{d^2 i_2}{dt^2} + \frac{1}{RC}\frac{di_2}{dt} + \frac{i_2}{LC} = \frac{v}{RLC}. \tag{7}$$

The technique for solving this equation will be shown in Chapter 3.

2.4.6 Mutual Inductance, M

If two electrical circuits are near each other, some of the flux set up by one circuit will link the second circuit. This is demonstrated in Fig. 2.21.

In Fig. 2.21 a core is shown so that the nature of the magnetic coupling between the two circuits may be illustrated. For the assumed positive direction of currents shown, the right hand rule relating current and magnetic flux shows that the flux linkages of each circuit is larger because of the flux set up by the other circuit. We observe also that there is flux which links only one of the circuits. Let

λ_{11} = flux linkages of circuit 1 set up by current i_1,
λ_{12} = flux linkages of circuit 1 set up by current i_2,
λ_{22} = flux linkages of circuit 2 set up by current i_2,
and λ_{21} = flux linkages of circuit 2 set up by current i_1.

Faraday's law of magnetic induction applied to circuit 1 results in an expression equivalent to Equation 2.12 except that linkages resulting from both i_1 and i_2 must be considered. Thus,

$$-v_1 + R_1 i_1 = -\frac{d\lambda}{dt} = -\frac{d}{dt}(\lambda_{11} + \lambda_{12}), \tag{2.25}$$

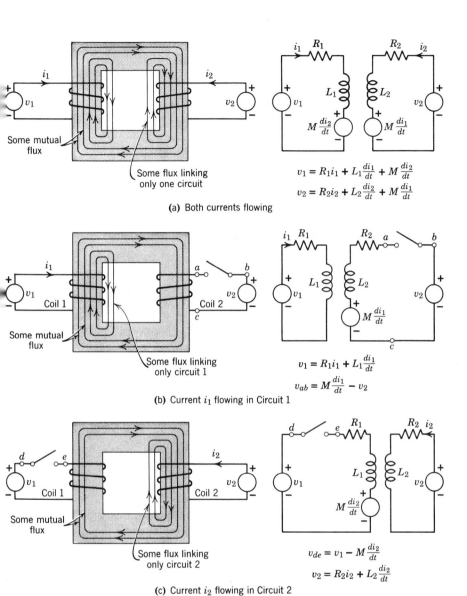

FIG. 2.21 Two circuits coupled magnetically, their electrical circuits, and their voltage equations.

the + sign between λ_{11} and λ_{12} being used because these flux linkages add for our assumed directions for the currents. Similarly, for circuit 2,

$$-v_2 + R_2 i_2 = -\frac{d}{dt}(\lambda_{22} + \lambda_{21}). \tag{2.26}$$

From the definition for self-inductance (Equation 2.9), let

$$L_1 = \frac{\lambda_{11}}{i_1}, \quad \text{and} \quad L_2 = \frac{\lambda_{22}}{i_2}.$$

Again define "mutual inductance," M_{12} and M_{21}, such that

$$M_{12} = \frac{\lambda_{12}}{i_2}, \quad \text{and} \quad M_{21} = \frac{\lambda_{21}}{i_1}.$$

Equations 2.25 and 2.26 become Kirchhoff's voltage equations,

$$v_1 = R_1 i_1 + L_1 \frac{di_1}{dt} + M_{12} \frac{di_2}{dt}, \tag{2.27}$$

and

$$v_2 = R_2 i_2 + L_2 \frac{di_2}{dt} + M_{21} \frac{di_1}{dt}. \tag{2.28}$$

The circuits for these equations are shown in Fig. 2.21a. The sign of the mutually induced voltage term is the same as the self-induced voltage term because the assumed positive directions for the currents is such that the flux linkages of each circuit is increased by current in the other circuit. If one of the currents in Fig. 2.21 had been assumed positive in the opposite direction, the net flux linkages would have been $\lambda_{11} - \lambda_{12}$ and $\lambda_{22} - \lambda_{21}$ and thus the sign of the mutually induced voltage term would have been opposite to the sign of the self induced term.

From an energy standpoint it can be shown that $M_{12} = M_{21} = M$ (see Appendix A). The magnitude of M may be varied by changing the physical position of one of the circuits with respect to the other. It is convenient to define a coefficient of coupling, k, as

$$k = \frac{M}{\sqrt{L_1 L_2}}.$$

This coefficient of coupling, k, will have a value less than one for physically realizable circuits (see Appendix A). However, with ferromagnetic cores it is possible to have values of k larger than 0.99.

Suppose there is no current actually flowing in one of the circuits. How may we then determine the polarity of the induced voltage term, $M(di/dt)$? One method is to **assume** a current and thus establish the proper polarity. Another method is to use a conservation of energy relationship known as Lenz's law which may be stated as follows. If the magnetic flux linking a circuit is changing with time the polarity of the induced voltage is such as **to tend to** cause current to flow in a direction which will oppose the change in flux. We demonstrate this with Fig. 2.22, which is essentially Fig. 2.21b redrawn. It is convenient to assume that i_1 is increasing positively with time since this makes di_1/dt numerically positive. The magnetic flux is proportional to the current and is therefore increasing positively with time in the direction shown. From the right hand rule relating current and magnetic flux, the current i_2 must **tend to** flow in the direction indicated in order to set up magnetic flux in a direction opposite to the change. This current i_2 causes the polarity across R_0 to be as indicated. The resistance R_0 need not be shown but is a convenient "crutch" to help us determine the polarity.

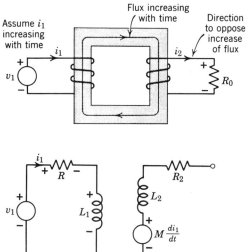

FIG. 2.22 Lenz's law applied to two circuits.

Figures 2.21 and 2.22 show the nature of the magnetic coupling of the two coils by indicating the way in which they are wound about a core. It is generally more convenient to indicate the coupling by the use of dot marking, the **convention being that if the direction of the current through the two coils is in the same sense with regard to the dot markings, the flux linkages are additive and the sign of the mutually induced voltage is the same as the self-induced voltage term.**

Figure 2.23 demonstrates how the dot markings may be used in determining the proper sign of the voltage term. Note that for i_1 in the direction shown the polarity of $M(di_1/dt)$ is independent of i_2. Furthermore, note that i_1 could have been indicated as positive in the opposite sense. If this were done in Fig. 2.23a the equation for the

Basic Concepts of Circuit Theory 57

circuits would be

$$-v_1 = R_1 i_1 + L_1 \frac{di_1}{dt} + M \frac{di_2}{dt}, \qquad (2.29)$$

$$0 = R_2 i_2 + L_2 \frac{di_2}{dt} + M \frac{di_1}{dt}. \qquad (2.30)$$

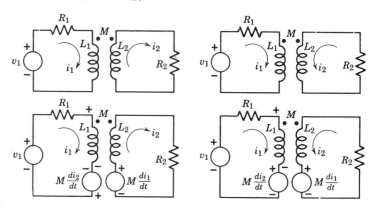

FIG. 2.23 Magnetically coupled coils with dot markings, equivalent circuits, and equations.

(a) One current into dot, other current out of dot.

(b) Both currents into dots.

$$v_1 = R_1 i_1 + L_1 \frac{di_1}{dt} - M \frac{di_2}{dt} \qquad v_1 = R_1 i_1 + L_1 \frac{di_1}{dt} + M \frac{di_2}{dt}$$

$$0 = R_2 i_2 + L_2 \frac{di_2}{dt} - M \frac{di_1}{dt} \qquad 0 = R_2 i_2 + L_2 \frac{di_2}{dt} + M \frac{di_1}{dt}$$

If more than two coils are mutually coupled it becomes necessary to use more than one type of dot. Fig. 2.24 shows three coils which are mutually coupled as indicated. The (●) dot marking shows the mutual M_{12} (between coils 1 and 2), the square (■) marking shows the mutual M_{13} (between coils 1 and 3), whereas the triangle (▲) marking shows the mutual M_{23} (between coils 2 and 3). Fig. 2.25 shows the coils of Fig. 2.24 with assumed positive directions for currents and polarities for voltages.

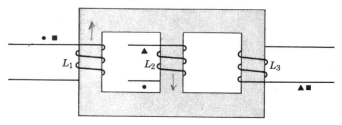

FIG. 2.24 Three coils showing magnetic coupling and dot markings.

If coil No. 1 is assumed to have a resistance R_1, coil No. 2 a resistance R_2, and coil No. 3 a resistance R_3, the voltage equations for the coils of Fig. 2.25 may be written:

$$v_1 = R_1 i_1 + L_1 \frac{di_1}{dt} - M'_{12} \frac{di_2}{dt} + M'_{13} \frac{di_3}{dt},$$

$$v_2 = R_2 i_2 + L_2 \frac{di_2}{dt} - M'_{21} \frac{di_1}{dt} + M'_{23} \frac{di_3}{dt},$$

and

$$v_3 = R_3 i_3 + L_3 \frac{di_3}{dt} + M'_{31} \frac{di_1}{dt} + M'_{32} \frac{di_2}{dt},$$

in which $M_{12} = M_{21}$, $M_{13} = M_{31}$, and $M_{23} = M_{32}$.

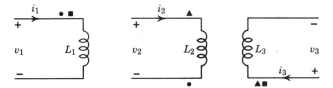

FIG. 2.25 Coils of Fig. 2.24 with assumed positive voltages and currents.

It would appear thus far that the quantity $M(di/dt)$ is a dependent voltage source since it is normally a function of a current in some other circuit. We prefer at this time not to consider these voltages as voltage sources, but to consider M as a circuit element.

Example 2.10

GIVEN: The circuit shown in Fig. E-10.1 which might be used to represent a two winding transformer. Let the resistance R_1 be 10 ohms (10 Ω); the resistance R_2 be 2 Ω; L_1 be 5 henries; L_2 be 2 henries and the mutual inductance M be 3 henries. The current, i, is specified by the waveform shown in Fig. E-10.2. No current flows in the other winding.

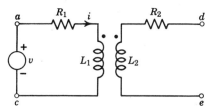

FIG. E-10.1

Basic Concepts of Circuit Theory 59

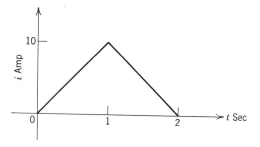

FIG. E-10.2

TO FIND: (a) The voltage v applied to the terminals a-c which produces this current, and (b) the resulting open circuit voltage at terminals d-e.

SOLUTION: First let us redraw the circuit showing the inclusion of the circuit elements to represent the mutual terms (Fig. E-10.3). Note that the polarity of $M(di/dt)$ is independent of a choice for current direction in L_2 and that there is no mutual voltage element in series with L_1 since no current is flowing in L_2. Nodes b and f are labeled on our circuit diagram for convenience in our solution. It should be noted that these nodes are available only on our fictitious representation of the actual transformer. Voltmeters or cathode ray oscilloscope elements could be attached at terminals a-c or d-e but not at a-b, b-c, d-f, or f-e.

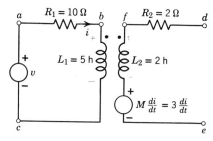

FIG. E-10.3

We shall start our solution by first sketching the voltages v_{ab}, v_{bc}, v_{ac}, and v_{de} from our knowledge of the terminal current-voltage relationships. These sketches are shown in Figs. E-10.4 to E-10.8 with an arrangement of waveforms to provide a quick at a glance comparison, (The voltage scales and analytical expressions are written in later, from the solutions which follow.) Having made our sketches as shown in Figs. E-10.4 to E-10.8, we can proceed with the analytical solution.

For the interval $0 < t < 1$, we can write $i = 10t$ amp. Then $v_{ab} = iR_1 = 10(10t) = 100t$ volts, and

$$v_{bc} = L_1 \frac{di}{dt} = 5 \frac{d}{dt}(10t) = 5(10) = 50 \text{ volts.}$$

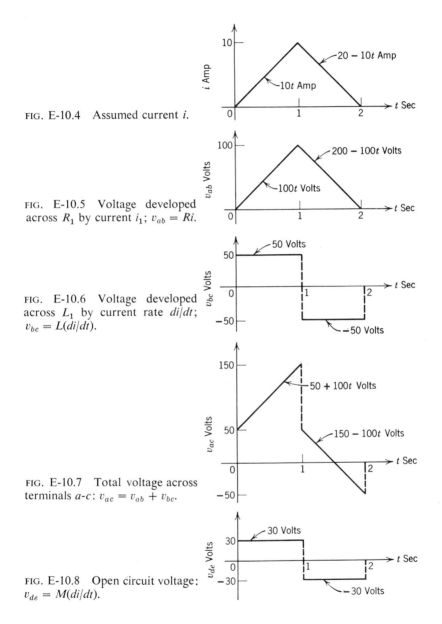

FIG. E-10.4 Assumed current i.

FIG. E-10.5 Voltage developed across R_1 by current i_1; $v_{ab} = Ri$.

FIG. E-10.6 Voltage developed across L_1 by current rate di/dt; $v_{bc} = L(di/dt)$.

FIG. E-10.7 Total voltage across terminals a-c: $v_{ac} = v_{ab} + v_{bc}$.

FIG. E-10.8 Open circuit voltage: $v_{de} = M(di/dt)$.

Basic Concepts of Circuit Theory 61

The voltage, $v_{ac} = v_{ab} + v_{bc} = 100t + 50$ volts.
For the interval, $1 < t < 2$,

$i = -10t + 20$ amp,

$v_{ab} = 10(-10t + 20) = -100t + 200$ volts,

$v_{bc} = 5\dfrac{d}{dt}(-10t + 20) = 5(-10) = -50$ volts,

and

$v_{ac} = -100t + 200 - 50 = -100t + 150$ volts.

For the interval $t > 2$

$i = 0$ amp,

$v_{ab} = 0$ volts,

and

$v_{bc} = 0$ volts.

Finally, for the waveform of voltage that would appear at terminals d–e,

$v_{de} = M\dfrac{di}{dt} = 3\dfrac{di}{dt}.$

For the time interval $0 < t < 1$,

$v_{de} = 3\dfrac{di}{dt} = 3\dfrac{d}{dt}(10t) = 30$ volts,

and for the time interval $1 < t < 2$,

$v_{de} = 3\dfrac{di}{dt} = 3\dfrac{d}{dt}(-10t + 20) = -30$ volts.

Example 2.11

GIVEN: The circuit of Fig. E-11.1 with the switch closed at $t = 0$, and with $i_1(0-) = 0$, $i_2(0-) = 0$.

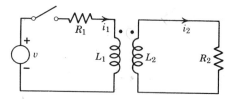

FIG. E-11.1

TO FIND: The equation whose solution gives i_1 for $t > 0$.
SOLUTION: Kirchhoff's voltage equations for the circuit are

$$L_1\dfrac{di_1}{dt} + R_1 i_1 - M\dfrac{di_2}{dt} = v, \qquad (1)$$

and

$$-M\frac{di_1}{dt} + L_2\frac{di_2}{dt} + R_2 i_2 = 0. \tag{2}$$

The variable i_2 may be eliminated but the techniques for doing so may not be obvious. We will introduce a method which is fairly general. Let us rewrite Equations 1 and 2.

$$\left(L_1\frac{d}{dt} + R_1\right)i_1 - M\frac{di_2}{dt} = v \tag{3}$$

and

$$-M\frac{di_1}{dt} + \left(L_2\frac{d}{dt} + R_2\right)i_2 = 0. \tag{4}$$

We interpret $\left(L_1\dfrac{d}{dt} + R_1\right)$ as an operator acting on i_1 such that

$$\left(L_1\frac{d}{dt} + R_1\right)i_1 = L_1\frac{di_1}{dt} + R_1 i.$$

If Equation 3 be operated on by the operator $\left(L_2\dfrac{d}{dt} + R_2\right)$ and Equation 4 be operated on by the operator $M(d/dt)$ the results are

$$\left(L_2\frac{d}{dt} + R_2\right)\left(L_1\frac{d}{dt} + R_1\right)i_1 - \left(L_2\frac{d}{dt} + R_2\right)M\frac{di_2}{dt} = \left(L_2\frac{d}{dt} + R_2\right)v, \tag{5}$$

and

$$-M^2\frac{d^2 i_1}{dt^2} + M\frac{d}{dt}\left(L_2\frac{d}{dt} + R_2\right)i_2 = 0. \tag{6}$$

The addition of Equations 5 and 6 eliminates i_2 and results in

$$\left(L_2\frac{d}{dt} + R_2\right)\left(L_1\frac{d}{dt} + R_1\right)i_1 - M^2\frac{d^2 i_1}{dt^2} = \left(L_2\frac{d}{dt} + R_2\right)v$$

or

$$L_1 L_2 \frac{d^2 i_1}{dt^2} + (R_2 L_1 + R_1 L_2)\frac{di_1}{dt} + R_2 R_1 i_1 - M^2\frac{d^2 i_1}{dt^2} = L_2\frac{dv}{dt} + R_2 v,$$

or

$$(L_1 L_2 - M^2)\frac{d^2 i_1}{dt^2} + (R_2 L_1 + R_1 L_2)\frac{di_1}{dt} + R_2 R_1 i_1 = L_2\frac{dv}{dt} + R_2 v. \tag{7}$$

The techniques for obtaining a solution to Equation 7 will be studied in Chapter 3.

2.5 INITIAL CONDITIONS

In order to solve the differential equations which arise in circuit theory it is essential that we determine the conditions that apply immediately after some switching operation. It is the purpose of this article to introduce us to the evaluation of these "initial conditions."

In discussing the elements of inductance and capacitance it was mentioned that we shall use time equal to zero as the time of some switching operation. It is presumed that the switching operation is completed in zero time, although this is not strictly true because of arcing which may occur at the switch terminals and because of the capacitance between switch terminals which changes from an undefined large value to a small value as the terminals are separated.

In physical reality electric and magnetic fields, and therefore voltages and currents must be continuous functions of time. This is because the velocity of propagation of energy is finite. In our circuit model we have made certain approximations in the interest of simplicity, and by our definitions we allow instantaneous changes in voltage and current through certain elements. Thus we allow instantaneous changes in current through a voltage source, resistance, or capacitance, and we allow instantaneous changes in voltage across a current source, resistance, or inductance. Usually we do not allow an instantaneous change in voltage across a capacitance or an instantaneous change in current through an inductance.

For the capacitance of Fig. 2.19b we repeat Equation 2.23

$$v_{ab} = v_{ab}(0-) + \frac{1}{C}\int_{0-}^{0+} i\, dt + \frac{1}{C}\int_{0+}^{t} i\, dt$$

$$= v_{ab}(0+) + \frac{1}{C}\int_{0+}^{t} i\, dt,$$

and note that if infinitely large currents (impulse currents) do not flow, the voltage across a capacitor can not change immediately. Sometimes it is useful to idealize our circuit elements so as to allow impulse currents to flow. This will be discussed in Chapter 9. Until then, it will be safe to use $v_{ab}(0-) = v_{ab}(0+)$.

For magnetic flux linkages we write

$$v = \frac{d\lambda}{dt},$$

or

$$\lambda = \int_0^t v\, dt + \lambda(0+),$$

and note that flux linkages must not change immediately unless there is a voltage impulse. This means that in a single inductive circuit the current can not change immediately since $\lambda = Li$. This limitation need not extend to several circuits mutually coupled. For example for two circuits we write

$$\lambda = L_1 i_1 \pm M i_2 = \int_0^t v \, dt + \lambda(0+).$$

It would appear possible to have i_1 and i_2 both change instantaneously so as to keep the flux linkages constant while having a finite value for v. It can be shown that this will occur only if the coupling coefficient, $k = M/\sqrt{L_1 L_2}$ is assumed equal to 1. Physically, it is not possible to have k exactly equal to 1[2]; from a circuit calculation standpoint it is sometimes convenient, if k is very close to 1, to say it is 1. Under this condition the currents i_1 and i_2 can change instantaneously. Otherwise the currents through an inductance can not change immediately without a voltage impulse. A discussion of impulses is postponed until Chapter 9.

Initial conditions are not restricted to currents and voltages at $t = 0+$. We will also need occasionally the derivatives of these quantities:

$$\frac{di}{dt}(0+), \frac{d^2 i}{dt^2}(0+), \frac{dv}{dt}(0+), \frac{d^2 v}{dt^2}(0+), \text{etc.}$$

The following will illustrate how initial quantities are determined.

The switch K in Fig. 2.26 is closed at time equal to zero, connecting a finite direct voltage source to a series RL circuit. The current $i(0-)$ is given as zero, and since the current through L cannot change immediately, $i(0+) = 0$. Kirchhoff's voltage equation which holds for $t > 0$ is

$$L\frac{di}{dt} + Ri = V. \qquad t > 0 \qquad (2.31)$$

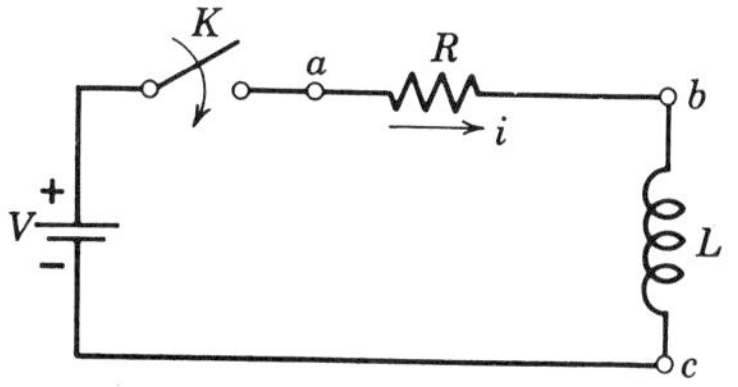

FIG. 2.26 A series RL circuit with direct voltage source.

At $t = 0+$,

$$L\frac{di}{dt}(0+) + Ri(0+) = V, \qquad t = 0+$$

and since $i(0+) = 0$,

$$\frac{di}{dt}(0+) = \frac{V}{L}.$$

---

[2] See Appendix A.

Note well that if $i(0+) = 0$, $(di/dt)(0+)$ is not necessarily zero.

To find $(d^2i/dt^2)(0+)$, we need to return to Equation 2.31 which holds for $t \geqslant 0$. This equation may be differentiated to yield

$$L\frac{d^2i}{dt^2} + R\frac{di}{dt} = 0, \quad t \geqslant 0 \tag{2.32}$$

Then at $t = 0+$,

$$L\frac{d^2i}{dt^2}(0+) + R\frac{di}{dt}(0+) = 0,$$

and

$$\frac{d^2i}{dt^2}(0+) = -\frac{R}{L}\frac{di}{dt}(0+) = -\frac{RV}{L^2}.$$

In a similar manner higher derivatives may be found.

Suppose that the voltage v_{ac} in Fig. 2.26 instead of being a constant, V, were equal to $V\cos(\omega t + \alpha)$ in which ω and α are constants.

Then the voltage equation becomes

$$L\frac{di}{dt} + Ri = V\cos(\omega t + \alpha). \quad t \geqslant 0. \tag{2.33}$$

at $t = 0+$

$$L\frac{di}{dt}(0+) + Ri(0+) = V\cos\alpha, \quad t = 0+$$

and

$$\frac{di}{dt}(0+) = \frac{V\cos\alpha}{L} \quad \text{since } i(0+) = 0.$$

Differentiation of Equation 2.33 yields

$$L\frac{d^2i}{dt^2} + R\frac{di}{dt} = -\omega V\sin(\omega t + \alpha), \quad t \geqslant 0$$

and thus

$$\frac{d^2i}{dt^2}(0+) = -\frac{R}{L}\frac{di}{dt}(0+) - \frac{\omega V}{L}\sin\alpha$$

$$= -\frac{RV\cos\alpha}{L^2} - \frac{\omega V}{L}\sin\alpha.$$

Example 2.12

GIVEN: The circuit shown in Fig. E-12.1 with an initial charge on the capacitor. The switch is closed at $t = 0$.

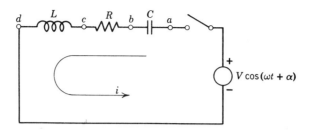

FIG. E-12.1

TO FIND: (a) The expressions for the initial conditions $\frac{di}{dt}(0+)$, $\frac{d^2i}{dt^2}(0+)$, $\frac{dv_{ab}}{dt}(0+)$, $\frac{dv_{bc}}{dt}(0+)$, and $\frac{dv_{cd}}{dt}(0+)$.

(b) The evaluation of these initial conditions if $V = 10$, $\omega = 2$, $\alpha = 60°$, $v_{ab}(0-) = 5$ volts, $R = 5\,\Omega$, $L = 2h$, and $C = \frac{1}{2}f$.

SOLUTION: Kirchhoff's voltage equation for the circuit is

$$L\frac{di}{dt} + Ri + \frac{1}{C}\int_0^t i\,dt + v_{ab}(0+) = V\cos(\omega t + \alpha). \quad (1)$$

This equation holds for $t \geqslant 0$. Since $i(0-) = 0$ because of the switch, and since no voltage impulse exists,

$$i(0-) = i(0+) = 0. \quad (2)$$

If we evaluate Equation 1 at $t = 0+$ we obtain,

$$L\frac{di}{dt}(0+) + Ri(0+) + 0 + v_{ab}(0+) = V\cos\alpha, \quad (3)$$

and, since $i(0+) = 0$,

$$\frac{di}{dt}(0+) = \frac{V\cos\alpha - v_{ab}(0+)}{L}. \quad (4)$$

why not?

Note that if $i(0+) = 0$, this **does not mean** that $(di/dt)(0+) = 0$.

In order to obtain an equation containing d^2i/dt^2, we must differentiate Equation 1.

$$L\frac{d^2i}{dt^2} + R\frac{di}{dt} + \frac{i}{C} = -\omega V\sin(\omega t + \alpha). \quad (5)$$

This equation also holds for $t \geqslant 0$, and we may evaluate this equation at $t = 0+$ to obtain,

$$L\frac{d^2i}{dt^2}(0+) + R\frac{di}{dt}(0+) + \frac{i(0+)}{C} = -\omega V\sin\alpha,$$

or

$$\frac{d^2i}{dt^2}(0+) = \frac{-\omega V\sin\alpha - R(di/dt)(0+)}{L}. \quad (6)$$

Basic Concepts of Circuit Theory 67

In order to determine $(dv_{ab}/dt)(0+)$, we note that

$$v_{ab} = \frac{1}{C}\int_0^t i\,dt + v_{ab}(0+)$$

or

$$i = C\frac{dv_{ab}}{dt}. \tag{7}$$

Equation 7 holds for $t \geqslant 0$ and thus

$$\frac{dv_{ab}}{dt}(0+) = \frac{i(0+)}{C}. \tag{8}$$

To evaluate $(dv_{bc}/dt)(0+)$, we note that $v_{bc} = Ri$, and therefore

$$\frac{dv_{bc}}{dt} = R\frac{di}{dt}.$$

Then

$$\frac{dv_{bc}}{dt}(0+) = R\frac{di}{dt}(0+). \tag{9}$$

Since

$$v_{cd} = L\frac{di}{dt}, \quad \frac{dv_{cd}}{dt} = L\frac{d^2i}{dt^2},$$

and thus

$$\frac{dv_{cd}}{dt}(0+) = L\frac{d^2i}{dt^2}(0+) \tag{10}$$

(b) For the constants given we may now evaluate the initial conditions.

From Equation 2, $i(0+) = 0$ amp. $\tag{11}$

From Equation 4,

$$\frac{di}{dt}(0+) = \frac{V\cos\alpha - v_{ab}(0+)}{L}$$

$$= \frac{10\cos 60° - 5}{2} = 0 \text{ amp/sec.} \tag{12}$$

From Equation 6,

$$\frac{d^2i}{dt^2}(0+) = \frac{-\omega V\sin\alpha - R(di/dt)(0+)}{L}$$

$$= \frac{-2(10)\sin 60° - 5(0)}{2}$$

$$= -8.67 \text{ amp/sec}^2. \tag{13}$$

From Equation 8,

$$\frac{dv_{ab}}{dt}(0+) = \frac{i(0+)}{C} = \frac{0}{(\frac{1}{2})}$$

$$= 0 \text{ volts/sec.} \tag{14}$$

From Equation 9,

$$\frac{dv_{bc}}{dt}(0+) = R\frac{di}{dt}(0+)$$

$$= 5(0) = 0 \text{ volts/sec.} \tag{15}$$

From Equation 10,

$$\frac{dv_{cd}}{dt}(0+) = L\frac{d^2i}{dt^2}(0+) = 2(-8.67)$$

$$= -17.3 \text{ volts/sec.} \tag{16}$$

Example 2.13

GIVEN: The circuit shown with $i_1(0-)$ and $v_{bo}(0-)$ given. Also the constants V, ω, α, L, R_1, R_2, R_3, and C are known.

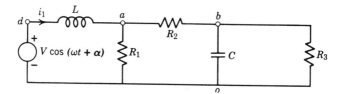

FIG. E-13.1

TO FIND: The technique for determining initial conditions for this circuit.

SOLUTION: Inspection of the circuit shows that the branch current method requires three unknown current whereas the node voltage method requires only two unknown voltages. We will therefore use the node voltage method and apply Kirchhoff's current law about nodes a and b. These equations are

$$\frac{1}{L}\int_0^t [v_{ao} - V\cos(\omega t + \alpha)] \, dt - i_1(0+) + \frac{v_{ao}}{R_1} + \frac{v_{ao} - v_{bo}}{R_2} = 0, \tag{1}$$

and

$$\frac{v_{bo} - v_{ao}}{R_2} + C\frac{dv_{bo}}{dt} + \frac{v_{bo}}{R_3} = 0. \tag{2}$$

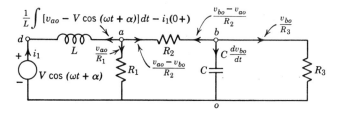

FIG. E-13.2

Since no impulses are present $i_1(0-) = i_1(0+)$ and $v_{bo}(0-) = v_{bo}(0+)$.

We note that $v_{ao}(0+)$ may be determined by evaluating Equation 1 at $t = 0$. Then

$$-i_1(0+) + \frac{v_{ao}(0+)}{R_1} + \frac{v_{ao}(0+)}{R_2} - \frac{v_{bo}(0+)}{R_2} = 0,$$

from which

$$v_{ao}(0+) = \frac{R_1[R_2 i_1(0+) + v_{bo}(0+)]}{R_1 + R_2}. \tag{3}$$

The currents in all the resistances at $t = 0+$ can readily be written now that $v_{ao}(0+)$ and $v_{bo}(0+)$ are known

Equation 2 may now be used to find $(dv_{bo}/dt)(0+)$. At $t = 0+$ this equation becomes

$$\frac{v_{bo}(0+)}{R_2} - \frac{v_{ao}(0+)}{R_2} + C\frac{dv_{bo}}{dt}(0+) + \frac{v_{bo}(0+)}{R_3} = 0, \tag{4}$$

and the expression for $(dv_{bo}/dt)(0+)$ may be written.

In order to determine $(dv_{ao}/dt)(0+)$ we must first differentiate Equation 1 to obtain

$$\frac{v_{ao}}{L} - \frac{V}{L}\cos(\omega t + \alpha) + \frac{1}{R_1}\frac{dv_{ao}}{dt} + \frac{1}{R_2}\frac{d}{dt}(v_{ao} - v_{bo}) = 0. \tag{5}$$

Then Equation 5 is evaluated at $t = 0+$ to give

$$\frac{v_{ao}(0+)}{L} - \frac{V\cos\alpha}{L} + \left(\frac{1}{R_1} + \frac{1}{R_2}\right)\frac{dv_{ao}}{dt}(0+) + \frac{1}{R_2}\frac{dv_{bo}}{dt}(0+). \tag{6}$$

$(dv_{ao}/dt)(0+)$ may be determined from Equation 6.

2.6 SUMMARY, A LIST OF CIRCUIT ELEMENTS, THEIR SYMBOLS, AND THEIR VOLT-AMPERE RELATIONSHIP

This chapter has stressed the dependence of circuit laws and elements on the laws and relations of field theory. Kirchhoff's voltage and current laws have been stated and the application of these laws to circuit solutions by the node voltage method and the branch current method have been demonstrated. Initial conditions have been investigated to prepare us for the methods of solving differential equations to be studied in Chapter 3.

The list of defined circuit elements together with their volt-ampere relationships is summarized below.

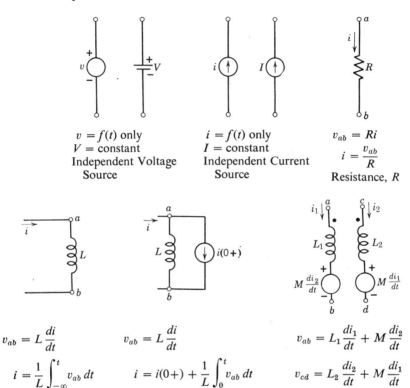

$v = f(t)$ only
$V =$ constant
Independent Voltage Source

$i = f(t)$ only
$I =$ constant
Independent Current Source

$v_{ab} = Ri$
$i = \dfrac{v_{ab}}{R}$
Resistance, R

$v_{ab} = L\dfrac{di}{dt}$
$i = \dfrac{1}{L}\int_{-\infty}^{t} v_{ab}\, dt$

$v_{ab} = L\dfrac{di}{dt}$
$i = i(0+) + \dfrac{1}{L}\int_{0}^{t} v_{ab}\, dt$

Self-inductance, L

$v_{ab} = L_1\dfrac{di_1}{dt} + M\dfrac{di_2}{dt}$
$v_{cd} = L_2\dfrac{di_2}{dt} + M\dfrac{di_1}{dt}$

Mutual inductance, M

$v_{ab} = \dfrac{1}{C}\int_{-\infty}^{t} i\, dt$
$i = C\dfrac{dv_{ab}}{dt}$

$v_{ab} = v_{ab}(0+) + \dfrac{1}{C}\int_{0}^{t} i\, dt$
$i = C\dfrac{dv_{ab}}{dt}$

Capacitance C

Basic Concepts of Circuit Theory 71

FURTHER READING

The student's college physics text will prove helpful for much of the material of this chapter. Also there is a good discussion of circuit elements, Kirchhoff's laws, and the circuit models of physical systems in Chapter 4 of *Electrical Engineering Science* by P. R. Clement and W. C. Johnson, McGraw-Hill, New York, 1960.

PROBLEMS

2.1 *a* For the elements shown power is being delivered to the element. Indicate with + and − marks the voltage polarity. If $p = 100$ watts and $i = 5$ amp, what is the magnitude of the voltage?

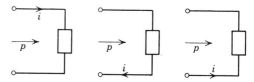

b For the elements shown, power is being taken from the element. Indicate with + and − marks the voltage polarity.

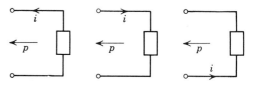

c For the direction of power, (power into $\xrightarrow{p}$ or power out of $\xleftarrow{p}$) shown, and the voltage polarities shown on each element, indicate the direction of current. If $p = 100$ watts and the voltage is 5 volts, what is the magnitude of the current?

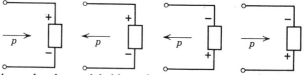

2.2 *a* For each of the nodes shown, label branch currents in terms of independent currents.

b Find the current i in Fig. (4) if, starting at the top and going clockwise, the currents are 1, 2, 3, 4, and 5 amperes respectively.

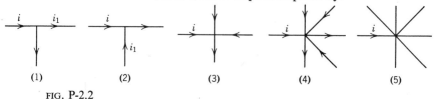

FIG. P-2.2

2.3 For each of the networks shown in Fig. P-2.3, all currents and voltages are unknown.
  a Label branch currents in terms of independent currents.
  b Label the nodes. How many independent voltages are there?

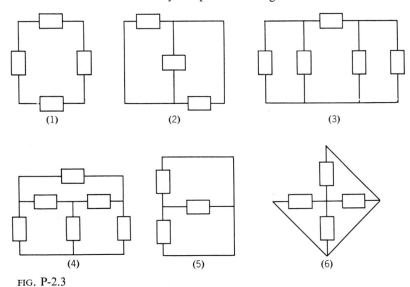

FIG. P-2.3

2.4 In the networks shown in Fig. P-2.4, assume the voltages across elements 1, 2, and 3 are known. Label the nodes. Write expressions for the voltages across the other elements in terms of the known voltages.

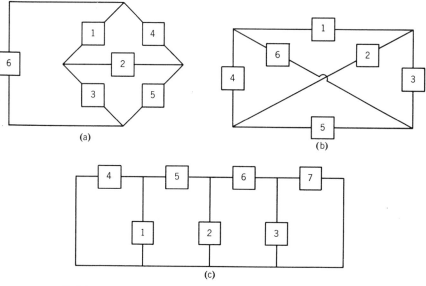

FIG. P-2.4

2.5 A certain No. 10 AWG (American Wire Gauge) copper wire is 1 meter long and has a diameter of 0.102 inches. The conductivity of copper is given as 5.8×10^7 mhos/meter at 20°C. Calculate the resistance of this wire at 20°C. If 1 ampere of d-c current flows in this wire, calculate the power absorbed as thermal power.
Partial answer: $P = 0.00325$ watts.

2.6 It is known that the resistance of copper varies with temperature (over the range of $-50°C$ to $250°C$) in accord with the following expression, $R_1 = R_2[1 + \alpha_2(T_1 - T_2)]$, where α_2 is the temperature coefficient of the material at T_2, and R_1 and R_2 are the resistances in ohms at the temperatures T_1 and T_2 respectively. From a handbook of tables we find that the temperature coefficient for copper at 20°C is 0.00393 per degree centigrade. For a certain No. 8 AWG copper wire the resistance at 50°C is found to be 10 ohms. Calculate the resistance of this wire at 150°C.
Answer: $R = 13.6 \, \Omega$.

2.7 The switch in the network shown in Fig. P-2.7 is closed at time $t = 0$. V, R_1, R_2, and R_3 are known. Using Kirchhoff's voltage and current laws in the form $\Sigma v = 0$ and $\Sigma i = 0$:
a Find the literal expressions for the voltages and currents by the node voltage method for $t > 0$.
b Find the literal expressions for the voltages and currents by the branch current method for $t > 0$.
c Calculate the numerical values for all currents and voltages if $V = 300$ volts, $R_1 = 10 \, \Omega$, $R_2 = 30 \, \Omega$, $R_3 = 15 \, \Omega$.
Partial answer: c, The magnitude of the current in R_2 is 8.33 amperes.

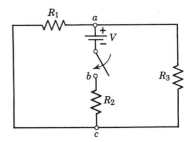

FIG. P-2.7

2.8 Repeat problem 2.7 for the circuit shown in Fig. P-2.8 [use same numerical values as for part c of Problem 2.7].
Partial answer: The magnitude of the current in R_1 is 15 amperes.

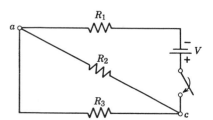

FIG. P-2.8

2.9 The switch in the network shown in Fig. P-2.9 is closed at time $t = 0$. Select a method (node-voltage or branch current) to solve for all currents and voltages. Give your reason for the method you have selected. Solve for all currents and voltages.
Partial answer: The magnitude of the current in R_4 is 5 amperes.

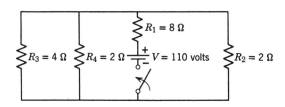

FIG. P-2.9

2.10 Repeat problem 2.9 for the network shown in Fig. P-2.10 (switch S_2 is open). $V = 140$ Volts, $R_1 = 3\Omega$, $R_2 = 6\Omega$, $R_3 = 2\Omega$, $R_4 = 1\Omega$, and $R_5 = 3\Omega$.

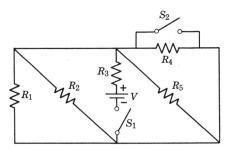

FIG. P-2.10

2.11 In the network of Fig. P-2.10 the switch S_2 is closed at some time after switch S_1 is closed. Explain what effect this will have on the circuit. Solve for all voltages and currents after switch S_2 is closed.
Partial answer: The current through the switch S_2 is 70 amperes.

2.12 *a* Write the expressions for i_1, i_2, and power for each element in Fig. P-2.12.
b Evaluate i_1, i_2, and the power for **each** element if $V = 5$ volts, $i = 5$ amp, and $R = 5\ \Omega$. Indicate for each element whether the direction of power is into or out of the element. Show that the total power supplied equals the total power absorbed.
Partial answer: Power is into the ideal voltage source V and is 20 watts.

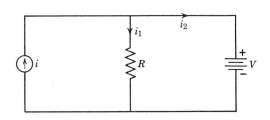

FIG. P-2.12

Basic Concepts of Circuit Theory 75

2.13 For the data of Example 2.3 (page 34).
  a Write the expressions for the voltage and power for the current source. Using the values of part b of the example, calculate and then sketch the voltage and power for one period. Note that the average power supplied by the source during one period is zero.
  b Write the expressions for the power absorbed by R_1 and sketch i_1, v_{ab}, and p_{R_1} for one period. Note that $v_{ab}i_1$ is never negative. Why is this so?

2.14 a Write the expressions for i_1, i_2, and power for *each* element in Fig. P-2.14.
  b Evaluate i_1, i_2 and the power for each element with $V = 100$ volts, $I = 10$ amperes, $R_1 = 5$ ohms, and

  1 $R_2 = 10$ ohms.

  2 $R_2 = 20$ ohms.

  3 $R_2 = 50$ ohms.

  Partial answer: b, 1, $i_2 = 10$ amperes.
  Partial answer: b, 2, Power is 500 watts into voltage source V.
  Partial answer: b, 3, Power is 200 watts into resistance R_2.

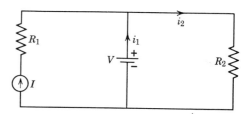

FIG. P-2.14

2.15 Use Kirchhoff's laws to prove that

$$i_1 = \frac{V - R_2 I}{R_1 + R_2}, \quad \text{and}$$

$$i_2 = \frac{V + R_1 I}{R_1 + R_2}.$$

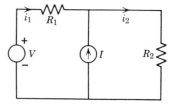

FIG. P-2.15

2.16 For each of the circuits shown, all sources and resistances are known quantities. Evaluate all unknown currents and voltages and find the power for each element indicating whether power is into or out of the element.

Partial answer to 2.16(b):

$P_{(5 \text{ amp current source})} = 317.5$ watts (out of)
$P_{(10 \text{ volt voltage source})} = 35.0$ watts (into)
$P_{(1\Omega \text{ resistance})} = 12.25$ watts (into)
$P_{(9\Omega \text{ resistance})} = 20.25$ watts (into)
$P_{(10\Omega \text{ resistance})} = 250.0$ watts (into)

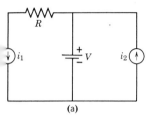

(a)

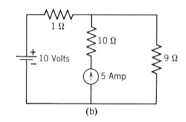

(b)

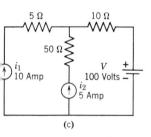

(c)

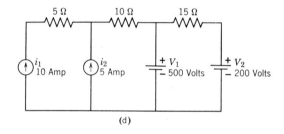

(d)

FIG. P-2.16

2.17 A sinusoidally varying current source, $i = 10 \cos 377t$ amperes, is applied to an inductance $L = 2$ mh (2 millihenries) as shown in Fig. P-2.17a. One cycle (or period) of the source current is shown in Fig. P-2.17b.
a Solve for the voltage v_{ab}.
b Sketch one cycle of the voltage.
c Determine the magnitude of the power and state whether it is into or out of the inductance at times $t = 1/120$ sec, $1/240$ sec, and $1/480$ sec.
Partial answer: *a*, $v_{ab} = -7.54 \sin 377t$ volts.
    c, At $t = 1/480$ sec $p = 37.5$ watts out of the inductance.

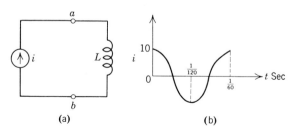

FIG. P-2.17   (a)   (b)

Basic Concepts of Circuit Theory   77

2.18 A sinusoidal voltage, $v = 100 \cos 377t$ volts, is applied to the terminals of an inductance of 2 millihenries as shown in Fig. P-2.18a. One cycle of the applied voltage, v, is shown in Fig. P-2.18b. $i(0+) = 0$.

   a From the voltage-current relationship for inductance sketch one cycle of the current i.

   b Determine the analytical expression for the current i and label your sketch of part *a* accordingly.

   c Sketch the power ($p = vi$) for one cycle and for one cycle determine the intervals of time in which the power is into the inductance and the intervals of time it is out of the inductance.

   d Determine the energy stored in the field of the inductance at $t = 1/120$ sec and $t = 1/240$ sec.

Partial answer: *b*, $i = 133 \sin 377t$ amp.

        d, Energy stored at time $t = 1/240$ sec is 17.7 joules.

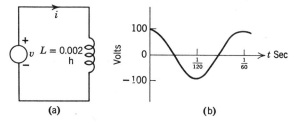

FIG. P-2.18

2.19 Given the circuit shown in Fig. P-2.19.

   a Calculate the voltages v_{ab}, v_{bc}, and v_{ac}.

   b Plot, carefully to scale, the voltages v_{ab}, v_{bc}, and v_{ac}.

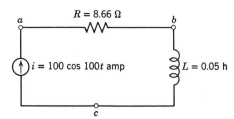

FIG. P-2.19

2.20 The switch in the circuit shown in Fig. P-2.20 is closed at time $t = 0$. $i_2(0+) = 0$. Calculate and plot to scale the currents i, i_1, and i_2.

Answer: $i_1 = +10 \cos 3.14t$ amp.

        $i_2 = +10 \sin 3.14t$ amp.

        $i = 14.14 \cos(3.14t - 45°)$ amp.

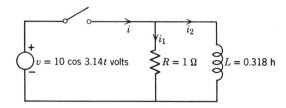

FIG. P-2.20

2.21 In each of the circuits shown in Fig. P-2.21 the switch is closed at time $t = 0$. All initial currents in the inductances are zero.

    a For circuits 1 and 2 write the equation whose solution will be the expression for the current i.

    b For circuits 3 and 4, write the equation whose solution will be the expression for the voltage v_{ab}.

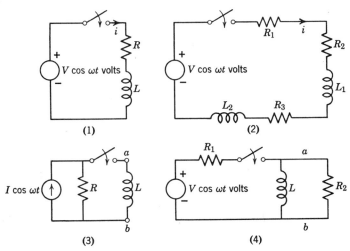

FIG. P-2.21

2.22 A voltage v, specified by Fig. P-2.22*b*, is applied to the terminals of an uncharged capacitance C as shown in Fig. P-2.22*a*.

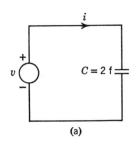

FIG. P-2.22*a*

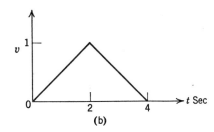

FIG. P-2.22*b*

a Determine and sketch the current i.
b Write the expression for power into the capacitance and sketch power versus time.
c Is power into or out of the capacitance at time $t = 1.5$ seconds and what is the magnitude of the power at this time?
d The energy stored in the electric field is $\int_{-\infty}^{t} p\, dt = \tfrac{1}{2}Cv^2$. Determine this energy at $t = 1.5$ seconds.
Partial answer At $t = 1.5$ sec power is into the capacitance and is 0.75 watts.

2.23 The waveform of an ideal voltage source is specified by Fig. P-2.23a. This voltage is applied to the circuit as shown in Fig. P-2.23b. $i_2(0-) = 0$.
a Sketch and determine the analytical expressions for (1) i_1, (2) i_2, (3) i_3, and (4) i. $R = 1\,\Omega$, $L = 1$ h, $C = 1$ f.
b Determine the energy in joules stored:
1 In the capacitance at $t = 3$ sec.
2 In the inductance at $t = 3$ sec.

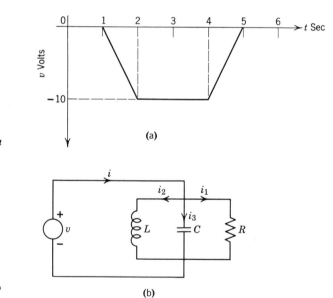

FIG. P-2.23a

FIG. P-2.23b

2.24 The current wave of Fig. P-2.24a is in the circuit of Fig. P-2.24b.
a Sketch v_{ab} versus time showing values at key points.
b Sketch v_{bc} versus time showing values at key points. Determine these values by using the slope of the i versus t curve.
c Sketch v_{cd} versus time showing values at key points. Assume $v_{cd}(0+) = 0$. Use the concept that $v_{cd} = v_{cd}(0+)$ plus $1/C$ times the area under the i versus t curve.
Partial answer: c, $v_{cd} = 150$ volts at $t = 0.4$ sec.

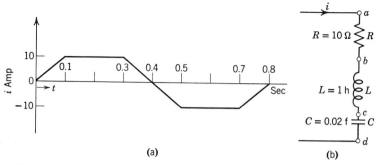

FIG. P-2.24

2.25 The switch in the circuit shown in Fig. P-2.25 is closed at time $t = 0$.
  a Label branch currents and write the necessary independent voltage equations in terms of the unknown currents for $t \gg 0$.
  b Find, by the node-voltage method, the equation whose solution will give v_{ab} for $t \gg 0$.

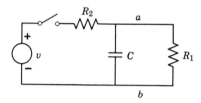

FIG. P-2.25

2.26 For the circuit of Fig. P-2.26, find, by the node-voltage method, the equation whose solution will give v_{ba} for $t \gg 0$.

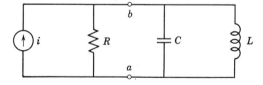

FIG. P-2.26

2.27 The switch in the circuit shown in Fig. P-2.27 is closed at time $t = 0$.
  a Determine, by the node-voltage method, the equation whose solution will give v_{mn}.

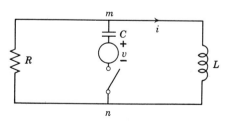

FIG. P-2.27

b Determine, by the branch current method, the equation whose solution will give the current *i*.

Answer: $b, L\dfrac{d^2i}{dt^2} + \dfrac{L}{RC}\dfrac{di}{dt} + \dfrac{i}{C} = \dfrac{dv}{dt}$.

2.28 Assume all switches are closed at time $t = 0$ in the circuit shown in Fig. P-2.28. Label independent currents and write the necessary number of voltage equations to solve for all of the currents for $t \geqslant 0$.

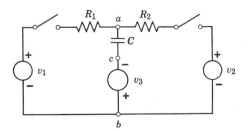

FIG. P-2.28

2.29 Use the node-voltage method for the circuit of Fig. P-2.28 to find the equation whose solution will give v_{ab} for $t \geqslant 0$.

2.30 Three coils are placed on a core as shown in Fig. P-2.30. Appropriately dot mark the terminals. Give your reasoning.

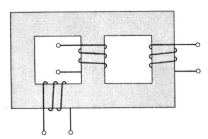

FIG. P-2.30

2.31 A current *i*, specified in Fig. P-2.31*b*, flows in one winding of a transformer represented by the circuit shown in Fig. P-2.31*a*. Sketch and then determine the analytical expression for:

a The waveform of the voltage v_{ab} which produces this current.

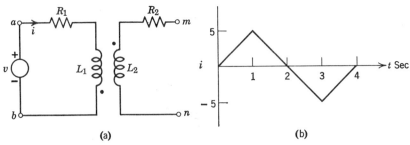

FIG. P-2.31 $R_1 = 3\Omega$, $R_2 = 2\Omega$, $L_1 = 4h$, $L_2 = 1h$, and $k = \tfrac{1}{2}$.

b The resulting open circuit voltage v_{mn}.
$R_1 = 3\Omega$, $R_2 = 2\Omega$, $L_1 = 4h$, $L_2 = 1h$, and $k = \frac{1}{2}$.

2.32 Two coils are magnetically coupled and properly dot-marked. Show by sketches and explanation that the relative polarities $L_1(di_1/dt)$ and $M(di_1/dt)$ are the same with respect to the dot marks. That is, if the voltage $L_1(di_1/dt)$ is of such polarity at an instant of time that the dot-marked end of this coil is + with respect to the unmarked end, then the polarity of $M(di_1/dt)$ is such that the dot-marked end of the second coil is also + with respect to its unmarked end.

2.33 In the circuit shown in Fig. P-2.33 the switch is closed at $t = 0$.
a Write the equation whose solution will give i for $t \geqslant 0$.
b Now assume that $V = 10$ volts, $R = 2$ ohms, $C = \frac{1}{2}$ farads and $v_{bc}(0-) = 0$. Calculate $i(0+)$, $(di/dt)(0+)$, and $(d^2i/dt^2)(0+)$.
c Same as b except $v_{bc}(0-) = -5$ volts.
Partial answer: b, $i(0+) = 5$ amperes; $(di/dt)(0+) = -5$ amperes/sec,
c, $i(0+) = 7.5$ amperes.

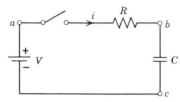

FIG. P-2.33

2.34 For a certain circuit the voltage equation is

$$\frac{1}{C}\int_0^t i\,dt + v_{ab}(0+) + Ri = V\cos(\omega t + \alpha)$$

The terms C, R, V, ω, α, and $v_{ab}(0+)$ are known constants.
a Determine the expression for $i(0+)$ in terms of the known constants.
b Determine the expression for $(di/dt)(0+)$ in terms of the known constants and $i(0+)$.

2.35 In the circuit shown in Fig. P-2.35 $i(0-) = 0$. Calculate

a, $\dfrac{di}{dt}(0+)$ b, $\dfrac{d^2i}{dt^2}(0+)$; c, $v_{ab}(0+)$; and d, $\dfrac{dv_{ab}}{dt}(0+)$

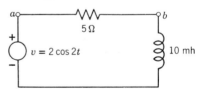

FIG. P-2.35

Answer:

a, $\dfrac{di}{dt}(0+) = 200$ amp/sec.

b, $\dfrac{d^2i}{dt^2}(0+) = -10^5$ amp/sec.^{2}

c, $v_{ab}(0+) = 0$ volts.

d, $\dfrac{dv_{ab}}{dt}(0+) = 10^3$ volts/sec.

2.36 The switch in the circuit shown in Fig. P-2.36 is closed at $t = 0$. $v_{bc}(0-) = 10$ volts, $i_1(0-) = 0$. Calculate:

a, $i(0+)$; b, $\dfrac{di}{dt}(0+)$; c, $v_{ac}(0+)$; and d, $\dfrac{dv_{ac}}{dt}(0+)$.

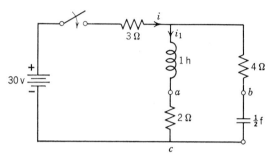

FIG. P-2.36

2.37 In Fig. P-2.37, $v_{bc}(0+) = -10^3$ volts, $i_2(0+) = 4$ amp. Calculate:

a, $i_1(0+)$; b, $\dfrac{di_2}{dt}(0+)$; c, $v_{ab}(0+)$; d, $\dfrac{dv_{ab}}{dt}(0+)$; e, $\dfrac{di_1}{dt}(0+)$; f, $\dfrac{dv_{bc}}{dt}(0+)$;

and g, $\dfrac{dv_{ac}}{dt}(0+)$.

Partial answer: a, $i_1(0+) = 2$ amperes.

b, $\dfrac{di_2}{dt}(0+) = -10^4$ amperes/sec.

f, $\dfrac{dv_{bc}}{dt}(0+) = -2 \times 10^6$ volts/sec.

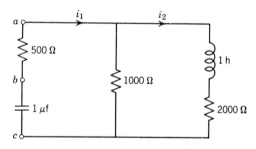

FIG. P-2.37

2.38 For the circuit shown in Fig. P-2.38, $M_{12} = 4$ h, $M_{23} = 2$ h. There is *no* coupling between coils L_1 and L_3. The switch is closed at $t = 0$.

Find: $v_{ab}(0+)$ and $\dfrac{dv_{ab}}{dt}(0+)$, $i(0-) = 0$

Partial answer: $v_{ab}(0+) = -10$ volts.

84

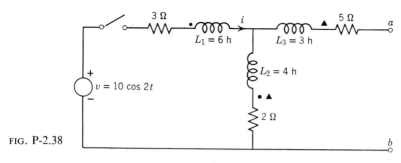

FIG. P-2.38

2.39 For the circuit shown in Fig. P-2.39 $i(0+) = 0$ and $v_{nm}(0+) = 0$. The switch is closed at time $t = 0$. $M_{12} = 2$ h. Find $v_{ab}(0+); \dfrac{dv_{ab}}{dt}(0+)$.

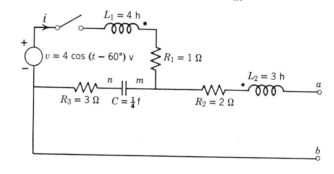

FIG. P-2.39

3 *The Solution of Linear Differential Equations with Constant Coefficients*

In Chapter 2 we found that our equilibrium equations for voltages and current were of the form

$$\frac{dx}{dt} + a_1 x + a_0 \int_0^t x\, dt = f_a(t), \qquad t \geqslant 0 \tag{3.1}$$

or of the form

$$\frac{d^2 x}{dt^2} + a_1 \frac{dx}{dt} + a_0 x = f_b(t), \qquad t \geqslant 0. \tag{3.2}$$

Equation 3.1 is an integro-differential equation which can be changed into the differential equation form of Equation 3.2 by differentiating once with respect to time. The $f(t)$'s are usually driving functions and the dependent variable is usually the response function. These equations are defined as linear equations if the coefficients a_1 and a_0 are not functions of the dependent variable x; the coefficients may be functions of the independent variable t. In our case, as in most linear systems, a_1 and a_0 are constants and the equations are linear equations with constant coefficients.

The principal properties of these equations are the following ones.

(1) These equations have a unique solution. This means that there is only one solution which will satisfy the equation and the initial conditions. Hence, any method by which we can secure a solution is permissible. A proof of this uniqueness in solution for a second order equation is given in Appendix B.

(2) Superposition applies to these equations. This means that if x_1 is the response to $f_1(t)$ and x_2 is the response to $f_2(t)$ then $(x_1 + x_2)$ is the response to $f_1(t) + f_2(t)$. The student can prove this for himself by substituting into either Equation 3.1 or Equation 3.2. This concept is especially useful in those cases in which there are several driving functions of different wave form; we may find the response to each driving function separately and add the responses to obtain the actual response. It should be noted that we must **always** include the response to a driving function of zero value. For purpose of illustration, if x_1 is the response to $f_a(t)$ for Equation 3.1, and x_2 is the response to $f_a(t) = 0$ for Equation 3.1, the complete response to Equation 3.1 is $x_1 + x_2$.

There are a number of methods for solving these equations. Some are given below.

(1) Classical method of differential equations.
(2) Laplace transformation method.
(3) Numerical integration—digital computers.
(4) Analogue simulation—analogue computers.

In this chapter we shall limit ourselves to the first two methods. We shall also generally limit ourselves to differential equations of second order or less. Example 3.11 will show that the techniques are identical for higher-order equations but for these more effort is usually needed to find the necessary roots.

In solving a differential equation we are integrating and should expect constants of integration. A second order differential equation will have two constants, a first order differential equation will have one constant. These constants must be evaluated from the initial conditions.

The first method we shall use is known as the "classical" method for solving differential equations. Our approach to this is "heuristic"; we shall find the solution by assuming the form of this solution.

3.1 CLASSICAL METHOD

3.1.1 Homogeneous Equation

The equation is said to be homogeneous if each term is a function of the dependent variable. For Equation 3.2 this means simply that the right hand side is zero, or

$$\frac{d^2x}{dt^2} + a_1 \frac{dx}{dt} + a_0 x = 0. \tag{3.3}$$

The solution x must be such that x and its derivatives are similar functions. It seems logical to try the exponential form and so we **assume**:

$$x = Ae^{mt}, \tag{3.4}$$

in which A and m are constants. Substitution of Equation 3.4 into Equation 3.3 yields

$$(m^2 + a_1 m + a_0)Ae^{mt} = 0,$$

having the only significant result that

$$m^2 + a_1 m + a_0 = 0. \tag{3.5}$$

Equation 3.5, called the auxiliary or characteristic equation for Equation 3.3, may be obtained immediately from Equation 3.3 by substituting m^2 for d^2/dt^2, and m for d/dt. Equation 3.5 is solved by means of the quadratic formula. Then

$$m = -\frac{a_1}{2} \pm \sqrt{\left(\frac{a_1}{2}\right)^2 - a_0}, \tag{3.6}$$

or

$$m_1 = -\frac{a_1}{2} + \sqrt{\left(\frac{a_1}{2}\right)^2 - a_0},$$

and

$$m_2 = -\frac{a_1}{2} - \sqrt{\left(\frac{a_1}{2}\right)^2 - a_0}.$$

There are generally two values for m or two solutions, $A_1 e^{m_1 t}$ and $A_2 e^{m_2 t}$, each of which satisfies Equation 3.3. Because this equation is linear, superposition applies and the sum of these solutions is also a solution. Therefore:

$$x = A_1 e^{m_1 t} + A_2 e^{m_2 t}, \tag{3.7}$$

and we are assured from our development that this equation is the unique solution to Equation 3.3 if the roots m_1 and m_2 are evaluated from Equation 3.6. The two constants A_1 and A_2 are the two constants of integration which are evaluated from initial conditions. There is one special case which needs separate attention. If $(a_1/2)^2 = a_0$, $m_1 = m_2 = m$, and we have a tentative solution, $x = A_1 e^{mt}$, with only one constant of integration. This means we must seek another solution; later in this chapter we will show that this second solution must be of the form, $A_2 t e^{mt}$.[1] So, for this case we write, as our solution,

$$x = e^{mt}(A_1 + A_2 t). \tag{3.8}$$

It can also be shown that for higher order equations in which there may be n equal values of m the solution becomes:

$$x = e^{mt}(A_1 + A_2 t + A_3 t^2 \ldots A_n t^{n-1}).$$

---

[1] An alternate approach is the "variation of parameter method" in which one assumes that the complete solution is the tentative solution with the constant, A_1, replaced by the variable parameter, $y(t)$, or $x = y(t)e^{mt}$. Substitution of this solution into Equation 3.3 yields $d^2y/dt^2 = 0$. Thus $dy/dt = A_2$, and $y = A_2 t + A_1$, or $x = e^{mt}(A_1 + A_2 t)$.

Although it appears as if there are only two types of solution, it is convenient to separate them into three cases depending on the evaluation of m from Equation 3.6.

Case I.

$$\left(\frac{a_1}{2}\right)^2 > a_0.$$

This means that m_1 and m_2 are negative real numbers. The solution is written,

$$x = A_1 e^{m_1 t} + A_2 e^{m_2 t}, \tag{3.7}$$

and is referred to as the overdamped case.

Case II.

$$\left(\frac{a_1}{2}\right)^2 = a_0.$$

This means $m_1 = m_2 = m$. The solution is written,

$$x = e^{mt}(A_1 + A_2 t) \tag{3.8}$$

and is referred to as the critically damped case.

Case III.

$$\left(\frac{a_1}{2}\right)^2 < a_0.$$

This means that m_1 and m_2 are complex conjugates.

$$m_1 = -\frac{a_1}{2} + \sqrt{\left(\frac{a_1}{2}\right)^2 - a_0} = -\alpha + j\beta, \tag{3.9}$$

$$m_2 = -\frac{a_1}{2} - \sqrt{\left(\frac{a_1}{2}\right)^2 - a_0} = -\alpha - j\beta, \tag{3.10}$$

and one form of the solution would be

$$x = e^{-\alpha t}(A_1 e^{j\beta t} + A_2 e^{-j\beta t}). \tag{3.11}$$

The imaginary unit of complex numbers is j, $\sqrt{-1}$. Electrical engineers use the symbol j instead of i because i is universally the symbol for current.

The most useful formula for dealing with j is the important relationship known as **Euler's formula**:

$$e^{j\phi} = \cos\phi + j\sin\phi. \tag{3.12}$$

If $\phi = \dfrac{\pi}{2} \pm n2\pi$, in which $n = 0, 1, 2, 3, 4 \ldots$ etc.,

$$e^{j(\pi/2 \pm n2\pi)} = j. \tag{3.13}$$

Also

$$j^2 = e^{j(\pi \pm n4\pi)} = -1. \tag{3.14}$$

For x to be a real function of time in Equation 3.11 the terms A_1 and A_2 must be complex conjugates. Rather than evaluate these terms it is simpler to expand the terms $e^{j\beta t}$ and $e^{-j\beta t}$ with Euler's formula. Equation 3.11 becomes

$$x = e^{-\alpha t}[A_1(\cos \beta t + j \sin \beta t) + A_2(\cos \beta t - j \sin \beta t)],$$

or

$$x = e^{-\alpha t}[(A_1 + A_2) \cos \beta t + j(A_1 - A_2) \sin \beta t]. \tag{3.15}$$

In order for x to be a real function of time, the quantities $(A_1 + A_2)$ and $j(A_1 - A_2)$ must be real. This requires that A_1 and A_2 be complex conjugates. Equation 3.15 is then rewritten:

$$x = e^{-\alpha t}[A_3 \cos \beta t + A_4 \sin \beta t], \tag{3.16}$$

in which A_3 and A_4 are real constants to be evaluated from initial conditions. We shall use Equation 3.16 for solutions to this case which is called the underdamped, or oscillatory case.

Example 3.1

GIVEN: The RC circuit shown, with a d-c voltage as the source or driving function. The switch is closed at $t = 0$.

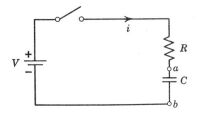

FIG. E-1.1

TO FIND: (a) The complete response i in literal form.
(b) A sketch of i vs. t if $v_{ab}(0-) = -20$ volts, $R = 5 \, \Omega$, $C = 0.1$ farads and $V = 30$ volts.

SOLUTION: (a) Kirchhoff's voltage equation is

$$Ri + \frac{1}{C}\int_0^t i \, dt + v_{ab}(0+) = V. \tag{1}$$

The integral in Equation 1 may be removed by differentiating once to give

$$R\frac{di}{dt} + \frac{1}{C}i = 0, \tag{2}$$

or

$$\frac{di}{dt} + \frac{1}{RC} i = 0. \tag{3}$$

We now assume the solution to Equation 3 to be $i = Ae^{mt}$, and obtain the characteristic equation, $m + 1/RC = 0$, or $m = -1/RC$. Hence

$$i = Ae^{-t/RC}. \tag{4}$$

Since Equation 3 is of the first order there is only one arbitrary constant A. This constant must be evaluated from the initial conditions of the circuit, which may be obtained from Equation 1. $v_{ab}(0-) = v_{ab}(0+)$ because the capacitor voltage cannot change instantaneously. Equation 1 at $t = 0+$ becomes

$$Ri(0+) + \frac{1}{C}(0) + v_{ab}(0+) = V,$$

or

$$i(0+) = \frac{V - v_{ab}(0+)}{R}.$$

From Equation 4, $i = Ae^{-t/RC}$, we may write

$$i(0+) = Ae^{-0/RC} = A = \frac{V - v_{ab}(0+)}{R}$$

Hence the complete response i of our circuit is

$$i = \frac{V - v_{ab}(0+)}{R} e^{-t/RC} \tag{5}$$

If no energy is initially stored in the electric field of the capacitor $v_{ab}(0-) = v_{ab}(0+) = 0$ and Equation 5 becomes

$$i = \frac{V}{R} e^{-t/RC}$$

Equation 5 shows that for our circuit the current increased from 0 to $[V - v_{ab}(0+)]/R$ amperes instantaneously at $t = 0$ and then decreased exponentially to zero.

It is sometimes convenient to speak of the time at which the exponent of our solution is numerically equal to 1 as the "time constant, t_0," of the circuit. Thus, $t_0/RC = 1$, or $t_0 = RC$ seconds. This represents the time for the exponential term to have changed from 1 to $1/e$ (approximately 0.37).

(*b*) Equation 5, evaluated for the constants given, becomes,

$$i = \frac{30 - (-20)}{5} e^{-t/0.5} = 10e^{-2t} \text{ amp.}$$

A sketch of i vs. t is shown in Fig. E-1.2. For our circuit $t_0 = (5)(0.1) = 0.5$ sec. Thus, at 0.5 sec the magnitude of i has decayed to $(0.37)(10)$ or 3.7 amp as shown.

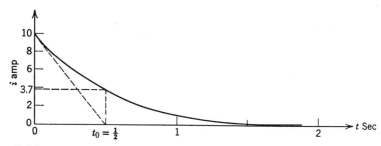

FIG. E-1.2

It is also of interest to note that as time becomes increasingly large, $t \to \infty$, the current becomes zero. This is because, for a capacitor in series with a d-c source, the capacitor voltage becomes equal to the source voltage at steady state (t very large).

Example 3.2

GIVEN: The circuit shown. $v_{ab}(0+) = 10$ volts. $i_2(0+) = 0$ amp.

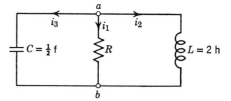

FIG. E-2.1

TO FIND: The complete response v_{ab}, if:

(a) $R = 2 \, \Omega$.

(b) $R = 1 \, \Omega$.

(c) $R = 1/2 \, \Omega$.

SOLUTION: The circuit may be redrawn, with the charged capacitor replaced by its equivalent circuit. Kirchhoff's current law applied at

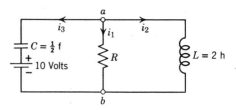

FIG. E-2.2

node a may be written, $i_1 + i_2 + i_3 = 0$. Written in terms of the voltage v_{ab},

$$i_1 = \frac{v_{ab}}{R}, \quad i_2 = \frac{1}{L}\int_0^t v_{ab}\, dt + i_2(0+)$$

and

$$i_3 = C\frac{dv_{ab}}{dt},$$

and hence,

$$\frac{v_{ab}}{R} + \frac{1}{L}\int_0^t v_{ab}\, dt + i_2(0+) + C\frac{dv_{ab}}{dt} = 0. \tag{1}$$

Differentiating Equation 1 gives

$$\frac{1}{R}\frac{dv_{ab}}{dt} + \frac{v_{ab}}{L} + C\frac{d^2v_{ab}}{dt^2} = 0 \tag{2}$$

Rearranging Equation 2 and making the coefficient of the highest ordered derivative unity by division by C gives

$$\frac{d^2v_{ab}}{dt^2} + \frac{1}{RC}\frac{dv_{ab}}{dt} + \frac{v_{ab}}{LC} = 0. \tag{3}$$

The characteristic equation of Equation 3 is

$$m^2 + \frac{1}{RC}m + \frac{1}{LC} = 0. \tag{4}$$

(a) $R = 2\,\Omega$. For this case Equation 4 becomes

$$m^2 + m + 1 = 0,$$

or

$$m = -\frac{1}{2} \pm j\frac{\sqrt{3}}{2}.$$

This is the underdamped or oscillatory case and the solution may be written

$$v_{ab} = e^{-\frac{1}{2}t}\left[A_3 \cos \frac{\sqrt{3}}{2}t + A_4 \sin \frac{\sqrt{3}}{2}t\right], \tag{5}$$

in which A_3 and A_4 are the real constants to be evaluated from initial conditions.

The application of the specified initial condition, $v_{ab}(0+) = 10$ volts, to Equation 5 results in

$$10 = e^0[A_3 \cos 0 + A_4 \sin 0],$$

or

$$10 = A_3.$$

Substituting the numerical value for A_3 and then differentiating Equation 5, we have

$$\frac{dv_{ab}}{dt} = e^{-\frac{1}{2}t}\left(-\frac{\sqrt{3}}{2}(10)\sin\frac{\sqrt{3}}{2}t + \frac{\sqrt{3}}{2}A_4\cos\frac{\sqrt{3}}{2}t\right)$$
$$-\tfrac{1}{2}e^{-\frac{1}{2}t}\left(10\cos\frac{\sqrt{3}}{2}t + A_4\sin\frac{\sqrt{3}}{2}t\right). \qquad (6)$$

If we now return to Equation 1 and substitute known values at $t = 0+$, and the numerical values of R, L, and C, we get

$$\frac{10}{2} + \frac{1}{2}(0) + (0) + \frac{1}{2}\frac{dv_{ab}}{dt}(0+) = 0, \qquad \text{at } t = 0+,$$

and hence

$$\frac{dv_{ab}}{dt}(0+) = -10.$$

Then from Equation 6 evaluated at $t = 0+$,

$$-10 = e^0\left(-\sqrt{3}\,5\sin 0 + \frac{\sqrt{3}}{2}A_4\cos 0\right) - \tfrac{1}{2}e^0(10\cos 0 + A_4\sin 0),$$

and

$$A_4 = -\frac{10}{\sqrt{3}}.$$

Thus our complete solution for v_{ab} with $R = 2\,\Omega$ becomes

$$v_{ab} = e^{-\frac{1}{2}t}\left[10\cos\frac{\sqrt{3}}{2}t - \frac{10}{\sqrt{3}}\sin\frac{\sqrt{3}}{2}t\right] \text{ volts}. \qquad (7)$$

or

$$v_{ab} = e^{-\frac{1}{2}t}\left[11.54\cos\left(\frac{\sqrt{3}}{2}t + 30°\right)\right] \text{ volts}.$$

This is the underdamped or oscillatory case. Fig. E-2.3 shows v_{ab} as a function of time. The function e^{-t} and $11.54\cos(\sqrt{3}/2t + 30°)$ are sketched separately so that the student may observe the effect of multiplying a sinusoidal function of time by a decaying exponential function.

(b) $R = 1\,\Omega$. Equation 4 becomes

$m^2 + 2m + 1 = 0$,

$(m + 1)^2 = 0$,

or

$m = -1$, and -1.

Our solution may then be written in the form

$$v_{ab} = e^{-t}(A_1 + A_2 t). \qquad (8)$$

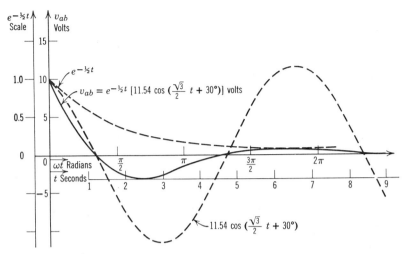

FIG. E-2.3 Plot of voltage v_{ab} for $R = 2\,\Omega$, oscillatory or underdamped case.

Evaluation of Equation 8 at $t = 0$ gives

$A_1 = 10$.

Using this value for A_1 and differentiating Equation 8, we find

$$\frac{dv_{ab}}{dt} = -e^{-t}(10 + A_2 t) + e^{-t}A_2. \tag{9}$$

From Equation 1 at $t = 0+$, $(dv_{ab}/dt)(0+) = -20$ volts/sec. For Equation 9 evaluated at $t = 0+$,

$$\frac{dv_{ab}}{dt}(0+) = -20 = -10 + A_2$$

and

$A_2 = -10$.

The complete solution for v_{ab} with $R = 1$ may then be written as

$$v_{ab} = 10e^{-t}(1 - t) \text{ volts}. \tag{10}$$

This is the critically damped case.

Figure E-2.4 shows v_{ab} as a function of time and also shows the components, $10e^{-t}$ and $-10te^{-t}$.

(c) $R = 1/2\,\Omega$. Equation 4 becomes

$m^2 + 4m + 1 = 0$,

$m = -2 \pm \sqrt{3}$, or

$m_1 = -0.27$, $m_2 = -3.73$.

The Solution of Linear Differential Equations with Constant Coefficients

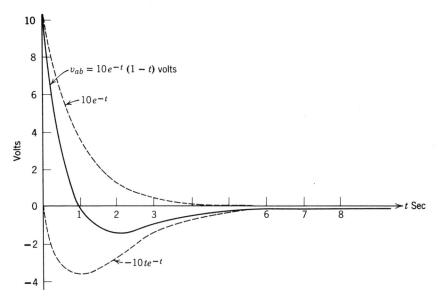

FIG. E-2.4 Plot of voltage v_{ab} (for $R = 1$ ohm), critically damped case.

Our solution is then of the form

$$v_{ab} = A_1 e^{-0.27t} + A_2 e^{-3.73t}. \tag{11}$$

Evaluation of Equation 11 at $t = 0+$ gives

$$10 = A_1 + A_2. \tag{12}$$

Taking the derivative of Equation 11, we have

$$\frac{dv_{ab}}{dt} = -0.27 A_1 e^{-0.27t} - 3.73 A_2 e^{-3.73t}. \tag{13}$$

From Equation 1, evaluated at $t = 0+$,

$$\frac{dv_{ab}}{dt}(0+) = -40.$$

Then from Equation 13 evaluated at $t = 0+$,

$$-40 = -0.27 A_1 - 3.73 A_2, \tag{14}$$

The simultaneous solution of Equations 12 and 14 gives

$$A_1 = -0.780 \quad \text{and} \quad A_2 = 10.8.$$

The complete solution for v_{ab} with $R = \tfrac{1}{2} \Omega$ is

$$v_{ab} = -0.78 e^{-0.27t} + 10.8 e^{-3.73t} \text{volts}.$$

This is the overdamped case.

Note that the time constant of our circuit for case *a* is 2 seconds; for case *b* it is 1 second, and for case *c* there are **two time constants**, 1/0.27 and 1/3.73 seconds.

Figure E-2.5 shows the components of v_{ab} and v_{ab} as functions of time.

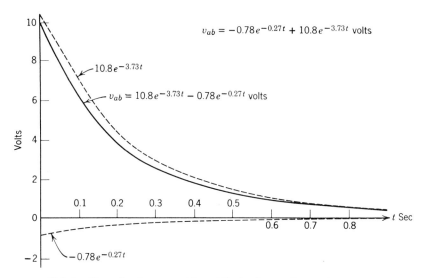

FIG. E-2.5 Plot of voltage v_{ab} ($R = 1/2$ ohm), overdamped case.

3.1.2 Nonhomogeneous Equation with Sinusoidal Driving Function (Classical Method)

We are limiting ourselves temporarily to driving functions or sources which vary sinusoidally with time. This is the most important form of driving function; the technique applied to this form can also be applied to most of the driving functions encountered in practice. Let Equation 3.2 be written

$$\frac{d^2x}{dt^2} + a_1 \frac{dx}{dt} + a_o x = B \cos(\omega t + \alpha) \qquad t > 0 \qquad (3.17)$$

in which B is the maximum value or amplitude of the sinusoidal function and $B \cos \alpha$ is the value of the driving function at $t = 0$.

The technique we shall use is to so operate on Equation 3.17 as to obtain a homogeneous equation for which we know the solution. Differentiating Equation 3.17 twice yields

$$\frac{d^2}{dt^2}\left(\frac{d^2x}{dt^2} + a_1 \frac{dx}{dt} + a_o x\right) = -\omega^2 B \cos(\omega t + \alpha) \qquad (3.18)$$

The Solution of Linear Differential Equations with Constant Coefficients 97

If we now multiply Equation 3.17 by ω^2 and add the resulting equation to Equation 3.18, we obtain

$$\left(\frac{d^2}{dt^2} + \omega^2\right)\left(\frac{d^2x}{dt^2} + a_1\frac{dx}{dt} + a_o x\right) = 0. \tag{3.19}$$

This technique is similar to that used in Example 2.11, page 62. Equation 3.19 is really a fourth-order equation; we have written it in the form shown in order that the roots can readily be found. For this homogeneous equation we again assume $x = Ae^{mt}$ and write the characteristic equation:

$$(m^2 + \omega^2)(m^2 + a_1 m + a_0) = 0. \tag{3.20}$$

The solution has four roots, two being the roots for the homogeneous equation, and the other two being $\pm j\omega$, a special case of complex conjugates (refer to Equation 3.16 and let $\alpha = 0$, $\beta = \omega$). The solution can be written (for real values of m_1 and m_2 and for $m_1 \neq m_2$):

$$x = A_1 e^{m_1 t} + A_2 e^{m_2 t} + K_1 \cos \omega t + K_2 \sin \omega t. \tag{3.21}$$

In order to distinguish the source response from the free response, the constants K_1 and K_2 are used here as coefficients for the source response whereas the constants A_1 and A_2 are used as coefficients for the free response. K_1 and K_2 may be evaluated by substituting Equation 3.21 into Equation 3.17. One need not substitute the entire Equation 3.21 since it is already known that use of $(A_1 e^{m_1 t} + A_2 e^{m_2 t})$ will result in a value of zero for the left side of the equation. [One special case is possible, with $m = \pm j\omega$. This means that the forced response has the same roots as the free response and we must use for the forced response, $t(K_1 \cos \omega t + K_2 \sin \omega t)$. This case is similar to that of the critically damped case of the previous article.]

The terms $(A_1 e^{m_1 t} + A_2 e^{m_2 t})$ are known as the free response, complementary function, or transient response. The remaining terms, those which satisfy the nonhomogeneous equation, are known as the particular integral, source response, forced response, or steady-state response. We shall refer to the solution of the homogeneous equation as the free response, and to the remaining terms as source response, forced response, or steady-state response. The complete response is the sum of these responses.

The reason that the particular integral or source response is often called the steady state response is that in almost all systems the free response approaches zero as time becomes large. This is true in any electric circuit in which there exists a resistance element having a nonzero value. The resistance element causes a reduction or attenuation of the free response as time progresses because some energy is dissipated as heat.

The steady-state solution to Equation 3.17, which has a driving function $B\cos(\omega t + \alpha)$, is $x_{ss} = K_1 \cos \omega t + K_2 \sin \omega t$. This solution applies to all finite values of ω, including $\omega = 0$, for which the steady-state solution is then the constant K_1. This means that the steady-state solution to a d-c driving function is a constant and that therefore steady-state values of voltages and currents will be constants. Thus, for direct current the steady-state voltage across an inductance must be zero since the current through the inductance is a constant. Also the steady state current through a capacitance must be zero since the voltage across the capacitance is a constant.

The technique used in this article to obtain the steady-state response will also apply to functions of time other than $\cos(\omega t + \alpha)$. Such functions of time include t^n, e^{at}, $e^{at} \cos(\omega t + \alpha)$, $t^n \cos(\omega t + \alpha)$, and $t^n e^{at} \cos(\omega t + \alpha)$, in which n is a positive integer and a is a constant. The steady-state solution will consist of the sum of a term of the same form as the driving function plus terms which include the forms of all the derivatives of the driving function. If the driving function is Bt^3 the steady-state solution will be $x_{ss} = K_1 t^3 + K_2 t^2 + K_3 t + K_4$. The method is referred to as "the method of undetermined coefficients."

Example 3.3

GIVEN: The circuit shown in Fig. E-3.1. $i(0+) = 1$ amp, $v_{cd}(0+) = 1$ volt.

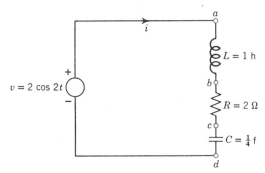

FIG. E-3.1

TO FIND: The complete response i.
SOLUTION: Kirchhoff's equation is

$$L\frac{di}{dt} + Ri + \frac{1}{C}\int_0^t i\, dt + v_{cd}(0+) = v \tag{1}$$

We differentiate Equation 1 and divide it by L to obtain the standard form,

$$\frac{d^2 i}{dt^2} + \frac{R}{L}\frac{di}{dt} + \frac{i}{LC} = \frac{1}{L}\frac{dv}{dt}. \tag{2}$$

Substitution of the values given for this problem results in

$$\frac{d^2i}{dt^2} + 2\frac{di}{dt} + 4i = -4\sin 2t. \quad (3)$$

We differentiate Equation 3 twice to obtain

$$\frac{d^2}{dt^2}\left(\frac{d^2i}{dt^2} + 2\frac{di}{dt} + 4i\right) = 16\sin 2t. \quad (4)$$

Multiplication of Equation 3 by the constant 4 and then addition of the resulting equation to Equation 4 gives

$$\left(\frac{d^2}{dt^2} + 4\right)\left(\frac{d^2i}{dt^2} + 2\frac{di}{dt} + 4i\right) = 0 \quad (5)$$

The roots are

$m_1 = j2$

$m_2 = -j2$

$m_3 = -1 + \sqrt{1-4} = -1 + j\sqrt{3}$

and

$m_4 = -1 - j\sqrt{3}.$

m_1 and m_2 are the roots of the source or steady-state response; m_3 and m_4 are the roots of the free response. We can write the complete solution as

$$i = \underbrace{K_1 \cos 2t + K_2 \sin 2t}_{i_{ss} \text{ nonhomog.}} + \underbrace{e^{-t}(A_1 \cos \sqrt{3}t + A_2 \sin \sqrt{3}t)}_{i_f \text{ homog.}}. \quad (6)$$

We use K_1 and K_2 as coefficients of the steady-state response, A_1 and A_2 as coefficients of the free response. The coefficients K_1 and K_2 are evaluated by substituting the source response, $K_1 \cos 2t + K_2 \sin 2t$, into Equation 3. (Recall that it is unnecessary to substitute the entire solution, as the free response must always result in a value of zero for the left side of the equation.)

We obtain

$$\frac{d^2}{dt^2}(K_1 \cos 2t + K_2 \sin 2t) + 2\frac{d}{dt}(K_1 \cos 2t + K_2 \sin 2t)$$

$$+ 4(K_1 \cos 2t + K_2 \sin 2t) = -4\sin 2t,$$

which results in

$$4K_2 \cos 2t - 4K_1 \sin 2t = -4 \sin 2t. \quad (7)$$

In order for this equation to hold at all values of time, the coefficients of like terms must be equal, or

$K_1 = 1$,

and

$K_2 = 0$.

Our solution for i becomes (from Equation 6),

$$i = \cos 2t + e^{-t}(A_1 \cos \sqrt{3}t + A_2 \sin \sqrt{3}t) \text{ amperes.} \tag{8}$$

The constants A_1 and A_2 are evaluated from the initial conditions. **It is very important to apply initial conditions to the complete solution ($i_{ss} + i_f$), and not to just one part of this solution.** The reason is that physically there is only one current, the complete solution, to which initial conditions apply. The separation of the current into two portions, i_{ss} and i_f, is for mathematical convenience only. The easiest initial conditions to use are $i(0+)$ and $(di/dt)(0+)$. $i(0+)$ is given as 1 ampere. To determine $(di/dt)(0+)$, we go to Equation 1 and substitute values to obtain

$$\frac{di}{dt} + 2i + 4 \int_0^t i \, dt + 1 = 2 \cos 2t. \tag{9}$$

Evaluation of Equation 9 at $t = 0+$ gives

$$\frac{di}{dt}(0+) + 2(1) + 4(0) + 1 = 2,$$

or

$$\frac{di}{dt}(0+) = -1 \text{ amp/sec.}$$

At $t = 0+$, Equation 8 becomes

$1 = 1 + 1(A_1)$,

or

$A_1 = 0$.

To apply the second initial condition we differentiate Equation 8, using the fact that $A_1 = 0$, to obtain

$$\frac{di}{dt} = -2 \sin 2t - e^{-t}A_2 \sin \sqrt{3}t + \sqrt{3}e^{-t}A_2 \cos \sqrt{3}t. \tag{10}$$

Then $(di/dt)(0+) = -1 = \sqrt{3}A_2$, or $A_2 = -1/\sqrt{3}$. Our complete solution may be written

$$i = \cos 2t - \frac{1}{\sqrt{3}} e^{-t} \sin \sqrt{3}t \text{ amperes.} \tag{11}$$

Figure E-3.2 shows the free response, the steady-state response and the complete response as functions of time. For this case the complete response approaches closely to the steady-state response in one cycle.

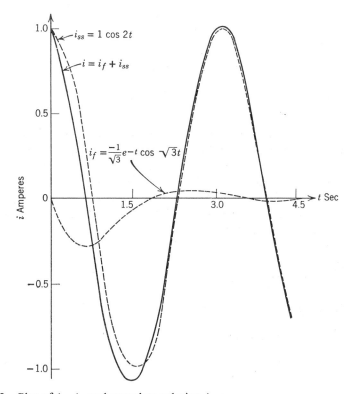

FIG. E-3.2 Plot of i_{ss}, i_f, and complete solution i.

3.2 LAPLACE TRANSFORMATION METHOD

The Laplace transformation method, as applied to electrical circuit transients, is essentially a means of transforming linear differential equations in the time domain to algebraic equations in the complex frequency (s) domain, and then, after suitable algebra, transforming back into the time domain to obtain the complete response. The pair of transforms is often written as shown below, the symbol, $\mathscr{L}$, being read "Laplace transform of" and the symbol, $\mathscr{L}^{-1}$, being read "inverse Laplace transform of."

$$\mathscr{L}f(t) = F(s) = \int_0^\infty f(t)e^{-st}\,dt \tag{3.22}$$

$$\mathscr{L}^{-1}F(s) = f(t) = \frac{1}{2\pi j}\int_{c-j\infty}^{c+j\infty} F(s)e^{st}\,ds. \tag{3.23}$$

The variable s is complex, $(\sigma + j\omega_s)$, and has the units of frequency, σ being called neper frequency and ω_s radian frequency.[2] The lower limit for the integral of the transform is zero because in circuit theory we are interested only in phenomena after time equals zero, the initial conditions serving as the starting point for the variables.

The student should not expect to see physical significance in s or in the Laplace transformation but should accept these as useful mathematical concepts. As one gains experience with the application of transform methods, one becomes so familiar with certain relations between the s domain and the time domain or between the s domain and the sinusoidal steady state domain that the s domain will appear to have physical significance. Until that time it is best to treat the transform as a mathematical operation.

The integral is improper because of the upper limit of ∞, and therefore we need to ask whether the transform exists for the functions we deal with. The answer is that the transform exists, or the integral converges, for functions of time ($t > 0$) which are single-valued, which have a finite number of finite discontinuities, and for which it is possible to find values of M, α, and T such that

$$|f(t)| < Me^{\alpha t}, \text{ for } t > T,$$

where M and T are positive real numbers and α is a real number. All of the functions which we encounter in ordinary linear circuit analysis meet the above conditions. Furthermore, there is one and only one $F(s)$ for a given $f(t)$, or the transform is unique.

It is relatively easy to secure the transform $F(s)$ from a given $f(t)$ and thus a table relating these quantities may be developed. The inverse operation, using Equation 3.23, is not easy, requiring a knowledge of functions of a complex variable. We need to ask whether a table developed from Equation 3.23 will correspond to a table developed from Equation 3.22. The answer is that there is an exact correpondence, for $t > 0$, wherever $f(t)$ is continuous. This is the only condition we are concerned with in our initial applications and so we shall use exclusively a table developed from Equation 3.22. We should expect a one-to-one correspondence since the solutions to our differential equations are known to be unique.

It is presumed that the student is taking calculus concurrently with the course in which this text is used. He will note that we have already introduced the solution to a specific and simple form of differential equations. We hope this will spur his interest in the mathematical course in differential equations which will be taken at a later time. He

---

[2] The subscript s is used with ω here to avoid confusion with ω in the trigonometric time functions, $\sin \omega t$ and $\cos \omega t$.

will note that we are also using conclusions obtained from the general areas of functions of a complex variable and transform calculus. We hope that the student will be motivated to take mathematical courses in these areas at such time as his background permits.

The Laplace transformation method has many advantages over the classical method, and these will become apparent as our use of these methods progress. There are a few instances in which the Laplace transform tends to hide some significant physical phenomena; we shall try to point these out.

3.2.1 Development of a Table of Transforms

We shall develop a portion of the short Table of Transforms shown on page 108. The student will find the remaining transforms given as exercises at the end of this chapter. Additional transforms will be developed as needed throughout this text.

From the definition of the Laplace transform, $\mathscr{L}[i(t)] = \mathscr{L}[i] = I(s)$ and $\mathscr{L}[v(t)] = \mathscr{L}[v] = V(s)$.

Since $\int [f_1(t) + f_2(t)]\,dt = \int f_1(t)\,dt + \int f_2(t)\,dt$, it follows that $\mathscr{L}[f_1(t) + f_2(t)] = F_1(s) + F_2(s)$.

For similar reasons it follows that

$$\mathscr{L}[\sum i = 0] = \sum I(s) = 0,$$
$$\mathscr{L}[\sum v = 0] = \sum V(s) = 0.$$

3.2.1.1 The Laplace Transform of a Constant (A). From Equation 3.22,

$$\mathscr{L}(A) = \int_0^\infty A e^{-st}\,dt = \frac{A e^{-st}}{-s}\bigg|_0^\infty = \frac{A e^{-\sigma t} e^{-j\omega_s t}}{-s}\bigg|_0^\infty = \frac{A}{s}.$$

For values of $\sigma > 0$ the integral converges. Fig. 3.1 shows how this may be illustrated, the integral converging for all values of s in the right half plane. Do we need to keep in mind that the transform is

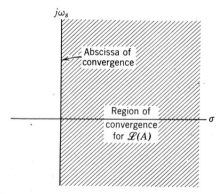

FIG. 3.1 Region of convergence for $\mathscr{L}(A)$.

apparently restricted to this region by the defining integral? In the case of the inverse transform, Equation 3.23, it is important that the constant c be in the region of convergence, or that c be greater than the abscissa of convergence. In all other cases of interest to us the restriction on the integral is not important. We shall therefore "extend by definition" the restricted region of $F(s)$ to include all of the s plane where $F(s)$ is well behaved. In this case $\mathscr{L}(A) = A/s$ is well behaved everywhere except at $s = 0$.

3.2.1.2 $\mathscr{L}(e^{-at})$. From Equation 3.22,

$$\mathscr{L}(e^{-at}) = \int_0^\infty e^{-at}e^{-st}\,dt = \int_0^\infty e^{-(s+a)t}\,dt$$

$$= \frac{-e^{-s(+a)t}}{s+a}\bigg|_0^\infty = \frac{-e^{-(\sigma+a)t}e^{-j\omega t}}{s+a}\bigg|_0^\infty$$

$$= \frac{1}{s+a}.$$

The integral converges for values $\sigma > -a$; this is shown graphically in Fig. 3.2. Since this restriction on the transform is important only in taking the inverse transform by Equation 3.23 (which we shall not do), we consider the transform $\mathscr{L}(e^{-at}) = 1/(s+a)$ to be defined over the entire s plane except at $s = -a$.

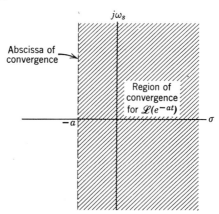

FIG. 3.2 Region of convergence for $\mathscr{L}(e^{-at})$.

3.2.1.3 $\mathscr{L}(te^{-at})$. From Equation 3.22,

$$\mathscr{L}(te^{-at}) = \int_0^\infty te^{-at}e^{-st}\,dt = \int_0^\infty te^{-(s+a)t}\,dt.$$

We shall integrate by parts, using the expression from the differential calculus:

$$d(uv) = u\,dv + v\,du.$$

Then
$$\int_{(1)}^{(2)} u \, dv = \int_{(1)}^{(2)} d(uv) - \int_{(1)}^{(2)} v \, du$$
or
$$\int_{(1)}^{(2)} u \, dv = uv \Big|_{(1)}^{(2)} - \int_{(1)}^{(2)} v \, du.$$

In this case we choose $u = t$ and $dv = e^{-(s+a)t} \, dt$. Then $du = dt$, $v = -e^{-(s+a)t}/(s+a)$, and

$$\mathcal{L}(te^{-at}) = -\frac{te^{-(s+a)t}}{s+a}\Big|_0^\infty + \int_0^\infty \frac{e^{-(s+a)t}}{s+a} \, dt$$

$$= 0 - \frac{e^{-(s+a)t}}{(s+a)^2}\Big|_0^\infty = \frac{1}{(s+a)^2}.$$

The term $te^{-(s+a)t}$ approaches zero as t approaches ∞ if $\sigma > -a$. This may be seen by considering the function te^{-mt} in which m is real and positive. This is sketched in Fig. 3.3.

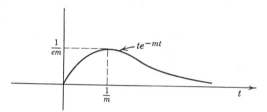

FIG. 3.3 Variation of te^{-mt} with t.

3.2.1.4 $\mathcal{L}[df(t)/dt]$. From the definition of the transform, Equation 3.22,

$$\mathcal{L}\left[\frac{df(t)}{dt}\right] = \int_0^\infty \frac{df(t)}{dt} e^{-st} \, dt.$$

We shall again integrate by parts, choosing

$u = e^{-st}$ and $dv = [df(t)/dt] \, dt$.

Then $du = -se^{-st} \, dt$, $v = f(t)$, and

$$\mathcal{L}\left[\frac{df(t)}{dt}\right] = e^{-st}f(t)\Big|_0^\infty + s\int_0^\infty f(t)e^{-st} \, dt = -f(0+) + sF(s).$$

Note that the quantity, $e^{-st}f(t)$, evaluated at $t = \infty$ is zero because for our functions $|f(t)| < Me^{\alpha t}$ and because there exist values of s for which $\sigma > \alpha$. We use $f(0+)$ instead of $f(0-)$ in order to avoid impulses at $t = 0$. In Chapter 9 we shall justify the use of $f(0-)$ values in order to include these singular functions.

2.1.5 $\mathscr{L}\left[\int_0^t f(t)\, dt\right]$. From Equation 3.22,

$$\mathscr{L}\left[\int_0^t f(t)\, dt\right] = \int_0^\infty \left[\int_0^t f(t)\, dt\right] e^{-st}\, dt.$$

Again we integrate by parts, letting

$$u = \int_0^t f(t)\, dt \quad \text{and} \quad dv = e^{-st}\, dt.$$

Then

$$du = f(t)\, dt, \quad v = \frac{e^{-st}}{-s}$$

and

$$\mathscr{L}\left[\int_0^t f(t)\, dt\right] = \frac{e^{-st}}{-s} \int_0^t f(t)\, dt \Bigg|_0^\infty$$
$$+ \frac{1}{s}\int_0^\infty f(t) e^{-st}\, dt = \frac{F(s)}{s}.$$

The reason that the quantity, $e^{-st}\int_0^t f(t)\, dt$, approaches zero as t approaches infinity follows from a proof that if $f(t)$ has a Laplace transform, $\int_0^t f(t)\, dt$ also has a Laplace transform.[3]

2.1.6 $\mathscr{L}(\cos \omega t)$. From Equation 3.22 $\mathscr{L}(\cos \omega t) = \int_0^\infty \cos \omega t\, e^{-st}\, dt$. From Euler's formula,

$$e^{j\omega t} = \cos \omega t + j \sin \omega t,$$

and

$$e^{-j\omega t} = \cos \omega t - j \sin \omega t,$$

or

$$\cos \omega t = \tfrac{1}{2}(e^{j\omega t} + e^{-j\omega t}).$$

Then

$$\mathscr{L}(\cos \omega t) = \frac{1}{2}\int_0^\infty e^{-(s-j\omega)t}\, dt + \frac{1}{2}\int_0^\infty e^{-(s+j\omega)t}\, dt$$
$$= \frac{1}{2}\left[\frac{1}{s-j\omega} + \frac{1}{s+j\omega}\right] = \frac{s}{s^2 + \omega^2}$$

2.1.7 $\mathscr{L}[e^{-at}\cos(\omega t + \phi)]$. Note that from Euler's formula,

$$\cos(\omega t + \phi) = \frac{e^{j(\omega t + \phi)} + e^{-j(\omega t + \phi)}}{2},$$

---

[3] See Murray F. Gardner and John L. Barnes, *Transients in Linear Systems*, Wiley, New York, 1942, Vol. 1, p. 129.

Table of Transforms

	$f(t)$	$F(s)$ for $s = \sigma + j\omega_s$
1.	i	$I(s)$
2.	v	$V(s)$
3.	$f_1(t) + f_2(t)$	$F_1(s) + F_2(s)$
4.	$\Sigma i = 0$	$\Sigma I(s) = 0$
5.	$\Sigma v = 0$	$\Sigma V(s) = 0$
6.	$Af(t)$	$AF(s)$
7.	A	$\dfrac{A}{s}$
8.	e^{-at}	$\dfrac{1}{s+a}$
9.	te^{-at}	$\dfrac{1}{(s+a)^2}$
10.	$\dfrac{t^{n-1}e^{-at}}{(n-1)!}$	$\dfrac{1}{(s+a)^n}$
11.	$\dfrac{df(t)}{dt}$	$sF(s) - f(0+)$
12.	$\displaystyle\int_0^t f(t)\,dt$	$\dfrac{F(s)}{s}$
13.	$\cos \omega t$	$\dfrac{s}{s^2 + \omega^2}$
14.	$\sin \omega t$	$\dfrac{\omega}{s^2 + \omega^2}$
15.	$\cos(\omega t + \phi)$	$\dfrac{s \cos \phi - \omega \sin \phi}{s^2 + \omega^2}$
16.	$\sin(\omega t + \phi)$	$\dfrac{s \sin \phi + \omega \cos \phi}{s^2 + \omega^2}$
17.	$e^{-at} \cos \omega t$	$\dfrac{s + a}{(s+a)^2 + \omega^2}$
18.	$e^{-at} \sin \omega t$	$\dfrac{\omega}{(s+a)^2 + \omega^2}$

and therefore

$$\mathcal{L}[e^{-at}\cos(\omega t + \phi)]$$

$$= \int_0^\infty e^{-at}\left[\frac{e^{j(\omega t+\phi)} + e^{-j(\omega t+\phi)}}{2}\right]e^{-st}\,dt$$

$$= \frac{1}{2}\int_0^\infty e^{-(s+a-j\omega)t}e^{j\phi}\,dt + \frac{1}{2}\int_0^\infty e^{-(s+a+j\omega)t}e^{-j\phi}\,dt$$

$$= \frac{e^{-(s+a-j\omega)t}}{-2(s+a-j\omega)}e^{j\phi}\bigg|_0^\infty - \frac{e^{-(s+a+j\omega)t}}{2(s+a+j\omega)}e^{-j\phi}\bigg|_0^\infty$$

$$= \frac{e^{j\phi}}{2(s+a-j\omega)} + \frac{e^{-j\phi}}{2(s+a+j\omega)}$$

$$= \frac{(s+a)(e^{j\phi}+e^{-j\phi}) + j\omega(e^{j\phi}-e^{-j\phi})}{2[(s+a)^2+\omega^2]}.$$

From Euler's formula,

$$\cos\phi = \frac{e^{j\phi}+e^{-j\phi}}{2} \quad \text{and} \quad \sin\phi = \frac{e^{j\phi}-e^{-j\phi}}{2j},$$

and therefore

$$\mathcal{L}[e^{-at}\cos(\omega t + \phi)] = \frac{(s+a)\cos\phi - \omega\sin\phi}{(s+a)^2+\omega^2}.$$

3.2.2 Application of Laplace Transforms to Circuit Equations

For each of the relations which we may write in the time domain, a similar relation may be written in the complex frequency (s) domain by means of the Laplace transform. We will demonstrate this first for the circuit of Fig. 3.4.

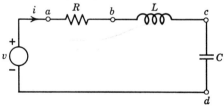

FIG. 3.4 A series R-L-C circuit with voltage source.

In the time domain Kirchhoff's voltage equation is

$$Ri + L\frac{di}{dt} + \frac{1}{C}\int_0^t i\,dt + v_{cd}(0+) = v, \qquad t > 0.$$

This equation holds for all time greater than zero; the Laplace transform of each term is presumed to hold for all values of s for which it is well behaved. This transformed equation is

$$RI(s) + LsI(s) - Li(0+) + \frac{1}{Cs}I(s) + \frac{v_{cd}(0+)}{s} = V(s). \qquad (3.24)$$

From Equation 3.24 a transform circuit may be sketched as shown in Fig. 3.5.

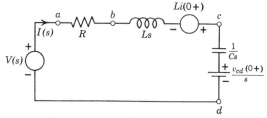

FIG. 3.5 Transform circuit for Fig. 3.4.

In this case the change from the time domain circuit to the transform circuit consists of changing v to $V(s)$, i to $I(s)$, R to R, L to Ls plus an independent transform voltage of $Li(0+)$ with polarity $-$ to $+$ in the direction of i for positive $i(0+)$, and C to $1/Cs$ with an independent transform voltage of $[v_{cd}(0+)]/s$, the polarity being the same as $v_{cd}(0+)$. The initial conditions (two of them in this case) appear as independent transform voltage sources. Observe that Equation 3.24 may be written:

$$I(s)\left[Ls + R + \frac{1}{Cs}\right] = V(s) + Li(0+) - \frac{v_{cd}(0+)}{s}. \qquad (3.25)$$

If we desired we could treat each term on the right hand side as a separate driving function and solve the problem by use of superposition.

The transform current $I(s)$ actually has the units of charge, since $I(s) = \int_0^\infty ie^{-st}\, dt$, and transform voltages $V(s)$, $Li(0+)$, and $[v_{cd}(0+)]/s$ actually have the units of volt seconds or flux linkages. However, we shall not call $I(s)$ charge or $V(s)$ flux linkage, but rather we shall continue to call $I(s)$ transform current and $V(s)$ transform voltage. The ratio of $V(s)$ to $I(s)$ has the units of ohms, as it should, and this unit applies to the terms Ls, R, $1/Cs$, and Ms.

Let us now consider in more detail the transform voltage-current relations for L and C. For L we have just demonstrated that for the equation, $v_{bc} = L(di/dt)$, which applies to Fig. 3.6a, the transform

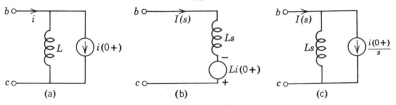

FIG. 3.6 Circuits and voltage-current relationship for inductance, L. (a) Time domain, $i = i(0+) + \frac{1}{L}\int_0^t v_{bc}\, dt$. (b) Transform domain, $V_{bc}(s) = LsI(s) - Li(0+)$. (c) Transform domain, $I(s) = [V_{bc}(s)]/Ls + [i(0+)]/s$.

equation is $V_{bc}(s) = LsI(s) - Li(0+)$ and the transform circuit is shown in Fig. 3.6b. Suppose we write the voltage current relation as

$$i = \frac{1}{L}\int_0^t v_{bc}\, dt + i(0+).$$

Then

$$I(s) = \frac{V_{bc}(s)}{Ls} + \frac{i(0+)}{s},$$

the circuit for which is shown in Fig. 3.6c. It is apparent that the two transform equations are identical and that the two circuits Fig. 3.6b and Fig. 3.6c are equivalent and may be used interchangeably. The transform circuit shown in Fig. 3.6c is the transform of the time domain circuit of Fig. 3.6a.

For a capacitor C, shown in Fig. 3.7a, the voltage-current relations are

$$v_{cd} = \frac{1}{C}\int_0^t i\, dt + v_{cd}(0+)$$

or

$$i = C\frac{dv_{cd}}{dt}.$$

The corresponding transform equations are

$$V_{cd}(s) = \frac{I(s)}{Cs} + \frac{v_{cd}(0+)}{s},$$

and

$$I(s) = CsV_{cd}(s) - Cv_{cd}(0+).$$

The last equation can be obtained by multiplying the previous equation by Cs. For these equations, two interchangeable circuits may be sketched as in Fig. 3.7b and c.

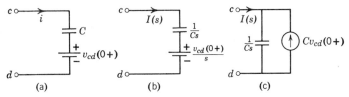

FIG. 3.7 Circuits and voltage-current relationship for capacitance, C.
(a) Time domain, $v_{cd} = v_{cd}(0+) + \frac{1}{C}\int_0^t i\, dt$. (b) Transform domain, $V_{cd}(s) = [I(s)]/Cs + [v_{cd}(0+)]/s$. (c) Transform domain, $I(s) = CsV_{cd}(s) - Cv_{cd}(0+)$.

In Table 3.1 is shown again the various elements defined, their voltage current relationship and their circuits in both the time domain and the complex frequency domain.

Table 3.1

Element	Time Domain	Complex Frequency Domain
Independent voltage source	v ⊙ ⎓ V $v = f(t)$ only $V =$ constant	$V(s)$ ⊙ ⎓ $\dfrac{V}{s}$
Independent current source	i ⊙ ⊙ I $i = f(t)$ only $I =$ constant	$I(s)$ ⊙ ⊙ $\dfrac{I}{s}$
Resistance, R	$v_{ab} = Ri$ $i = \dfrac{v_{ab}}{R}$	$V_{ab}(s) = RI(s)$ $I(s) = \dfrac{V_{ab}(s)}{R}$
Self inductance, L	$v_{ab} = L\dfrac{di}{dt}$ $i = \dfrac{1}{L}\displaystyle\int_0^t v_{ab}\, dt + i(0+)$	$V_{ab}(s) = LsI(s) - Li(0+)$ $I(s) = \dfrac{V_{ab}(s)}{Ls} + \dfrac{i(0+)}{s}$

(*continued*)

Table 3.1 (*Continued*)

Mutual inductance, M

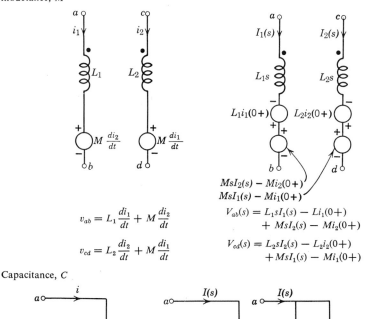

$$v_{ab} = L_1 \frac{di_1}{dt} + M \frac{di_2}{dt}$$

$$v_{cd} = L_2 \frac{di_2}{dt} + M \frac{di_1}{dt}$$

$$V_{ab}(s) = L_1 s I_1(s) - L_1 i_1(0+) + M s I_2(s) - M i_2(0+)$$

$$V_{cd}(s) = L_2 s I_2(s) - L_2 i_2(0+) + M s I_1(s) - M i_1(0+)$$

Capacitance, C

$$v_{ab} = \frac{1}{C} \int_0^t i\, dt + v_{ab}(0+)$$

$$i = C \frac{dv_{ab}}{dt}$$

$$V_{ab}(s) = \frac{I(s)}{Cs} + \frac{v_{ab}(0+)}{s}$$

$$I(s) = C s V_{ab}(s) - C v_{ab}(0+)$$

3.2.3 Solution of Circuits by Means of the Laplace Transform

We have shown that the Laplace transform changes differential equations into algebraic equations. The algebraic equations will have several $I(s)$'s or $V(s)$'s as unknowns, and a solution for some particular unknown can be obtained by the rules of algebra. We need now to transform our particular unknown (a function of s) into the time domain. We shall find that the function of s is the ratio of two polynomials in s. Instead of using Equation 3.23 to obtain the inverse Laplace transform we shall use partial fraction expansion and the table of transforms. Before showing a complete solution of an electrical circuit we will demonstrate the partial fraction expansion of terms commonly encountered.

The Solution of Linear Differential Equations with Constant Coefficients 113

Example 3.4

GIVEN: The quotient of polynomials

$$V(s) = \frac{s+3}{s^2 + 5s + 4} \tag{1}$$

TO FIND: The partial fraction expansion.
SOLUTION: We first factor the denominator to give

$$V(s) = \frac{s+3}{(s+1)(s+4)}. \tag{2}$$

The expansion is then

$$\frac{s+3}{(s+1)(s+4)} = \frac{A}{s+1} + \frac{B}{s+4}. \tag{3}$$

Multiplying Equation 3 by $s+1$ we have

$$\frac{(s+1)(s+3)}{(s+1)(s+4)} = \frac{(s+1)A}{s+1} + \frac{(s+1)B}{s+4}, \tag{4}$$

and upon canceling common factors,

$$\frac{s+3}{s+4} = A + \frac{s+1}{s+4}B. \tag{5}$$

The term s is an algebraic term that may have any value. If we let $s = -1$ the coefficient of B in Equation 5 becomes zero, and we may evaluate A. Thus

$$A = \frac{s+3}{s+4}\bigg|_{s=-1} = \frac{-1+3}{-1+4} = \frac{2}{3}.$$

B may then be evaluated in exactly the same manner.[4] Thus

$$B = \frac{s+3}{s+1}\bigg|_{s=-4} = \frac{-4+3}{-4+1} = +\frac{1}{3}.$$

Then

$$V(s) = \frac{\frac{2}{3}}{s+1} + \frac{\frac{1}{3}}{s+4}.$$

Example 3.5

GIVEN: The quotient of polynomials.

$$I(s) = \frac{2s+2}{(s+2)^2(s+3)}. \tag{1}$$

---

[4] This method is referred to as Heaviside's expansion method.

TO FIND: The partial fraction expansion.

SOLUTION: We note that there is a repeated root $(s + 2)^2$ in the denominator of Equation 1, and therefore the partial fraction expansion becomes

$$I(s) = \frac{2(s + 1)}{(s + 2)^2(s + 3)} = \frac{A}{s + 2} + \frac{B}{(s + 2)^2} + \frac{C}{s + 3} \tag{2}$$

We may evaluate the constants B and C by the same method as in Example 3.4. Thus

$$B = \frac{2(s + 1)}{(s + 3)}\bigg|_{s=-2} = -2$$

and

$$C = \frac{2(s + 1)}{(s + 2)^2}\bigg|_{s=-3} = -4$$

To evaluate A we substitute the known values of B and C into Equation 2, multiply both sides by the entire denominator $(s + 2)^2(s + 3)$, and equate like terms. Thus,

$$2s + 2 = A(s + 2)(s + 3) - 2(s + 3) - 4(s + 2)^2$$

or

$$2s + 2 = s^2(A - 4) + s(5A - 2 - 16) + 6A - 6 - 16.$$

Since this equation holds for all values of s, the equation must hold separately for the terms in s^2, the terms in s and the constants. Then

$$0 = A - 4,$$
$$2 = 5A - 18,$$

and

$$2 = 6A - 22,$$

from which $A = 4$.

Thus

$$I(s) = +\frac{4}{s + 2} - \frac{2}{(s + 2)^2} - \frac{4}{s + 3}.$$

Note that the inverse transform of $4/(s + 2)$ is $4e^{-2t}$ while the inverse transform of $2/(s + 2)^2$ is $2te^{-2t}$. This justifies the form of solution assumed for two equal roots in the classical method, page 88.

Example 3.6

GIVEN: The quotient of polynomials,

$$I(s) = \frac{11s + 1}{(s + 1)(s^2 + 4)}. \tag{1}$$

TO FIND: The partial fraction expansion.
SOLUTION: This ratio of polynomials may be expanded as

$$\frac{11s + 1}{(s + 1)(s^2 + 4)} = \frac{A}{s + 1} + \frac{Bs + C}{s^2 + 4}. \qquad (2)$$

A is evaluated in the manner of previous examples.

$$A = \frac{11s + 1}{(s^2 + 4)}\bigg|_{s=-1} = \frac{-11 + 1}{5} = -2.$$

With the value of A substituted into Equation 2, both sides of Equation 2 are multiplied by the denominator $(s + 1)(s^2 + 4)$ to give

$$11s + 1 = -2(s^2 + 4) + (Bs + C)(s + 1),$$

or

$$11s + 1 = s^2(B - 2) + s(B + C) + C - 8.$$

Then

$$C = +9, \qquad B = +2,$$

and

$$I(s) = \frac{-2}{s + 1} + \frac{2s}{s^2 + 4} + \frac{9}{s^2 + 4}.$$

Example 3.7

GIVEN: The circuit shown in Fig. E-7.1. The switch is closed at time $t = 0$. The capacitor has an initial charge $v_{ab}(0-) = -20$ volts.

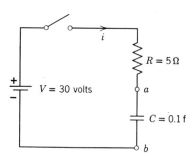

FIG. E-7.1

TO FIND: The complete solution i, employing the methods of the Laplace Transform, and to compare the answer with the solution by the classical method of Example 3.1.

SOLUTION: Kirchhoff's voltage equation is

$$Ri + \frac{1}{C}\int_0^t i\, dt + v_{ab}(0+) = V, \qquad (1)$$

with

$$v_{ab}(0-) = v_{ab}(0+) = -20.$$

116

Taking the Laplace transform of Equation 1 gives the literal transform equation,

$$RI(s) + \frac{I(s)}{Cs} + \frac{v_{ab}(0+)}{s} = \frac{V}{s},$$

or

$$I(s)\left[R + \frac{1}{Cs}\right] = \frac{V - v_{ab}(0+)}{s}. \tag{2}$$

Rearrangement of Equation 2 gives

$$I(s) = \frac{V - v_{ab}(0+)}{s\left(R + \frac{1}{Cs}\right)} = \frac{V - v_{ab}(0+)}{R\left(s + \frac{1}{RC}\right)}. \tag{3}$$

With numerical values Equation 3 becomes

$$I(s) = \frac{30 - (-20)}{5(s+2)} = \frac{10}{s+2}.$$

From our Table of Transforms (page 108) we can then write the time domain solution $i = 10e^{-2t}$ amp, which is the same as the response found by classical methods in Example 3.1. Note here that the initial conditions are incorporated in the beginning of the solution by the transform method rather than near the end of the solution as in the classical method.

Example 3.8

GIVEN: The circuit shown in Fig. E-8.1. Steady-state exists with the switch open. At time $t = 0$ the switch is closed.

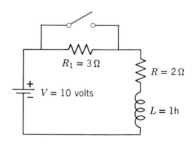

FIG. E-8.1

TO FIND: The complete response for the current i for $t \geqslant 0$, by the Laplace transform method.

SOLUTION: For our circuit $i(0+)$ must equal $i(0-)$ (there are no voltage impulses present). To find $i(0-)$, and thus $i(0+)$, we must solve the circuit for i with the switch open. Since it is stated that the circuit is in the steady-state with the switch open we need only find

The Solution of Linear Differential Equations with Constant Coefficients

the steady-state solution to determine $i(0-)$. This can be readily accomplished as follows. Our previous experience in finding the complete response of a circuit shows us that the steady-state response is always of the same form as the source. Thus, if the source is a constant the steady-state response will also be a constant. Therefore, Kirchhoff's voltage equation for our circuit for $t < 0$ (switch open) is

$$(R + R_1)i + L\frac{di}{dt} = V. \tag{1}$$

The source response i_{ss} is found by substituting $i_{ss} = K$ into Equation 1, which gives

$$(R + R_1)i_{ss} + L\frac{di_{ss}}{dt} = V, \tag{2}$$

or

$$(R + R_1)K + L\frac{d}{dt}K = V. \tag{3}$$

The derivative of a constant is 0, $L\frac{di_{ss}}{dt} = 0$, and thus our solution for Equation 3 is

$$i_{ss} = K = \frac{V}{R + R_1} = i(0-) = i(0+). \tag{4}$$

We now write Kirchhoff's voltage equation for our circuit for $t > 0$. For $t > 0$ (switch closed)

$$Ri + L\frac{di}{dt} = V. \tag{5}$$

The Laplace transform of Equation 5 is

$$RI(s) + LsI(s) - Li(0+) = \frac{V}{s},$$

or

$$(Ls + R)I(s) = \frac{V}{s} + L\left(\frac{V}{R + R_1}\right). \tag{6}$$

Substitution of values into Equation 6 gives

$$(s + 2)I(s) = \frac{10}{s} + 2 \tag{7}$$

Then

$$I(s) = \frac{2s + 10}{s(s + 2)}.$$

By partial fraction expansion,

$$I(s) = \frac{2s + 10}{s(s + 2)} = \frac{K}{s} + \frac{A}{s + 2}$$

Then
$$K = \left.\frac{2s + 10}{s + 2}\right|_{s=0} = \frac{10}{2} = 5,$$
and
$$A = \left.\frac{2s + 10}{s}\right|_{s=-2} = \frac{-4 + 10}{-2} = -3,$$
Thus
$$I(s) = \frac{5}{s} - \frac{3}{s + 2}.$$

From our Table of Transforms the complete response i is

$$i = 5 - 3e^{-2t} \text{ amp.} \tag{8}$$

We can check our solution to see if it satisfies the initial and final values for i:

$$i(0+) = \frac{V}{R + R_1} = \frac{10}{5} = 2 = 5 - 3e^0 = 2, \tag{9}$$

which checks.

Also,

$$i_{ss} = i(t \to \infty) = 5 - 0 = 5. \tag{10}$$

Substitution of Equation 10 into Equation 5 shows that our solution satisfies the original equation under steady state conditions.

A more direct solution for our problem could have been achieved by noting that a Laplace transform equivalent circuit can be drawn directly from the information arrived at in Table 3.1 (page 112). This circuit is shown in Fig. E-8.2. Kirchhoff's voltage equation in terms of the Laplace variable, s, from Fig. E-8.2 is

$$RI(s) + LsI(s) - Li(0+) = V(s)$$
$$2I(s) + sI(s) - 2 = \frac{10}{s},$$
$$(s + 2)I(s) = \frac{10}{s} + 2. \tag{11}$$

Equation 11 is the same as Equation 7.

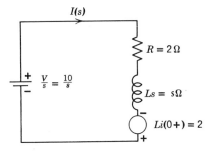

FIG. E-8.2

The Solution of Linear Differential Equations with Constant Coefficients 119

Example 3.9

GIVEN: The circuit shown in Fig. E-9.1. $i(0+) = 1$ amp, $v_{cd}(0+) = 1$ volt. (This is the same circuit as that of Example 3.3.)

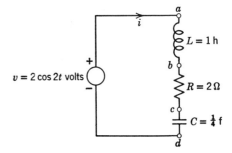

FIG. E-9.1

TO FIND: The complete response i employing the methods of the Laplace transform, and to compare this solution with the solution of Example 3.3.

SOLUTION: Here we will take a direct approach and draw the transform circuit as shown in Fig. E-9.2. Kirchhoff's voltage equation for the circuit written in terms of the Laplace transform variable (s) is

$$sI(s) - 1 + 2I(s) + \frac{4}{s}I(s) + \frac{1}{s} = \frac{2s}{s^2 + 2^2}, \tag{1}$$

or

$$I(s)\left[s + 2 + \frac{4}{s}\right] = \frac{2s}{s^2 + 2^2} + 1 - \frac{1}{s},$$

and

$$I(s) = \frac{1}{s + 2 + \frac{4}{s}}\left[\frac{2s}{s^2 + 2^2} + 1 - \frac{1}{s}\right]. \tag{2}$$

Simplification of (2) gives us

$$I(s) = \frac{s}{s^2 + 2s + 4}\left[\frac{s^3 + s^2 + 4s - 4}{s(s^2 + 2^2)}\right]. \tag{3}$$

The partial fraction expansion may be written as

$$\frac{s^3 + s^2 + 4s - 4}{[(s + 1)^2 + \sqrt{3}^2][s^2 + 2^2]} = \frac{A_1 s + A_2}{(s + 1)^2 + \sqrt{3}^2} + \frac{K_1 s + K_2}{s^2 + 2^2}. \tag{4}$$

Note that the denominator of Equation 4 was formed by changing the term $s^2 + 2s + 4$ to $(s + 1)^2 + \sqrt{3}^2$. This is done that we may have the denominator term in our partial fraction expansion in a recognizable form of the Table of Laplace Transforms.

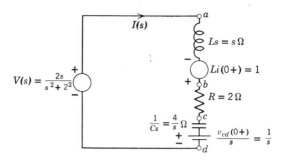

FIG. E-9.2

To evaluate the constants, we multiply both sides of Equation 4 by the denominator to give

$$s^3 + s^2 + 4s - 4 = (A_1 s + A_2)(s^2 + 2^2) + (K_1 s + K_2) \times [(s+1)^2 + \sqrt{3}^2].$$

By performing the indicated multiplication and gathering terms, we have

$$s^3 + s^2 + 4s - 4 = s^3(A_1 + K_1) + s^2(A_2 + 2K_1 + K_2) \\ + s(4A_1 + 4K_1 + 2K_2) + 4A_2 + 4K_2.$$

Equating the coefficients of like terms gives us

$$A_1 + K_1 = 1 \tag{5}$$
$$A_2 + 2K_1 + K_2 = 1 \tag{6}$$
$$4A_1 + 4K_1 + 2K_2 = 4 \tag{7}$$
$$4A_2 + 4K_2 = -4 \tag{8}$$

Simultaneous solution of the Equations 5, 6, 7, and 8 yields

$$A_1 = 0, \quad A_2 = -1, \quad K_1 = +1, \quad K_2 = 0.$$

Then

$$I(s) = \frac{-1}{(s+1)^2 + \sqrt{3}^2} + \frac{s}{s^2 + 2^2}. \tag{9}$$

To make the first term of Equation 9 satisfy the Table of Transforms, numerator and denominator of the first term are multiplied by $\sqrt{3}$ and Equation 9 becomes

$$I(s) = \frac{\sqrt{3}(-1)}{\sqrt{3}[(s+1)^2 + \sqrt{3}^2]} + \frac{s}{s^2 + 2^2} \tag{10}$$

From the Table of Transforms, Equation 10 then becomes

$$i = \frac{-1}{\sqrt{3}} e^{-t} \sin \sqrt{3}t + \cos 2t \tag{11}$$

and we see that our complete solution, Equation 11, is the same as the solution arrived at by the classical method of Example 3.3.

The Solution of Linear Differential Equations with Constant Coefficients

Example 3.10

GIVEN: The circuit shown in Fig. E-10.1, which is the circuit of Example 3.2. $i_2(0+) = 0$

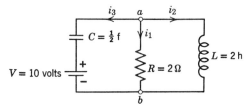

FIG. E-10.1

TO FIND: The complete response v_{ab}, by the method of the Laplace transform.

SOLUTION: The transform circuit of Fig. E-10.1 is shown in Fig. E-10.2.

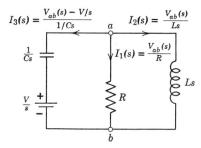

FIG. E-10.2

Kirchhoff's current law applied at junction a of Fig. E-10.2 gives us

$$I_3(s) + I_2(s) + I_1(s) = 0$$

or

$$\frac{V_{ab}(s) - \frac{V}{s}}{\frac{1}{Cs}} + \frac{V_{ab}(s)}{R} + \frac{V_{ab}(s)}{Ls} = 0. \quad (1)$$

Then

$$V_{ab}(s)\left[Cs + \frac{1}{R} + \frac{1}{Ls}\right] = CV$$

or

$$V_{ab}(s) = \frac{RLCVs}{RLCs^2 + Ls + R} = \frac{Vs}{s^2 + \frac{s}{RC} + \frac{1}{LC}}$$

Substituting numerical values, we have

$$V_{ab}(s) = \frac{10s}{s^2 + s + 1}. \quad (2)$$

122

If we complete the square in the denominator of Equation 2 we have

$$V_{ab}(s) = \frac{10s}{\left(s + \frac{1}{2}\right)^2 + \left(\frac{\sqrt{3}}{2}\right)^2}. \tag{3}$$

Consulting our Table of Transforms we see that the form of Equation 3 must be modified to suit transform pair No. 17. This can be accomplished simply by adding and subtracting 5 from the numerator to give

$$V_{ab}(s) = \frac{10s + 5 - 5}{\left(s + \frac{1}{2}\right)^2 + \left(\frac{\sqrt{3}}{2}\right)^2} = \frac{10(s + \frac{1}{2})}{\left(s + \frac{1}{2}\right)^2 + \left(\frac{\sqrt{3}}{2}\right)^2}$$

$$- \frac{5}{\left(s + \frac{1}{2}\right)^2 + \left(\frac{\sqrt{3}}{2}\right)^2}. \tag{4}$$

The first term of Equation 4 is now in proper form to utilize transform pair No. 17. The second term of (4) may be made to satisfy transform pair No. 18 by multiplying numerator and denominator by $\sqrt{3}/2$ to give

$$- \frac{5}{\sqrt{3}/2} \left[\frac{\sqrt{3}/2}{(s + \frac{1}{2})^2 + (\sqrt{3}/2)^2} \right].$$

Thus

$$V_{ab}(s) = \frac{10\left(s + \frac{1}{2}\right)}{\left(s + \frac{1}{2}\right)^2 + \left(\frac{\sqrt{3}}{2}\right)^2} - \frac{10\left(\frac{\sqrt{3}}{2}\right)}{\sqrt{3}\left[\left(s + \frac{1}{2}\right)^2 + \left(\frac{\sqrt{3}}{2}\right)^2\right]}$$

and

$$v_{ab} = 10e^{-t/2}\left[\cos\frac{\sqrt{3}}{2}t - \frac{1}{\sqrt{3}}\sin\frac{\sqrt{3}}{2}t\right] \text{ volts}.$$

This is identical to the solution obtained by the classical method.

It is the purpose of the following example to demonstrate the solution of a slightly more complex circuit, a circuit requiring the solution of two simultaneous equations and one whose differential equation is of third order. We will see that as our circuits grow in complexity, so do their solutions, and that the method of the Laplace transform has advantages over the classical method.

Example 3.11

GIVEN: The circuit shown in Fig. E-11.1. The switch is closed at time $t = 0$. No energy is initially stored in any of the elements. (Initial conditions are zero.)

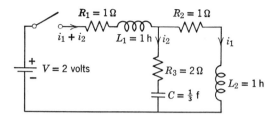

FIG. E-11.1

TO FIND: The complete solution for the current i_1 by (a) the classical method, and (b) the Laplace transform method.

SOLUTION: (a) **Solution by classical method.** Kirchhoff's voltage equations, written around the paths of the independent currents are

$$i_1 + \frac{di_1}{dt} + i_2 + \frac{di_2}{dt} + i_1 + \frac{di_1}{dt} = 2, \tag{1}$$

and

$$i_1 + \frac{di_1}{dt} + i_2 + \frac{di_2}{dt} + 2i_2 + 3\int_0^t i_2\, dt = 2. \tag{2}$$

Gathering terms and rewriting Equations 1 and 2 we obtain

$$2\frac{di_1}{dt} + 2i_1 + \frac{di_2}{dt} + i_2 = 2 \tag{3}$$

and

$$\frac{di_1}{dt} + i_1 + \frac{di_2}{dt} + 3i_2 + 3\int_0^t i_2\, dt = 2. \tag{4}$$

Differentiating Equation 4 gives

$$\frac{d^2 i_1}{dt} + \frac{di_1}{dt} + \frac{d^2 i_2}{dt^2} + 3\frac{di_2}{dt} + 3i_2 = 0. \tag{5}$$

Equations 3 and 5 are rewritten as

$$2\left(\frac{d}{dt} + 1\right)i_1 + \left(\frac{d}{dt} + 1\right)i_2 = 2 \tag{6}$$

and

$$\left(\frac{d^2}{dt^2} + \frac{d}{dt}\right)i_1 + \left(\frac{d^2}{dt^2} + 3\frac{d}{dt} + 3\right)i_2 = 0. \tag{7}$$

The variable i_2 is eliminated from Equations 6 and 7 by first multiplying Equation 6 by $[d^2/dt^2 + 3(d/dt) + 3]$ and Equation 7 by $[d/dt + 1]$ to obtain the equations,

$$\left[\frac{d^2}{dt^2} + 3\frac{d}{dt} + 3\right]2\left(\frac{d}{dt} + 1\right)i_1 + \left[\frac{d^2}{dt^2} + 3\frac{d}{dt} + 3\right]\left(\frac{d}{dt} + 1\right)i_2$$
$$= \left[\frac{d^2}{dt^2} + 3\frac{d}{dt} + 3\right]2, \tag{8}$$

and

$$\left[\frac{d}{dt}+1\right]\left(\frac{d^2}{dt^2}+\frac{d}{dt}\right)i_1 + \left[\frac{d}{dt}+1\right]\left(\frac{d^2}{dt^2}+3\frac{d}{dt}+3\right)i_2 = 0. \tag{9}$$

Then Equation 9 is subtracted from Equation 8 and the indicated operations are performed to obtain

$$\left(2\frac{d^3}{dt^3}+6\frac{d^2}{dt^2}+6\frac{d}{dt}+2\frac{d^2}{dt^2}+6\frac{d}{dt}+6-\frac{d^3}{dt^3}\right.$$
$$\left.-\frac{d^2}{dt^2}-\frac{d^2}{dt^2}-\frac{d}{dt}\right)i_1 = 6, \tag{10}$$

or

$$\left(\frac{d^3}{dt^3}+6\frac{d^2}{dt^2}+11\frac{d}{dt}+6\right)i_1 = 6 \tag{11}$$

We now differentiate Equation 11 to reduce the right-hand side to zero. This results in

$$\frac{d}{dt}\left(\frac{d^3}{dt^3}+6\frac{d^2}{dt^2}+11\frac{d}{dt}+6\right)i_1 = 0. \tag{12}$$

The roots of the characteristic equations of Equation 12 are

$m = 0$ for the source response

and the roots of $m^3 + 6m^2 + 11m + 6 = 0$ for the free response.

We are now confronted with having to find the roots of the third-order equation $m^3 + 6m^2 + 11m + 6 = 0$. There are indeed many ways this may be done.[5] We shall use the method of synthetic division. We arrange the coefficients of our equation in descending order and, after 6 trials obtain -1 as one root.

$$\begin{array}{r} 1 + 6 + 11 + 6 \underline{/1} \\ -1 - 5 - 6 \\ \hline 1 + 5 + 6 \end{array}$$

The remainder $1 + 5 + 6$ may now be used to write the quadratic equation $m^2 + 5m + 6 = 0$, which is seen to have the factors $(m + 3)$ and $(m + 2)$. Thus the roots of our equation are

$m_1 = -1$,
$m_2 = -2$,
and
$m_3 = -3$.

---

[5] See Ley, Lutz, and Rehberg, *Linear Circuit Analysis*, pp. 216–227, McGraw-Hill, New York, 1956.

The solution may be written as

$$i_1 = Ke^{0t} + A_1 e^{-t} + A_2 e^{-2t} + A_3 e^{-3t} \text{ amp.} \tag{13}$$

Substitution of the source response $i_1 = K$ into Equation 11 gives

$$6K = 6,$$

or

$$K = 1.$$

Thus $i_{1ss} = 1$ amp, which can be verified by inspection of the circuit.

The constants A_1, A_2, and A_3 must be determined from initial conditions as follows.

(1) At time $t = 0+$, $i_1(0+) = 0$, from the given initial conditions, and thus

$$i_1(0+) = 0 = 1 + A_1 + A_2 + A_3. \tag{14}$$

(2) Differentiating Equation 13 gives

$$\frac{di_1}{dt} = -A_1 e^{-t} - 2A_2 e^{-2t} - 3A_3 e^{-3t} \tag{15}$$

From Equation 1 we get

$$\frac{di_2}{dt} = 2 - 2i_1 - 2\frac{di_1}{dt} - i_2 \tag{16}$$

Substituting di_2/dt from Equation 16 into Equation 2 gives

$$i_1 + \frac{di_1}{dt} + i_2 + 2 - 2i_1 - 2\frac{di_1}{dt} - i_2 + 2i_2 + 3\int_0^t i_2 \, dt = 2. \tag{17}$$

Since $i_1(0+) = 0$ and $i_2(0+) = 0$, Equation 17 evaluated for time $t = (0+)$ gives

$$\frac{di_1}{dt}(0+) = 0. \tag{18}$$

Then

$$\frac{di_1}{dt}(0+) = 0 = -A_1 - 2A_2 - 3A_3. \tag{19}$$

(3) Differentiating Equation 13 twice gives

$$\frac{d^2 i_1}{dt^2} = A_1 e^{-t} + 4A_2 e^{-2t} + 9A_3 e^{-3t} \tag{20}$$

Differentiating Equations 1 and 2 gives

$$2\frac{di_1}{dt} + 2\frac{d^2 i_1}{dt^2} + \frac{di_2}{dt} + \frac{d^2 i_2}{dt^2} = 0 \tag{21}$$

and

$$\frac{di_1}{dt} + \frac{d^2i_1}{dt^2} + 3\frac{di_2}{dt} + \frac{d^2i_2}{dt^2} + 3i_2 = 0. \tag{22}$$

Solving Equation 21 for d^2i_2/dt^2 and substituting the result into Equation 22 gives

$$\frac{d^2i_1}{dt^2} + \frac{di_1}{dt} - 2\frac{di_2}{dt} - 3i_2 = 0. \tag{23}$$

Evaluation of Equation 23 for time $t = 0+$ yields,

$$\frac{d^2i_1}{dt^2}(0+) = 2\frac{di_2}{dt}(0+),$$

and from Equation 1 evaluated at time $t = 0+$, $\frac{di_2}{dt}(0+) = 2$.

Thus

$$\frac{d^2i_1}{dt^2}(0+) = 4,$$

and we may now write

$$\frac{d^2i_1}{dt^2}(0+) = 4 = A_1 + 4A_2 + 9A_3. \tag{24}$$

Solution by determinants of Equations 14, 19, and 24 gives

$$A_1 = \frac{\begin{vmatrix} -1 & 1 & 1 \\ 0 & -2 & -3 \\ 4 & 4 & 9 \end{vmatrix}}{\begin{vmatrix} 1 & 1 & 1 \\ -1 & -2 & -3 \\ 1 & 4 & 9 \end{vmatrix}} = \frac{-1(-18+12) - 0(9-4) + 4(-3+2)}{1(-18+12) + 1(9-4) + 1(-3+2)} = \frac{2}{-2} = -1,$$

and similarly

$$A_2 = \frac{\begin{vmatrix} 1 & -1 & 1 \\ -1 & 0 & -3 \\ 1 & 4 & 9 \end{vmatrix}}{-2} = -1,$$

and

$$A_3 = \frac{\begin{vmatrix} 1 & 1 & -1 \\ -1 & -2 & 0 \\ 1 & 4 & 4 \end{vmatrix}}{-2} = +1.$$

The complete solution may now be written as

$$i_1 = 1 - e^{-t} - e^{-2t} + e^{-3t} \text{ amp}. \tag{25}$$

(b) **Solution by Laplace transform method.** The Laplace transform circuit is drawn as shown in Fig. E-11.2. The Kirchhoff's voltage equations are

$$(s+1)I_2(s) + (2s+2)I_1(s) = \frac{2}{s}, \tag{26}$$

and

$$(s + 3/s + 3)I_2(s) + (s+1)I_1(s) = \frac{2}{s}. \tag{27}$$

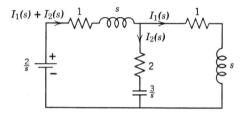

FIG. E-11.2

Solving Equations 26 and 27 for $I_1(s)$ by determinants, we get

$$I_1(s) = \frac{\begin{vmatrix} s+1 & \frac{2}{s} \\ \left(\frac{s^2+3s+3}{s}\right) & \frac{2}{s} \end{vmatrix}}{\begin{vmatrix} s+1 & 2s+2 \\ \frac{s^2+3s+3}{s} & s+1 \end{vmatrix}} = \frac{+2(2s+3)}{s(s^3+6s^2+11s+6)}. \tag{28}$$

The roots of the cubic in the denominator may be determined in the same manner as was used for the classical method. Then by partial fraction expansion we may write

$$I_1(s) = \frac{+2(2s+3)}{s(s+1)(s+2)(s+3)}$$

$$= \frac{K}{s} + \frac{A_1}{s+1} + \frac{A_2}{s+2} + \frac{A_3}{s+3}.$$

Evaluating the constants, we obtain

$$I_1(s) = \frac{1}{s} - \frac{1}{s+1} - \frac{1}{s+2} + \frac{1}{s+3}. \tag{29}$$

The inverse Laplace transform of Equation 29 gives us the final solution,

$$i_1 = 1 - e^{-t} - e^{-2t} + e^{-3t} \text{ amp}. \tag{30}$$

We can readily see that the two methods of solution are of about the same difficulty with one major exception. In the classical method, as the order of the equations gets higher, more derived initial conditions are required to evaluate the constants. This, as we have just seen, entails considerable effort and difficulty, which are avoided in the Laplace transform solution.

3.3 SUMMARY

Both the classical and Laplace transform methods of solving linear differential equations have been studied. The classical method has the advantage of being in the time domain and forcing us to relate to the physical phenomena for which the circuit is a model. It has the disadvantage of requiring appreciable algebra and calculus to evaluate the constants of integration, the amount of work increasing with the order of the equation. The Laplace transform changes the differential equations to algebraic equations, thus simplifying the mathematics. Also, in the transform method, initial conditions are introduced immediately and thus less effort is usually required to obtain a complete solution, this effect being particularly noticeable for higher order equations. The transform method has the disadvantage of being more removed from the physical world.

One feature of the transform method that is particularly useful is that a transform circuit may be sketched in which initial conditions are always included. We shall show in Chapter 4 that such a circuit may often be simplified by reduction techniques; such simplification is not possible in the time domain.

Transform voltage and current have the units of volt seconds and ampere seconds respectively.

FURTHER READING

Nearly all introductory texts on circuit analysis include the classical method, although some do not introduce this until after sinusoidal steady state has been studied. Some include the classical method but not the Laplace transform method. In the classical method some authors use the letters p or D to represent the operator d/dt. We have not used this notation because our experience has shown that the student tends to confuse the operator with the transform variable; also there is little to be gained with this additional nomenclature. In the Laplace transform method some writers use the letters p or λ to represent the complex frequency variable; however, in recent literature the letter s is almost always used.

Several good texts which present both classical and Laplace transform methods are *Network Analysis* by M. E. VanValkenburg, Prentice-Hall, New York, 1955, and *Transient Analysis in Electrical Engineering* by S. Fich, Prentice-Hall, New York, 1951. A more advanced text is *Linear Circuit Analysis* by Ley, Lutz, and Rehberg, McGraw-Hill, New York, 1956. For graduate level work we recommend *Linear Transient Analysis*, Vols. I and II, by E. Weber, New York, 1954–1956. Also at the graduate level is the classic text dealing solely with the Laplace transform method, *Transients in Linear Systems*, by Murray F. Gardner and John L. Barnes, New York, 1942.

With regard to methods for finding the roots of a polynomial, good discussions are given by D. F. Tuttle in *Network Synthesis*, Vol. 1, Appendix A., Wiley, New York, 1958; and by Ley, Lutz, and Rehberg in *Linear Circuit Analysis*, pp. 216–227, McGraw-Hill, New York, 1956. Also mathematical texts on numerical methods usually include the subject of finding roots.

PROBLEMS

3.1 For the differential equation,

$$\frac{d^2x}{dt^2} + a_1 \frac{dx}{dt} + a_0 x = f_1(t), \tag{1}$$

assume that x_0 is the response to the homogeneous equation, or

$$\frac{d^2x_0}{dt^2} + a_1 \frac{dx_0}{dt} + a_0 x_0 = 0.$$

Assume that x_1 is the response to the driving function $f_1(t)$, such that,

$$\frac{d^2x_1}{dt^2} + a_1 \frac{dx_1}{dt} + a_0 x_1 = f_1(t).$$

Show that $x = x_0 + x_1$ satisfies Equation 1.

3.2 By substituting Equation 3.8 into Equation 3.3 show that Equation 3.8 is a solution under a certain condition for m. Show that this condition is that $m = -(a_1/2)$, which occurs when $(a_1/2)^2 = a_0$.

3.3 In Example 3.1 assume that Equation 3 was differentiated to obtain

$$\frac{d^2i}{dt^2} + \frac{1}{RC}\frac{di}{dt} = 0.$$

This results in the roots $m = 0$, and $m = -1/RC$. Therefore, a solution may be written as

$$i = A_1 e^0 + A_2 e^{-t/RC} = A_1 + A_2 e^{-t/RC}.$$

Show that to satisfy Equation 3, $A_1 = 0$, and that, therefore, the complete solution was obtained in Example 3.1.

3.4 In part c of Example 3.2 the solution is written in the form of Equation 13 as

$$v_{ab} = A_1 e^{-0.27t} + A_2 e^{-3.73t},$$

or

$$e^{-2t}[A_1 e^{1.73t} + A_2 e^{-1.73t}].$$

By using identities relating exponential and hyperbolic functions show that this equation may be written

$$v_{ab} = e^{-2t}[A_3 \cosh 1.73t + A_4 \sinh 1.73t].$$

3.5 In the circuit of Fig. P-3.5a, $i(0-) = 10$ amp.
 a Find the response i.
 b Complete the table of Fig. P-3.5b and plot the curve of i vs. t.
 c Calculate the time constant.
 Answer: $a, i = 10e^{-2t}$ amperes.

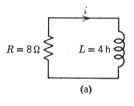

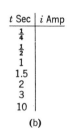

t Sec	i Amp
$\frac{1}{4}$	
$\frac{1}{2}$	
1	
1.5	
2	
3	
10	

(a) (b)

FIG. P-3.5

3.6 a For a solution of the form $x = Ae^{-t/t_0}$ show that if the slope at $t = 0+$, $dx/dt(0+)$, were maintained for $t > 0$, the quantity x would go to zero in one time constant t_0.
 b Now show that the nature of this exponential function is such that if the slope is maintained at any value of t, x would become zero in one time constant. Note that if

$$x(t_1) = Ae^{-t_1/t_0}, \quad \frac{dx}{dt}(t_1) = -\frac{1}{t_0} Ae^{-t_1/t_0}.$$

3.7 Fig. P-3.7 shows an RLC series circuit with an ideal voltage source. The switch is closed at time $t = 0$. $v_{ab}(0-) = 0$.
 a Calculate the complete response i.
 b Make a sketch (roughly to scale) of i vs. t.
 c Repeat parts a and b for $R = 2$ ohms.
 d Repeat parts a and b for $R = 2\sqrt{3}$ ohms.
 Answer: $a, i = 8(e^{-t} - e^{-3t})$ amp; $c, i = 11.3 e^{-t} \sin \sqrt{2}t$ amp; and $d, i = 16te^{-\sqrt{3}t}$ amp.

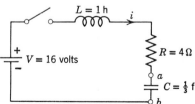

FIG. P-3.7

The Solution of Linear Differential Equations with Constant Coefficients 131

3.8 Repeat Problem 3.7a with (a) $v_{ab}(0-) = +32$ volts, (b) $v_{ab}(0-) = +16$ volts.

3.9 For each part of Example 3.2 determine i_1, i_3, and i_2. Show that each expression satisfies the initial conditions for $i_1(0+)$, $i_3(0+)$, and $i_2(0+)$ respectively.

3.10 In part a of Example 3.2, Equation 5, with the real constants A_3 and A_4, was used instead of the form demonstrated by Equation 3.11 which uses the complex constants A_1 and A_2. The purpose of this problem is to demonstrate to the student the advantages of such a change. So, solve for A_1 and A_2 by applying the proper initial conditions and demonstrate that your answer is identical to Equation 7, which was obtained by solving for the real constants A_3 and A_4.

3.11 a Using the technique of Section 3.1.2, show that the differential equation

$$\frac{d^2x}{dt^2} + a_1 \frac{dx}{dt} + a_0 x = e^{-\beta t},$$

has the characteristic equation, $(m + \beta)(m^2 + a_1 m + a_0) = 0$.
b Using the technique of Section 3.1.2, show that the differential equation

$$\frac{d^2x}{dt^2} + a_1 \frac{dx}{dt} + a_0 x = bt$$

has the characteristic equation $m^2(m^2 + a_1 m + a_0) = 0$.

3.12 For the circuit shown in Fig. P-3.12, prove that the complete response of the current is

$$i = A_1 e^{-(Rt/L)} + K_1 \cos \omega t + K_2 \sin \omega t,$$

in which

$$K_1 = \frac{VR}{\omega^2 L^2 + R^2}, \quad K_2 = \frac{\omega L V}{\omega^2 L^2 + R^2},$$

and

$$A_1 = i(0+) - \frac{RV}{\omega^2 L^2 + R^2}.$$

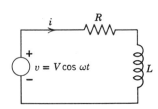

FIG. P-3.12

3.13 The switch in the circuit shown in Fig. P-3.13 is closed at time $t = 0$. $i(0-) = 0$. Find and sketch the free response, i_f, the steady-state response, i_{ss}, and the complete response, $i_f + i_{ss}$.
Partial answer: $i_{ss} = 16.5 \cos(2t - 76°)$ amp.

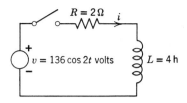

FIG. P-3.13

3.14 In the circuit shown in Fig. P-3.14, the switch is closed at $t = 0$. $v_{ab}(0-) = 0$, $v = 100 \cos t/10$ volts, $R = 10 \, \Omega$, and $C = 1$ f.
  a Calculate the complete response i.
  b Sketch, roughly to scale on the same set of axes, the free response, the steady-state response, and the complete response.
  c Calculate and sketch the complete response i if $v = 100 \cos(t/10 - 45°)$ volts.
  d Calculate and sketch the complete response i if $v = 100 \cos(t/10 + 45°)$ volts.
  Answer: c, $i = 7.07 \cos t/10$ amp.

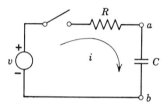

FIG. P-3.14

3.15 A charged capacitor, C_1, a resistor and an uncharged capacitor, C_2, are arranged as shown in Fig. P-3.15. The switch is closed at $t = 0$.
  a Find the complete response i in literal form.
  b What is the time constant in literal form?
  c If $C_1 = 0.5$ farads, $C_2 = 0.125$ farads, $R = 5$ ohms, and $v_{ab}(0-) = 50$ volts, calculate the complete response i, and the time constant t_0.

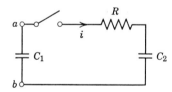

FIG. P-3.15

3.16 Solve Example 3.3, page 99, for
  a $R = 4$ ohms.
  b $R = 5$ ohms.
  Answer: b, $i = 1/15[17e^{-4t} - 8e^{-t} + 6 \cos 2t]$ amp.

3.17 The switch in Fig. P-3.17 is closed at $t = 0$ with $v_{ab}(0-) = 0$ and $i(0-) = 0$.
  a Determine the differential equation for the circuit with i as the dependent variable.

The Solution of Linear Differential Equations with Constant Coefficients   133

b Solve for i with $R = 5$ ohms, $L = 1$ henry, and $C = 0.1$ farad.
c After the switch has been closed a long time the switch is opened. Consider this a separate problem and assume $t = 0$ when the switch is opened. Solve for i.
d Sketch curve of i vs. t.
Answer: c, $i = 2 \cos 3.16t$ amp.

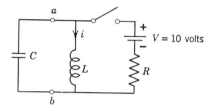

FIG. P-3.17

3.18 Determine the Laplace transform of each of the following.

a $f(t) = t$.
b $f(t) = \sin \omega t$.
c $f(t) = \cos(\omega t + \phi)$.
d $f(t) = e^{-at} \sin \omega t$.

3.19 Expand the following by partial fractions.

a $\dfrac{1}{s(s+1)}$

b $\dfrac{3s}{s^2 + 5s + 6}$

c $\dfrac{s+1}{s^2 + 2s}$

d $\dfrac{s^2 + 3s + 9}{s(s^2 + 6s + 9)}$

e $\dfrac{5s^2 + 3s + 11}{(s^2 + 1)(s^2 + 4)}$

3.20 Find the inverse Laplace transformation $i(t)$.

a $I(s) = \dfrac{s+1}{s^2 + 2s}$

b $I(s) = \dfrac{1}{s^2(s+2)}$

c $I(s) = \dfrac{s+1}{s^2 + 2s + 2}$

d $I(s) = \dfrac{s}{s^2 + 4s + 6}$

e $I(s) = \dfrac{5s^2 + 27s + 10}{s(s^2 + 6s + 10)}$

f $I(s) = \dfrac{1}{s^2 + 4s + 1}$

Answers: a, $i = \tfrac{1}{2}(1 + e^{-2t})$ amp.
c, $i = e^{-t} \cos t$ amp.
d, $i = e^{-2t} (\cos \sqrt{2}t - \sqrt{2} \sin \sqrt{2}t$ amp.

3.21 For the circuit shown in Fig. P-3.21 find the complete response i using the Laplace transform method. $v_{ab}(0+) = 100$ volts.

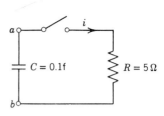

FIG. P-3.21

3.22 Use the Laplace transform method to solve Example 3.2 (p. 92) for (a) $R = 1\ \Omega$, and (b) $R = \frac{1}{2}\ \Omega$.

3.23 The circuit shown in Fig. P-3.23 is in the steady-state with the switch closed. At time $t = 0$, the switch is opened.
  a Find the complete response of the current i for $t > 0$ by the classical method.
  b Find the complete response of the current i for $t > 0$ by the Laplace transform method.
  c Make a sketch of i roughly to scale.
  d What is the time constant of the circuit? Find the value of i for $t = 0.02$ sec, $t = 0.08$ sec, and $t = 0.1$ sec.
Partial answer: $i = 4 + 6e^{-50t}$ amp.

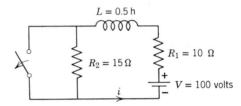

FIG. P-3.23

3.24 Find the complete solution v_{ab} of the circuit shown in Fig. P-3.24 by the following two methods. (The switch is closed at $t = 0$ and $v_{ab}(0+) = 5$ volts.)
  a The classical method.
  b The Laplace transform method.
  c From the solution for v_{ab} determine the current through the capacitor as a function of time. Show that your solution satisfies the conditions for $i_c(0+)$.
Partial answer: $v_{ab} = 2 + 3e^{-20t}$ volts.

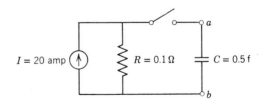

FIG. P-3.24

3.25 Repeat Problem 3.7 using the Laplace transform method.

3.26 The circuit shown in Fig. P-3.26 is in the steady-state with the switch closed. At time $t = 0$ the switch is *opened*.

a Let $R_0 = 10$ ohms. Find the complete solution for the current i for $t > 0$ by the following methods.
  (*1*) The classical method.
  (*2*) The Laplace transform method.
b Repeat *a*, but let $R_0 = 18$ ohms.
c Repeat *a*, but let $R_0 = 23$ ohms.
d Sketch the i versus time curve.
  (Show values at key points.)
Partial answers: *a*, $i = 11.2^{-6t} \cos(8t + 26.6°)$ amp.
  b, $i = 10e^{-10t}(1 - 8t)$ amp.
  c, $i = 12e^{-20t} - 2e^{-5t}$ amp.

FIG. P-3.26

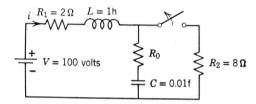

3.27 Find the complete response of the current i for the circuit shown in Fig. P-3.27 by the Laplace transform method. Assume $i(0+) \neq 0$. Check answer with Prob. 3.12.

FIG. P-3.27

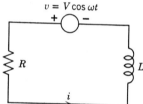

3.28 The switch in the circuit shown in Fig. P-3.28 is closed at time $t = 0$.
a Write Kirchhoff's voltage equation in literal form in the time domain.
b Take the Laplace transform of the equation of part *a*.
c From the equation of part *b* draw a Laplace transform equivalent circuit.
d Find the Laplace transform current $I(s)$ (using numerical values).
e From $I(s)$ find i.
f Show that your solution satisfies the initial and final values of i.
Partial answer: $i = 2 - 2e^{-t}$ amp.

FIG. P-3.28

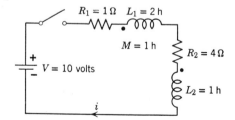

136

3.29 Solve problem 3.17b and c by the Laplace transform method.

3.30 Solve problem 3.14c by the Laplace transform method.

3.31 In Fig. P-3.31 the switch is closed at $t = 0$. $i(0-) = 0$ and $i_2(0-) = 0$. $L_1 = 4$ henries, $L_2 = 1$ henry, $R_1 = 12$ ohms, and $R_2 = 3$ ohms.
a Using both the classical method and the Laplace transform method solve for i_1 and i_2 for $k = 0.5$.
b Sketch the results of part *a* on the same graph.
c Solve for i_2 for $k = 0.95$.
d Sketch i_2 for parts *a* and *c* on the same graph.
Partial answer: $a, i_1 = 2 - e^{-2t} - e^{-6t}$ amp.

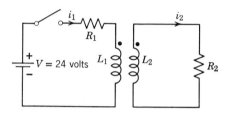

FIG. P-3.31

3.32 In the circuit shown in Fig. P-3.32 the switch is closed at $t = 0$.
a By both the classical and Laplace transform methods determine the complete response for i.
b Sketch, versus time, the free response, the steady-state response to the 8-volt source, the steady-state response to the a-c source, and the complete response.
Answer: $i = -4 + 4.46 \cos(2t - 26.6°)$ amperes.

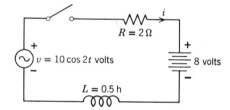

FIG. P-3.32

3.33 The circuit shown is in the steady-state with the switch open. The switch is closed at $t = 0$.
a Find the complete solution for the current i by the classical method.
b Find the complete solution for the current i by the Laplace transform method.

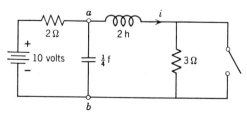

FIG. P-3.33

The Solution of Linear Differential Equations with Constant Coefficients 137

3.34 The circuit shown in Fig. P-3.34 is in the steady-state with the switch closed. At time $t = 0$ the switch is opened.

a Find the complete solution for the current i by the classical method.

b Find the complete solution for the current i by the Laplace transform method.

Answer: $i = V/R \cos(1/\sqrt{LC})t$ amp.

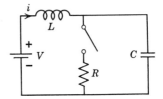

FIG. P-3.34

3.35 The circuit shown in Fig. P-3.35 is in the steady-state with the switch open. The switch is closed at time $t = 0$.

a Apply Kirchhoff's current law at node m and write the time domain equation for determining the voltage v_{mn} for all $t \geqslant 0$.

b Take the Laplace transform of the equation of part a.

c Find the complete solution for v_{mn} if $R_1 = R_2 = 1\,\Omega$.

Partial answer: c, $v_{mn} = 10e^{-t} \sin t$ volts.

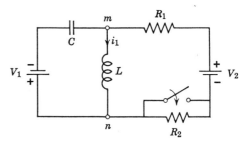

FIG. P-3.35 $L = 1$ h, $C = \frac{1}{2}$ f, $V_2 = 10$ volts, and $V_1 = 5$ volts.

3.36 The circuit shown in Fig. P-3.36 is in the steady-state with the switch open. At time $t = 0$ the switch is closed.

a Use the Laplace transform method to find the complete response for the current i.

b Use the Laplace transform method to find the complete response for the voltage v_{ab}.

Partial answer: $i = 10 - 5e^{-3t}$ amperes.

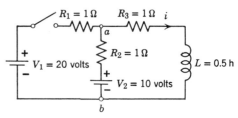

FIG. P-3.36

3.37 In the circuit shown in Fig. P-3.37 the charge on the capacitance C at time $t = 0$ is zero. At time $t = 0$ the switch is closed to position 1. At time $t = 1$ seconds the switch is moved to position 2 and 1 second later is moved again to position 1. Find the complete response of the current i by the following methods.
a The classical method.
b The Laplace transform method.
Make a sketch of i.

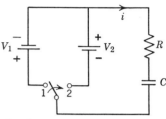

FIG. P-3.37 $R = 1\,\Omega$, $C = 1$ f, $V_1 = V_2 = 1$ volt.

3.38 The circuit shown in Fig. P-3.38 is in the steady-state with the switch open. At time $t = 0$ the switch is closed. Use the Laplace transform method to find the complete response of v_{ab}. $v_{ma}(0-) = 0$.
Answer: $v_{ab} = 1 + \frac{1}{5}(\cos t - 3 \sin t) - \frac{1}{5}e^{-t}(\cos t - 2 \sin t)$ volts.

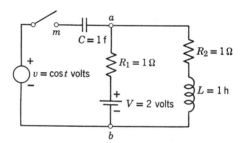

FIG. P-3.38

3.39 The circuit shown in Fig. P-3.39 is in the steady-state with the switch open. At time $t = 0$, the switch is closed. Find the complete solution for the current i by the following methods.
a The classical method.
b The Laplace transform method.
Answer: $i = \frac{1}{3}(10 - 4e^{-2t} \cos t)$ amp.

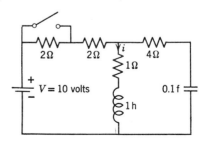

FIG. P-3.39

3.40 In the circuit of Example 3.11, Fig. E-11.1 (page 124), change R_3 to 1 ohm and C to 1 farad. Find the complete solution for the current i_1 by the methods below.
  a The classical method.
  b The Laplace transform method.
  Answer: $i_1 = 1 - 2e^{-t} + e^{-2t}$ amperes.

3.41 The circuit in Fig. P-3.41 is initially at rest (all initial conditions are zero). The switch is closed at time $t = 0$. Find the complete solution for the currents i and i_1.

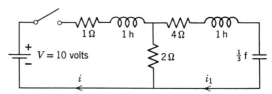

FIG. P-3.41

Partial answer: $i = 3.33 + 10.1e^{-t} - 13.1e^{-1.35t} - 0.362e^{-6.65t}$ amp.

4 Network Functions, Equivalent Circuits, and General Network Methods

In Chapter 3 we were introduced to the classical and Laplace transform methods for solving linear differential equations with constant coefficients. These methods were used on relatively simple circuits in which the driving functions were restricted to sinusoidal functions of time. The Laplace transform method changed the differential equations to algebraic equations which showed all initial conditions as voltage or current sources. In these equations appeared terms such as Ls, R, $1/Cs$, and Ms with the dimensions of ohms and the reciprocal of these terms with the dimensions of reciprocal ohms or mhos.

Because the Laplace transform has changed our equations into algebraic equations it becomes possible to define certain functions, called network functions, which are independent of the voltages and currents. It then becomes possible to replace a network with a simpler network which is equivalent as far as its terminal effects are concerned. It is the purpose of this chapter to introduce network functions and then to apply these in the simplification of circuits and in general methods for circuit analysis.

4.1 TRANSFORM NETWORK FUNCTIONS

Transform network functions are functions of those circuit elements which are exclusive of independent sources. **Initial conditions must be zero.** Thus network functions consist of combinations of terms such as Ls, R, $1/Cs$ and Ms plus terms which arise from dependent sources (see Section 9.9). Incidentally, the elements L, R, C, and M are considered passive elements, and networks which contain only these elements are referred to as passive networks. Independent and dependent sources are generally referred to as active elements; the term, active network, is however often reserved for networks which contain dependent sources and which are capable of amplification.

Although network functions are **not** functions of the voltages and currents which are the independent driving functions or the response functions, the network functions are best defined as the ratio of a response function to a driving function. Network functions are usually categorized as either driving point immittances or transfer functions.

4.1.1 Transform Driving Point Impedance and Admittance (Immittance)

Transform driving point impedance applies to a network having only one independent source and is the ratio of the transform voltage across the source to the transform current through the source, the positive reference direction for the current being from − to + through the source. Initial conditions are zero. Transform impedance has the units of ohms, the symbol $Z(s)$, and is defined for the above conditions as

$$Z(s) = \frac{V(s)}{I(s)}. \tag{4.1}$$

In Fig. 4.1 is a network in which the source or driving function is shown but the remainder of the network is only outlined. Such a network in which only two terminals are accessible is referred to as a one

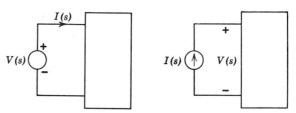

FIG. 4.1 Reference quantities for driving point transform impedance, $Z(s) = V(s)/I(s)$.

terminal-pair network or a one-port network. The term "port" comes from the terminology of wave propagation and implies the opening through which energy may flow into or out of a system.

The reciprocal of transform driving point impedance is defined as transform driving point admittance, has the symbol $Y(s)$, and the units of reciprocal ohms or mhos. Thus

$$Y(s) = \frac{I(s)}{V(s)}. \tag{4.2}$$

If we wish to state "impedance and/or admittance" we use the coined word, immittance. If we wish to picture either impedance or admittance in a diagram we use the jagged line,⋀⋀, which we have used to depict resistance, R.

Figure 4.2a shows a transform circuit consisting of R in series with Ls. The transform driving point impedance may be calculated from Kirchhoff's voltage law.

$$V(s) = LsI(s) + RI(s),$$

or

$$\frac{V(s)}{I(s)} = Z(s) = Ls + R.$$

The transform driving point admittance is

$$\frac{I(s)}{V(s)} = Y(s) = \frac{1}{Ls + R}.$$

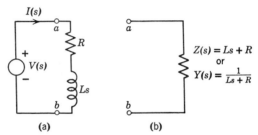

FIG. 4.2 Driving point immittance of an RL series circuit.

The transform immittances are not functions of either the voltage or current. It is therefore permissible to speak of the transform immittance of a network or a branch of a network without showing the voltage or current. Thus in Fig. 4.2b the transform driving point impedance, $Z(s)$, is pictured without showing voltage or current. This means that each branch of a network may be characterized by a single transform driving point immittance, and if several branches are in series or parallel it is customary to say that their transform driving point immittances are in series or parallel.

The words "transform driving point immittances" are unwieldy so we shall have it understood that immittances are driving point immittances unless specified otherwise. Also the word "transform" may be omitted where such omission causes no confusion.

4.1.2 Transform Transfer Functions

Transform transfer function, $T(s)$, applies to a network in which there is only one independent source and initial conditions are zero and is the

ratio of a transform response function, $R(s)$, to the transform source or driving function, $D(s)$, the response function being restricted to such functions that $T(s)$ will not be a driving point immittance. The response function $R(s)$ is **always** in the numerator so that

$$T(s) = \frac{R(s)}{D(s)}. \qquad (4.3)$$

Since $R(s)$ and $D(s)$ may be either a transform current or a transform voltage, the transform transfer function, $T(s)$ may have the units of ohms or mhos in which case it is called a transform transfer immittance. Also its units may be dimensionless, in which case it is called a transform gain function.

Figure 4.3 shows a transform network with three branches and a single driving source, $V(s)$. The following transfer functions may be determined:

$$\frac{I_2(s)}{V(s)}, \frac{I_1(s)}{V(s)}, \frac{V_{ab}(s)}{V(s)}, \quad \text{and} \quad \frac{V_{bc}(s)}{V(s)}.$$

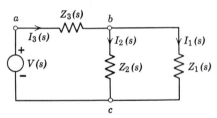

FIG. 4.3 Network to demonstrate transform transfer functions.

The first two of these are transfer admittances, the last two are transfer gain functions. The ratio, $[I_3(s)]/[V(s)]$, is **not** a transfer function by definition, since it is a driving point admittance.

4.2 EQUIVALENT CIRCUITS AND NETWORK REDUCTION

With the transform circuit it becomes possible to replace immittances in series or in parallel by a single immittance which is **equivalent as far as external effects are concerned**. This replacement reduces the number of nodes or branches in the network and thus reduces the number of unknowns which must be considered simultaneously.

4.2.1 Transform Immittances in Series

Suppose that there are n branches or one-port networks connected in series, the transform driving point impedances being known as $Z_1(s), Z_2(s), Z_3(s) \cdots Z_n(s)$, the circuit being shown in Fig. 4.4a.

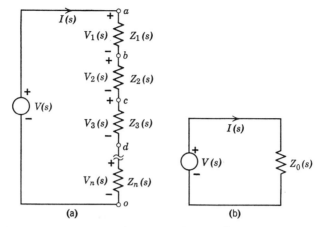

FIG. 4.4 Transform impedances in series and equivalent circuit.

Kirchhoff's transform voltage law requires

$$V(s) = I(s)[Z_1(s) + Z_2(s) + Z_3(s) + \cdots + Z_n(s)] = I(s)Z_0(s) \qquad (4.4)$$

or the equivalent transform impedance is

$$Z_0(s) = Z_1(s) + Z_2(s) + Z_3(s) + \cdots + Z_n(s), \qquad (4.5)$$

and the equivalent circuit may be sketched as in Fig. 4.4b. Note that in using $Z_0(s)$ to represent several impedances in series we have "lost" the nodes between the individual impedances. In Fig. 4.4a these are the nodes b, c, d, etc. We can always "recover" these nodes by replacing the circuit of Fig. 4.4b with its equivalent circuit of Fig. 4.4a. If the node voltage method of solution were used it is apparent that the transform circuit would have fewer unknown voltages than the time domain circuits. This represents one of the advantages of the Laplace transform method over the classical method of solution.

Suppose there are three transform impedances in series:

$$Z_1(s) = R_1 + L_1 s + \frac{1}{C_1 s},$$

$$Z_2(s) = R_2 + L_2 s + \frac{1}{C_2 s},$$

and

$$Z_3(s) = R_3 + L_3 s + \frac{1}{C_3 s}.$$

Network Functions, Equivalent Circuits, and General Network Methods

Then

$$Z_0(s) = Z_1(s) + Z_2(s) + Z_3(s),$$
$$= R_1 + R_2 + R_3 + (L_1 + L_2 + L_3)s + \left(\frac{1}{C_1} + \frac{1}{C_2} + \frac{1}{C_3}\right)\frac{1}{s},$$
$$= R_0 + L_0 s + \frac{1}{C_0 s}, \tag{4.6}$$

or

$$R_0 = R_1 + R_2 + R_3, \tag{4.7}$$
$$L_0 = L_1 + L_2 + L_3, \tag{4.8}$$
$$\frac{1}{C_0} = \frac{1}{C_1} + \frac{1}{C_2} + \frac{1}{C_3}. \tag{4.9}$$

For resistances or inductances in series the equivalent resistance or inductance is equal to the sum of the resistances or inductances, while for capacitances in series the reciprocal of the equivalent capacitance is equal to the sum of the reciprocals of the individual capacitances. The latter infers that the equivalent capacitance in farads will be less than any one of the individual capacitances.

For transform impedances in series the ratio of the transform voltage across one transform impedance to the total transform voltage can readily be written. For example, in Fig. 4.4a the ratio, $[V_2(s)]/[V(s)]$, is

$$\frac{V_2(s)}{V(s)} = \frac{Z_2(s)I(s)}{Z_0(s)I(s)} = \frac{Z_2(s)}{Z_0(s)}, \tag{4.10}$$

or

$$V_2(s) = \frac{Z_2(s)}{Z_0(s)} V(s). \quad \leftarrow \text{For transform impedances is series} \tag{4.11}$$

This method is referred to as the "voltage ratio" method.

Example 4.1

GIVEN: The circuit of Fig. E-1.1 with no voltage on capacitances. The switch is closed at $t = 0$. $v = 10$ volts, $C_1 = 1$ f, $R_1 = 1\ \Omega$. $C_2 = 4$ f, and $R_2 = 4\ \Omega$.

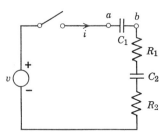

FIG. E-1.1

TO FIND: (a) The R_0 and C_0 which may be used to replace the given series system.

(b) The current i by means of the concept of transform driving point impedance.

(c) The voltage v_{ab} by means of the concept of transform transfer function.

SOLUTION: (a) The transform circuit is shown in Fig. E-1.2. The equivalent transform impedance is found by adding the individual impedances. Thus

$$Z_0(s) = \frac{1}{C_1 s} + R_1 + \frac{1}{C_2 s} + R_2 = \frac{1}{C_0 s} + R_0, \tag{1}$$

in which

$$\frac{1}{C_0} = \frac{1}{C_1} + \frac{1}{C_2} = 1 + \frac{1}{4}, \quad \text{or} \quad C_0 = 0.8 \text{ f}$$

and

$$R_0 = R_1 + R_2 = 1 + 4 = 5 \, \Omega.$$

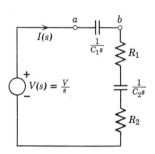

FIG. E-1.2 Using transform driving point impedance

(b) Since $Z_0(s) = \dfrac{V(s)}{I(s)}$,

$$I(s) = \frac{V(s)}{Z_0(s)} = \frac{V/s}{R_0 + 1/C_0 s} = \frac{C_0 V}{R_0 C_0 s + 1} = \frac{V/R_0}{s + 1/R_0 C_0}. \tag{2}$$

Then

$$i = \frac{V}{R_0} e^{-t/R_0 C_0},$$

and with values inserted,

$$i = \tfrac{10}{5} e^{-t/4} = 2 e^{-0.25 t} \text{ amp.} \tag{3}$$

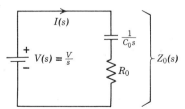

FIG. E-1.3

(c) The transform transfer function or gain function, $[V_{ab}(s)]/[V(s)]$, is determined by referring to Fig. E-1.2.

$$\frac{V_{ab}(s)}{V(s)} = \frac{\frac{1}{C_1 s} I(s)}{Z_0(s) I(s)} = \frac{\frac{1}{C_1 s}}{R_0 + \frac{1}{C_0 s}} = \frac{\frac{1}{R_0 C_1}}{s + \frac{1}{R_0 C_0}}. \tag{4}$$

Then

$$V_{ab}(s) = \frac{\frac{1}{R_0 C_1} V(s)}{s + \frac{1}{R_0 C_0}} = \frac{\frac{V}{R_0 C_1}}{s\left(s + \frac{1}{R_0 C_0}\right)}.$$

Expansion by partial fractions results in

$$V_{ab}(s) = \frac{\frac{V C_0}{C_1}}{s} - \frac{\frac{V C_0}{C_1}}{s + \frac{1}{R_0 C_0}}.$$

Thus

$$v_{ab} = \frac{V C_0}{C_1}(1 - e^{-t/R_0 C_0}). \tag{5}$$

Substitution of numbers into Equation 5 results in

$$v_{ab} = \frac{10(0.8)}{1}(1 - e^{-0.25t}) = 8(1 - e^{-0.25t}) \tag{6}$$

Equation 6 shows that as $t \to \infty$, $v_{ab} \to 8$ volts. This implies that the voltage across the larger capacitor, C_2, would approach 2 volts. We would expect this since both capacitors have the same change in charge, Δq, and since $\Delta q = C \Delta v$.

4.2.2 Transform Immittances in Parallel

Suppose that there are n branches or one-port networks connected in parallel as in Fig. 4.5, the driving point transform impedances being

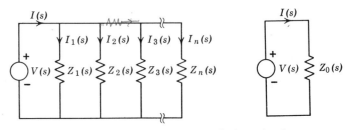

FIG. 4.5 Transform admittances in parallel and equivalent circuit.

known as $Z_1(s), Z_2(s), \ldots, Z_n(s)$. Kirchhoff's transform current law requires

$$I(s) = I_1(s) + I_2(s) + \ldots + I_n(s)$$
$$= V(s)\left[\frac{1}{Z_1(s)} + \frac{1}{Z_2(s)} + \ldots + \frac{1}{Z_n(s)}\right] = V(s)\frac{1}{Z_0(s)}$$
$$= V(s)[Y_1(s) + Y_2(s) + \ldots + Y_n(s)] = V(s)Y_0(s), \quad (4.12)$$

or

$$Y_0(s) = Y_1(s) + Y_2(s) + \ldots + Y_n(s). \quad (4.13)$$

Thus for transform admittances connected in parallel the equivalent admittance is the sum of the individual admittances. This is a mathematical equivalence as far as external effects are concerned. As a general rule we should not think of $Z_0(s)$ as physically realizable and consisting of an actual single element or several elements in series. There are exceptions to this, the primary exception occurring if the impedances $Z_1(s), Z_2(s)$, etc. are related such that

$$Z_1(s) = k_2 Z_2(s) = k_3 Z_3(s) = \ldots = k_n Z_n(s), \quad (4.14)$$

in which k_n is a positive real constant and $n = 1, 2, 3$, etc.

For the special case of $Z_1(s) = Z_2(s) = \ldots = Z_n(s)$

$$Y_0(s) = nY_1(s)$$

or

$$Z_0(s) = \frac{Z_1(s)}{n}. \quad (4.15)$$

For the special case of two impedances in parallel,

$$Y_0(s) = \frac{1}{Z_0(s)} = \frac{1}{Z_1(s)} + \frac{1}{Z_2(s)} = \frac{Z_1(s) + Z_2(s)}{Z_1(s)Z_2(s)},$$

or

$$Z_0(s) = \frac{Z_1(s)Z_2(s)}{Z_1(s) + Z_2(s)}. \quad (4.16)$$

Suppose that each transform impedance were purely resistive, that is $Z_1(s) = R_1$, $Z_2(s) = R_2$, etc. Then

$$\frac{1}{R_0} = \frac{1}{R_1} + \frac{1}{R_2} + \ldots + \frac{1}{R_n}. \quad (4.17)$$

Suppose that each transform impedance were purely inductive, that is $Z_1(s) = L_1 s$, $Z_2(s) = L_2 s$, etc. Then

$$\frac{1}{L_0} = \frac{1}{L_1} + \frac{1}{L_2} + \ldots + \frac{1}{L_n}. \quad (4.18)$$

Suppose that each transform impedance were purely capacitive; that is $Z_1(s) = 1/C_1 s$, $Z_2(s) = 1/C_2 s$, etc. Then

$$C_0 = C_1 + C_2 + \ldots + C_n. \tag{4.19}$$

From Equations 4.17, 4.18, and 4.19, we can determine the equivalent element to replace resistances, inductances, or capacitances in parallel.

The ratio of the transform current in one branch to the total transform current can readily be found. From Kirchhoff's voltage law,

$$\begin{aligned} V(s) &= Z_0(s)I(s) = Z_1(s)I_1(s) \\ &= Z_2(s)I_2(s) = Z_n(s)I_n(s). \end{aligned} \tag{4.20}$$

From Equation 4.20 the ratio of $[I_1(s)]/[I(s)]$ is

$$\frac{I_1(s)}{I(s)} = \frac{Z_0(s)}{Z_1(s)} = \frac{Y_1(s)}{Y_0(s)}. \tag{4.21}$$

This is referred to as the "current ratio" method.

If there are only two immittances in parallel, $Y_0(s) = Y_1(s) + Y_2(s)$, and

$$\begin{aligned} \frac{I_1(s)}{I(s)} = \frac{Y_1(s)}{Y_1(s) + Y_2(s)} &= \frac{\dfrac{1}{Z_1(s)}}{\dfrac{1}{Z_1(s)} + \dfrac{1}{Z_2(s)}} \\ &= \frac{Z_2(s)}{Z_1(s) + Z_2(s)}. \end{aligned} \tag{4.22}$$

Equations 4.21 and 4.22 are useful in determining the branch currents after the total current is determined.

The use of equivalent circuits does not permit us to solve problems which we could not solve otherwise. In fact one could argue that theoretically we need only the node voltage and loop current methods and that for a given network, some systematic method of eliminating unknowns from the equations of this network would serve in the same manner as the use of equivalent circuits. However, in actual practice the use of equivalent circuits often permit us to solve problems faster and with less chance of error. This result is probably caused by the facts that the calculations for each change are relatively simple and that the sketches of the various circuits give us a visual aid which tends to reduce errors.

Example 4.2

GIVEN: The circuit of Fig. E-2.1.
TO FIND: The current $I_2(s)$ by means of network reduction methods.

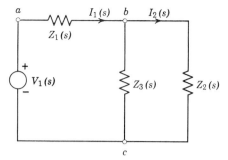

FIG. E-2.1

SOLUTION: The impedances $Z_2(s)$ and $Z_3(s)$ are in parallel and thus may be replaced by an equivalent impedance, $Z_0(s)$, equal to $\dfrac{Z_2(s)Z_3(s)}{Z_2(s) + Z_3(s)}$ (see Equation 4.16). This is sketched in Fig. E-2.2. Note

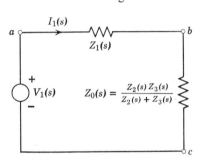

FIG. E-2.2

that the number of branches has been reduced and for Fig. E-2.2 there is no current $I_2(s)$ or $[I_1(s) - I_2(s)]$. $I_1(s)$ may readily be calculated since $Z_1(s)$ and $Z_0(s)$ are in series.

$$I_1(s) = \frac{V_1(s)}{Z_1(s) + Z_0(s)} = \frac{V_1(s)}{Z_1(s) + \dfrac{Z_2(s)Z_3(s)}{Z_2(s) + Z_3(s)}} \tag{1}$$

If we had wanted the voltage $V_{bc}(s)$, note that this can be written by the voltage ratio method (see Equation 4.10).

$$\frac{V_{bc}(s)}{V_1(s)} = \frac{Z_0(s)}{Z_1(s) + Z_0(s)},$$

or

$$V_{bc}(s) = \frac{Z_0(s)V_1(s)}{Z_1(s) + Z_0(s)}. \tag{2}$$

In order to determine $I_2(s)$ we refer to Fig. E-2.1 and note that the ratio of $[I_2(s)]/[I_1(s)]$ may be written (see Fig. 4.5 and Equation 4.22):

$$\frac{I_2(s)}{I_1(s)} = \frac{Z_3(s)}{Z_2(s) + Z_3(s)},$$

or

$$I_2(s) = \frac{Z_3(s)I_1(s)}{Z_2(s) + Z_3(s)}$$

$$= \frac{Z_3(s)V_1(s)}{Z_1(s)Z_2(s) + Z_1(s)Z_3(s) + Z_2(s)Z_3(s)} \quad (3)$$

Equation 3 could be interpreted as giving a transfer function $[I_2(s)]/[V_1(s)]$, or

$$T(s) = \frac{I_2(s)}{V_1(s)}$$

$$= \frac{Z_3(s)}{Z_1(s)Z_2(s) + Z_1(s)Z_3(s) + Z_2(s)Z_3(s)}. \quad (4)$$

We have defined the transform network functions on the basis that there be only one independent source. This does not mean that these functions can not be used for circuits in which several sources exist. For these circuits superposition applies. The following example demonstrates the use of superposition.

Example 4.3

GIVEN: The circuit of Fig. E-3.1.

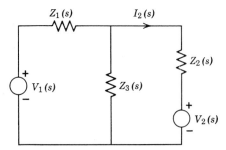

FIG. E-3.1

TO FIND: The solution for the current $I_2(s)$ by the use of superposition.

SOLUTION: By the principle of superposition, the current $I_2(s)$ is the algebraic sum of the currents which result from each voltage source acting separately. The two circuits are shown in Figs. E-3.2 and E-3.3.

For Fig. E-3.2, $I_2'(s)$ may be written down from the previous example.

$$I_2'(s) = \frac{Z_3(s)V_1(s)}{Z_1(s)Z_2(s) + Z_1(s)Z_3(s) + Z_2(s)Z_3(s)}. \quad (1)$$

For Fig. E-3.3 we note that $Z_1(s)$ and $Z_3(s)$ are in parallel, giving an equivalent impedance of

$$\frac{Z_1(s)Z_3(s)}{Z_1(s) + Z_3(s)},$$

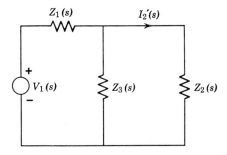

FIG. E-3.2

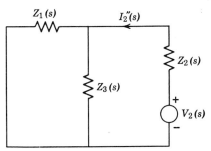

FIG. E-3.3

thus,

$$I_2''(s) = \frac{V_2(s)}{Z_2(s) + \frac{Z_1(s)Z_3(s)}{Z_1(s) + Z_3(s)}}. \tag{2}$$

Since $I_2(s) = I_2'(s) - I_2''(s)$,

$$I_2(s) = \frac{Z_3(s)V_1(s) - [Z_1(s) + Z_3(s)]V_2(s)}{Z_1(s)Z_2(s) + Z_1(s)Z_3(s) + Z_2(s)Z_3(s)}. \tag{3}$$

The advantages of network reduction are most obvious in resistive networks, networks in which there are no capacitances or inductances. In such networks the network functions are all real numbers and the response functions have the same form as the driving functions. Actually there is no advantage in using the Laplace transform for resistive circuits since the original equations in the time domain are algebraic and not differential equations.

As an illustration, for Example 4.3 assume that $V_1 = 100$ volts, $V_2 = 40$ volts, $Z_1(s) = 5$ ohms, $Z_2(s) = Z_3(s) = 10$ ohms. Fig. E-3.4 corresponds to Fig. E-3.2.

The parallel combination of R_2 and R_3 has an equivalent resistance of 5 ohms and thus

$$I_1' = \frac{100}{5+5} = 10 \text{ amp,}$$

and $I_2' = 5$ amp by current ratio.

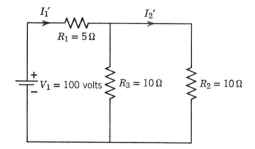

FIG. E-3.4

Figure E-3.5 corresponds to Fig. E-3.3. The parallel combination of R_1 and R_3 has an equivalent resistance of $[5(10)]/15 = 3.33$ ohms. Thus

$$I_2'' = \frac{40}{10 + 3.33} = 3 \text{ amp.}$$

The actual current, $I_2 = I_2' - I_2'' = 5 - 3 = 2$ amp.

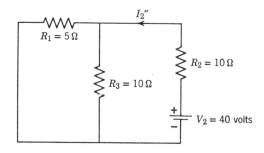

FIG. E-3.5

Example 4.4

GIVEN: The circuit of Fig. E-4.1 with $i = 10 \cos \omega t$ amperes.
TO FIND: The expressions for a group of independent voltages.

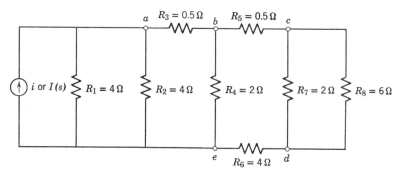

FIG. E-4.1

154

SOLUTION: We select e as the reference node so the independent unknown voltages are v_{ae}, v_{be}, v_{ce}, and v_{de}. If the node voltage method were used we would need to write four independent equations.

The resistances R_1 and R_2 are in parallel and so are the resistances R_7 and R_8. We calculate the equivalent resistances and sketch a new circuit shown in Fig. E-4.2.

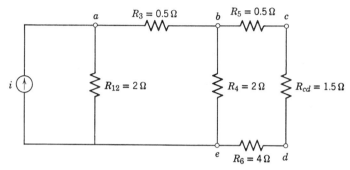

FIG. E-4.2

The resistances R_5, R_{cd}, and R_6 are in series, the equivalent resistance of the combination being 6 ohms. The 6-ohm resistance is in parallel with $R_4 = 2$ ohms, the equivalent resistance of the combination being 1.5 ohms. Thus Fig. E-4.3 may be sketched as shown.

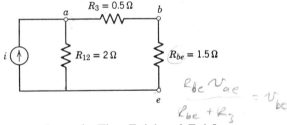

FIG. E-4.3

The next steps in the solution are shown in Figs. E-4.4 and E-4.5. From Fig. E-4.5,
$$v_{ae} = R_{ae}i = 10 \cos \omega t \text{ volts.}$$

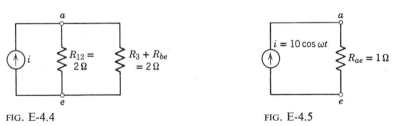

FIG. E-4.4   FIG. E-4.5

Network Functions, Equivalent Circuits, and General Network Methods   155

From Fig. E-4.3, using voltage ratio, we obtain

$$v_{be} = \frac{R_{be}v_{ae}}{R_{be} + R_3} = \frac{1.5}{2}(10 \cos \omega t) = 7.5 \cos \omega t \text{ volts.}$$

From Fig. E-4.2 we use voltage ratio to obtain

$$v_{ce} = \frac{(R_{cd} + R_6)v_{be}}{R_5 + R_{cd} + R_6} = \left(\frac{5.5}{6}\right) 7.5 \cos \omega t$$

$$= 6.88 \cos \omega t \text{ volts.}$$

$$v_{de} = \frac{R_6 v_{ce}}{R_6 + R_{cd}} = \frac{4}{5.5}(6.88 \cos \omega t) = 5 \cos \omega t \text{ volts.}$$

If we had used $I(s) = 10s/(s^2 + \omega^2)$ instead of $i = 10 \cos \omega t$ the voltages would have become

$$V_{ae}(s) = \frac{10s}{s^2 + \omega^2},$$

$$V_{be}(s) = \frac{7.5s}{s^2 + \omega^2}, \text{ etc.}$$

4.2.3 Wye-Delta Equivalence

Although impedances in series and in parallel are very common, other types of connections are needed. Two of these are known as the wye and delta connection of three impedances.

Three transform impedances may be connected to three nodes as shown in Fig. 4.6. In either connection there are three quantities and

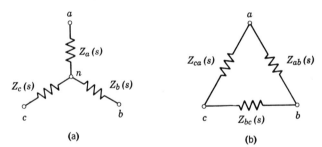

FIG. 4.6 Wye-delta impedances. (a) Wye (star) connection. (b) Delta (mesh) connection.

therefore it should be possible to replace the wye with the delta or vice versa as far as the effect on the remainder of the network is concerned. Note that in the wye connection there is an additional node, n, which does not appear in the delta connection, whereas in the delta connection there is one more independent current than in the wye connection. In order for the two networks to be equivalent as far as the three terminals a, b, and c are concerned, any test made at these terminals must give the same results for either network.

We shall first determine the relationships which permit us to calculate the wye transform impedances from the delta transform impedances. We label independent unknown transform currents as in Fig. 4.7 and then write transform voltage equations around these current paths. The rest of the circuit is not shown.

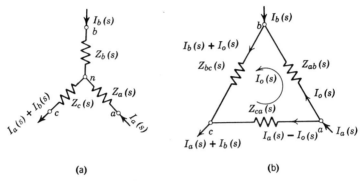

FIG. 4.7 Branch-current labeling for delta-wye equivalence.

For the wye connection of Fig. 4.7a, the independent transform voltage equations are

$$V_{ab}(s) = Z_a(s)I_a(s) - Z_b(s)I_b(s) \tag{4.23}$$

$$V_{bc}(s) = Z_b(s)I_b(s) + Z_c(s)[I_a(s) + I_b(s)]$$
$$= [Z_b(s) + Z_c(s)]I_b(s) + Z_c(s)I_a(s). \tag{4.24}$$

For the delta connection of Fig. 4.7b, the independent transform voltage equations are

$$V_{ab}(s) = Z_{ab}(s)I_0(s), \tag{4.25}$$

$$V_{bc}(s) = Z_{bc}(s)[I_b(s) + I_0(s)], \tag{4.26}$$

and

$$Z_{ab}(s)I_0(s) + Z_{bc}(s)[I_b(s) + I_0(s)] - Z_{ca}(s)[I_a(s) - I_0(s)] = 0. \tag{4.27}$$

The equations for the delta connection have one more unknown, $I_0(s)$, than the equations for the wye connection. One way of showing the conditions for equivalence is to eliminate the transform current, $I_0(s)$, from Equations 4.25, 4.26, and 4.27.

From Equation 4.27,

$$I_0(s) = \frac{Z_{ca}(s)I_a(s) - Z_{bc}(s)I_b(s)}{Z_{ab}(s) + Z_{bc}(s) + Z_{ca}(s)}. \tag{4.28}$$

Substitution of Equation 4.28 into Equations 4.25 and 4.26 results in

$$V_{ab}(s) = Z_{ab}(s)I_0(s) \quad (4.29)$$
$$= \frac{Z_{ab}(s)Z_{ca}(s)I_a(s) - Z_{ab}(s)Z_{bc}(s)I_b(s)}{Z_{ab}(s) + Z_{bc}(s) + Z_{ca}(s)}$$

and

$$V_{bc}(s) = Z_{bc}(s)[I_b(s) + I_0(s)]$$
$$= \frac{[Z_{ab}(s)Z_{bc}(s) + Z_{bc}(s)Z_{ca}(s)]I_b(s) + Z_{bc}(s)Z_{ca}(s)I_a(s)}{Z_{ab}(s) + Z_{bc}(s) + Z_{ca}(s)} \quad (4.30)$$

Now compare Equation 4.23 with Equation 4.29 and compare Equation 4.24 with Equation 4.30. In order for the equations to be identical,

$$Z_a(s) = \frac{Z_{ab}(s)Z_{ca}(s)}{Z_{ab}(s) + Z_{bc}(s) + Z_{ca}(s)},$$

$$Z_b(s) = \frac{Z_{bc}(s)Z_{ab}(s)}{Z_{ab}(s) + Z_{bc}(s) + Z_{ca}(s)}, \quad (4.31)$$

and

$$Z_c(s) = \frac{Z_{bc}(s)Z_{ca}(s)}{Z_{ab}(s) + Z_{bc}(s) + Z_{ca}(s)}.$$

Equations 4.31 permit calculations of the wye impedances if the delta impedances are known. Note that an impedance of the wye connection is equal to the product of the two delta impedances connected to the same node divided by the sum of the delta impedances. If the delta impedances are all equal,

$$Z(s)_{\text{wye}} = \frac{Z(s)_{\text{delta}}}{3}.$$

In order to determine how the delta impedances may be calculated from the wye impedances we shall start over again, using admittance terminology and writing current equations about the nodes (Fig. 4.8). Our reference node is labeled o and is in the part of the network not shown.

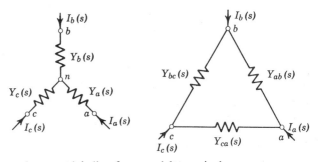

FIG. 4.8 Nodes and current labeling for wye-delta equivalence.

For the wye connection,

$$I_a(s) = Y_a(s)V_{an}(s) = Y_a(s)[V_{ao}(s) - V_{no}(s)], \quad (4.32)$$

$$I_b(s) = Y_b(s)[V_{bo}(s) - V_{no}(s)], \quad (4.33)$$

$$I_c(s) = Y_c(s)[V_{co}(s) - V_{no}(s)], \quad (4.34)$$

and

$$I_a(s) + I_b(s) + I_c(s) = 0,$$

or

$$V_{no}(s)[Y_a(s) + Y_b(s) + Y_c(s)] - Y_a(s)V_{ao}(s)$$
$$- Y_b(s)V_{bo}(s) - Y_c(s)V_{co}(s) = 0. \quad (4.35)$$

For the delta connection,

$$I_a(s) = Y_{ab}(s)[V_{ao}(s) - V_{bo}(s)] + Y_{ca}(s)[V_{ao}(s) - V_{co}(s)] \quad (4.36)$$

$$I_b(s) = Y_{ab}(s)[V_{bo}(s) - V_{ao}(s)] + Y_{bc}(s)[V_{bo}(s) - V_{co}(s)] \quad (4.37)$$

$$I_c(s) = Y_{bc}(s)[V_{co}(s) - V_{bo}(s)] + Y_{ca}(s)[V_{co}(s) - V_{ao}(s)] \quad (4.38)$$

In order to prove equivalence we shall eliminate $V_{no}(s)$ from Equations 4.32, 4.33, 4.34, and 4.35. We shall use a systematic elimination technique such as we would employ if we were solving for the four voltages, the currents and admittances being known. However, we shall stop after eliminating $V_{no}(s)$.

For our purposes we therefore rewrite the equations in standard form for solving simultaneous equations, placing Equation 4.35 first. For the wye connection,

$$[Y_a(s) + Y_b(s) + Y_c(s)]V_{no}(s) - Y_a(s)V_{ao}(s)$$
$$- Y_b(s)V_{bo}(s) - Y_c(s)V_{co}(s) = 0 \quad (4.39)$$

$$- Y_a(s)V_{no}(s) + Y_a(s)V_{ao}(s) = I_a(s) \quad (4.40)$$

$$- Y_b(s)V_{no}(s) + Y_b(s)V_{bo}(s) = I_b(s) \quad (4.41)$$

$$- Y_c(s)V_{no}(s) + Y_c(s)V_{co}(s) = I_c(s) \quad (4.42)$$

The transform voltage $V_{no}(s)$ may be eliminated by the technique of making linear combinations of the equations so that the coefficients of $V_{no}(s)$ in the last three equations is zero. If Equation 4.39 is multiplied by $[Y_a(s)]/[Y_a(s) + Y_b(s) + Y_c(s)]$ and the resulting equation added to Equation 4.40 we obtain a new equation which will be used instead of Equation 4.40. This same technique is applied to Equations

4.41 and 4.42 and we obtain three equations in three unknowns.

$$\frac{Y_a(s)[Y_b(s) + Y_c(s)]V_{ao}(s)}{Y_a(s) + Y_b(s) + Y_c(s)} - \frac{Y_a(s)Y_b(s)V_{bo}(s)}{Y_a(s) + Y_b(s) + Y_c(s)}$$
$$- \frac{Y_a(s)Y_c(s)V_{co}(s)}{Y_a(s) + Y_b(s) + Y_c(s)} = I_a(s) \quad (4.43)$$

$$- \frac{Y_a(s)Y_b(s)V_{ao}(s)}{Y_a(s) + Y_b(s) + Y_c(s)} + \frac{Y_b(s)[Y_a(s) + Y_c(s)]V_{bo}(s)}{Y_a(s) + Y_b(s) + Y_c(s)}$$
$$- \frac{Y_b(s)Y_c(s)V_{co}(s)}{Y_a(s) + Y_b(s) + Y_c(s)} = I_b(s) \quad (4.44)$$

$$- \frac{Y_a(s)Y_c(s)V_{ao}(s)}{Y_a(s) + Y_b(s) + Y_c(s)} - \frac{Y_b(s)Y_c(s)V_{bo}(s)}{Y_a(s) + Y_b(s) + Y_c(s)}$$
$$+ \frac{Y_c(s)[Y_a(s) + Y_b(s)]V_{co}(s)}{Y_a(s) + Y_b(s) + Y_c(s)} = I_c(s). \quad (4.45)$$

Now compare Equation 4.43 with Equation 4.36, Equation 4.44 with Equation 4.37, and Equation 4.45 with Equation 4.38. In order for the equations to be identical,

$$Y_{ab}(s) = \frac{Y_a(s)Y_b(s)}{Y_a(s) + Y_b(s) + Y_c(s)}$$

$$Y_{bc}(s) = \frac{Y_b(s)Y_c(s)}{Y_a(s) + Y_b(s) + Y_c(s)} \quad (4.46)$$

and

$$Y_{ca}(s) = \frac{Y_c(s)Y_a(s)}{Y_a(s) + Y_b(s) + Y_c(s)}.$$

Equations of (4.46) permit calculation of the delta admittances if the wye admittances are known. An admittance of the delta connection is equal to the product of the two wye admittances connected to the same nodes divided by the sum of the wye admittances. If the wye admittances are all equal,

$$Y(s)_{\text{delta}} = \frac{Y(s)_{\text{wye}}}{3}$$

Note that the derivations for equivalence between wye and delta immittances apply only if there are no sources in any of the branches. Keep in mind also that although changing from a wye to a delta or vice-versa may be helpful and may represent a step in the most straightforward solution, there is no assurance that this is so. One should consider several approaches to a solution before committing oneself to one of these.

Example 4.5

GIVEN: The bridge circuit of Fig. E-5.1.

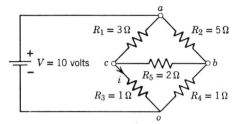

FIG. E-5.1

TO FIND: The current i by means of wye-delta equivalence.

SOLUTION: There are several choices possible. The delta formed by R_1, R_2, and R_5 can be changed to a wye. The wye formed by R_2, R_4, and R_5 could be changed into a delta. Also the delta formed by R_3, R_4, and R_5 could be changed to a wye, but this would result in losing the branch for which we desire the current. We select the first of these as representing a reasonably good choice and calculate the equivalent resistances from Equation 4.31. The resultant circuit is shown in Fig. E-5.2, a new node, d, appearing for the wye connection. Figs. E-5.3 and E-5.4 show the additional reduction of the circuit. Using voltage ratio, we obtain

$$v_{do} = \left(\frac{0.89}{1.5 + 0.89}\right)10 = 3.72 \text{ volts.}$$

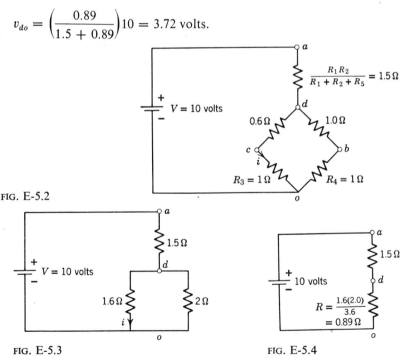

FIG. E-5.2

FIG. E-5.3

FIG. E-5.4

Then, from Fig. E-5.3,

$$i = \frac{v_{do}}{1.6} = \frac{3.72}{1.6} = 2.33 \text{ amp.}$$

4.2.4 Equivalent Circuits for Sources

An actual source can often be represented approximately by a circuit consisting of an ideal source and the elements resistance, inductance, and capacitance. In the transform domain such a combination can always be replaced by an ideal transform voltage source in series with a transform impedance. (The proof of this theorem is given in Appendix E, applications are discussed in the next Section.) We shall now prove that an ideal transform voltage source in series with a transform impedance can be replaced by an equivalent circuit consisting of an ideal transform current source in parallel with a transform impedance. This equivalence applies only to terminal effects.

Figure 4.9a shows a transform voltage source in series with a transform impedance $Z(s)$ which supplies a transform current $I_1(s)$ at a transform voltage $V_{ab}(s)$ to a circuit not shown. The equation relating $V_{ab}(s)$ and $I_1(s)$ is

$$V_{ab}(s) = V(s) - Z(s)I_1(s). \tag{4.47}$$

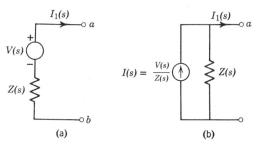

FIG. 4.9 Equivalent circuit for sources. (a) Voltage source. (b) Equivalent current source.

If we divide this equation by $Z(s)$ we obtain

$$\frac{V_{ab}(s)}{Z(s)} = \frac{V(s)}{Z(s)} - I_1(s) = I(s) - I_1(s). \tag{4.48}$$

A circuit with transform current source, $I(s) = [V(s)]/[Z(s)]$, and parallel transform impedance $Z(s)$ satisfies this equation and is shown in Fig. 4.9b. Thus the two circuits shown in Fig. 4.9 are equivalent as far as external effects are concerned. They are not equivalent internally.

Our proof is an extension of our work in Chapter 3 with initial conditions. We should recall that an inductance and its initial condition may be represented by either of the circuits shown in Fig. 4.10a and that a capacitance and its initial condition may be represented by either of the circuits shown in Fig. 4.10b.

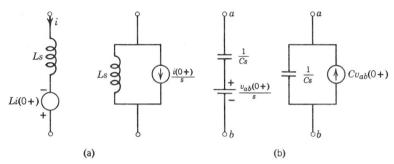

(a) (b)

FIG. 4.10 Equivalent circuits for elements with initial energy storage. (a) Inductance. (b) Capacitance.

Example 4.6

GIVEN: The circuit of Fig. E-6.1, which is the same as for Example 4.2, page 151.

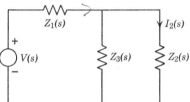

FIG. E-6.1

TO FIND: The current $I_2(s)$ by replacing the voltage source and series impedance with a current source and parallel impedance.

SOLUTION: The voltage source and series impedance $Z_1(s)$ may be replaced by the current source, $V(s)/Z_1(s)$, and the parallel impedance $Z_1(s)$ as indicated in Fig. E-6.2. The current $I_2(s)$ may be obtained by several methods. One method is to use "current ratio" on the circuit of Fig. E-6.2.

$$I_2(s) = \left(\frac{Y_2(s)}{Y_1(s) + Y_2(s) + Y_3(s)}\right)\frac{V(s)}{Z_1(s)}$$

$$= \left(\frac{\frac{1}{Z_2(s)}}{\frac{1}{Z_1(s)} + \frac{1}{Z_2(s)} + \frac{1}{Z_3(s)}}\right)\frac{V(s)}{Z_1(s)}$$

$$= \frac{Z_3(s)V(s)}{Z_1(s)Z_2(s) + Z_1(s)Z_3(s) + Z_2(s)Z_3(s)} \tag{1}$$

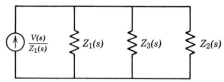

FIG. E-6.2

Another method is to replace the parallel combination of $Z_1(s)$ and $Z_3(s)$ by a single impedance as shown in Fig. E-6.3, and then change the current source circuit to a voltage source circuit as shown in Fig. E-6.4. From Fig. E-6.4,

$$I_2(s) = \frac{\dfrac{Z_3(s)V(s)}{Z_1(s) + Z_3(s)}}{Z_2(s) + \dfrac{Z_1(s)Z_3(s)}{Z_1(s) + Z_3(s)}}$$

$$= \frac{Z_3(s)V(s)}{Z_1(s)Z_2(s) + Z_1(s)Z_3(s) + Z_2(s)Z_3(s)} \qquad (2)$$

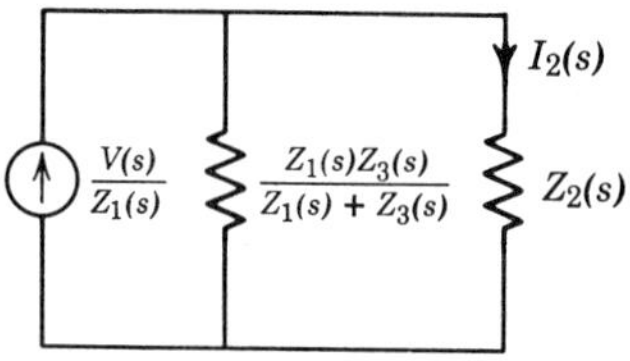

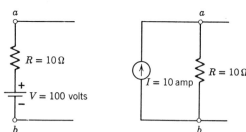

FIG. E-6.3 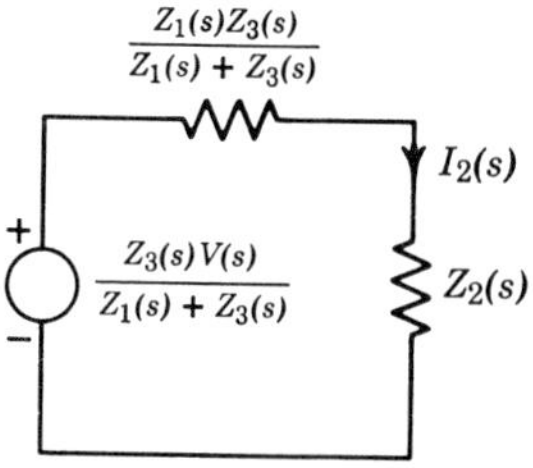   FIG. E-6.4

With equivalent circuits for sources it is easy to show that the circuits are not equivalent internally. Consider the d-c voltage source with series resistance indicated in Fig. 4.11a and the equivalent current source of Fig. 4.11b.

FIG.4.11   Source equivalents for direct current.

These two circuits are equivalent as far as external effects are concerned. However, this is not true for internal effects. If no external connections are made to the terminals a–b, there is no current flow and no energy dissipated in the voltage source circuit. However, a current of 10 amperes exists and 1000 watts of power is being transmitted in the current source circuit.

Suppose an actual source is idealized to the extent that we assume the series impedance to be zero for the voltage source equivalent circuit. Then the equivalent current source circuit would lack value since it would appear to be of infinite magnitude but in parallel with a short circuit! As we stray from reality in order to simplify, it should not surprise us to find that penalties will be paid.

4.2.5 Thévenin's Theorem

A theorem named after Leon Thévenin is often useful in reducing a complex circuit to a simpler one. This theorem, which is proved in Appendix E, may be stated as follows.

Any one terminal pair (one port) network which is linear and which may have any number of independent and dependent transform sources (as long as the dependent sources are not functions of quantities outside the network) may be replaced by a transform voltage source in series with a transform impedance. The transform voltage source is the voltage across the terminal pair when these are open circuited and the transform impedance is the ratio of this transform voltage to the transform current which flows between these terminals when short-circuited.

For our networks, which thus far have no dependent sources, the Thévenin's transform impedance is the driving point transform impedance of the one port network.

We demonstrate this with the example shown in Fig. 4.12.

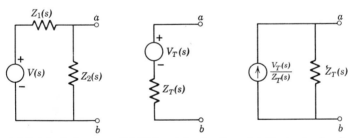

(a) Original circuit.   (b) Thévenin's circuit.   (c) Norton's circuit.

FIG. 4.12 Circuit with Thévenin's and Norton's equivalent circuits.

We find Thévenin's transform voltage, $V_T(s)$, by determining $V_{ab}(s)$ in Fig. 4.12 with terminals a and b open-circuited. By inspection,

$$V_T(s) = V_{ab}(s) = \frac{Z_2(s)V(s)}{Z_1(s) + Z_2(s)}. \tag{4.49}$$

The short-circuit transform current flowing from terminal a to terminal b through the short is $[V(s)]/[Z_1(s)]$. Therefore, Thévenin's transform impedance, $Z_T(s)$, is

$$Z_T(s) = \frac{V_T(s)}{\frac{V(s)}{Z_1(s)}} = \frac{Z_1(s)Z_2(s)}{Z_1(s) + Z_2(s)}. \qquad (4.50)$$

It may be easier in this case to find $Z_T(s)$ by determining the driving point transform impedance at the terminals a and b. In accord with the definition, the internal independent transform sources are made zero and we determine the transform impedance as consisting of $Z_1(s)$ in parallel with $Z_2(s)$. Thus

$$Z_T(s) = \frac{Z_1(s)Z_2(s)}{Z_1(s) + Z_2(s)}. \qquad (4.51)$$

Thévenin's equivalent circuit is shown in Fig. 4.12b. We have already shown in Section 4.2.4 that a transform voltage source in series with a transform impedance may be replaced by a transform current source in parallel with the transform impedance. This latter circuit is usually referred to as Norton's equivalent circuit. Norton's theorem, which is closely related to Thevenin's theorem, states that a one terminal pair network may be replaced by a transform current source in parallel with a transform impedance. The transform current source is the current between these terminals when short-circuited. The transform impedance is Thévenin's impedance.

Example 4.7

GIVEN: The circuit of Fig. E-7.1 with $v_{mb}(0+) = 0$. The switch is closed at $t = 0$.

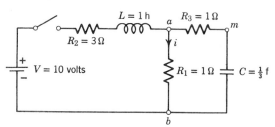

FIG. E-7.1

TO FIND: The current, i, through the branch having just the 1 Ω resistance using the Laplace transform and Thévenin's theorem.

SOLUTION: The Laplace transform circuit with the 1 Ω resistance removed is shown in Fig. E-7.2. We find Thévenin's transform voltage $V_T(s)$ by determining $V_{ab}(s)$.

$$V_T(s) = V_{ab}(s) = \frac{10}{s} - (s + 3)I_0(s),$$

166

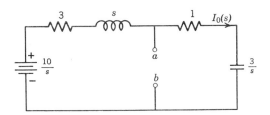

FIG. E-7.2

where

$$I_0(s) = \frac{10/s}{4 + s + \frac{3}{s}} = \frac{10}{(s+3)(s+1)}.$$

Thus

$$V_T(s) = \frac{10}{s} - \frac{(s+3)(10)}{(s+3)(s+1)} = \frac{10}{s(s+1)}.$$

To determine $Z_T(s)$ we shall find the driving point transform impedance between terminals a and b with all sources reduced to zero. Thus

$$Z_T(s) = \frac{(s+3)(3/s+1)}{s+3+\frac{3}{s}+1} = \frac{(s+3)(s+3)}{(s+3)(s+1)} = \frac{s+3}{s+1}.$$

Fig. E-7.3 shows the Thévenin's equivalent circuit connected to the resistance R_1.

$$I(s) = \frac{V_T(s)}{Z_T(s) + 1} = \frac{\frac{10}{s(s+1)}}{\frac{s+3}{s+1} + 1} = \frac{5}{(s)(s+2)}.$$

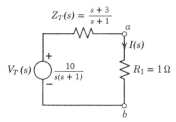

FIG. E-7.3

Expanding $I(s)$ by partial fractions and taking the inverse transform gives

$i = \frac{5}{2}(1 - e^{-2t})$ amp.

Example 4.8

GIVEN: The resistive bridge circuit shown in Fig. E-8.1
TO FIND: The current, i, in R_5 by means of Thévenin's Theorem.

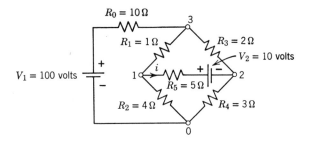

FIG. E-8.1

SOLUTION: We need to find V_T and R_T of the circuit shown in Fig. E-8.2.

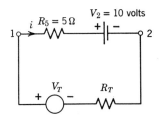

FIG. E-8.2

Thévenin's voltage, V_T, is the voltage, v_{12}, of the original circuit when branch 1–2 is open circuited.

The circuit is redrawn with branch 1–2 temporarily removed as shown in Fig. E-8.3.

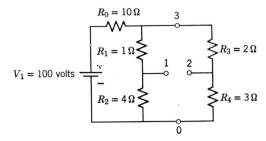

FIG. E-8.3

We shall first determine Thévenin's open-circuit voltage, v_{12}. The resistance, $R_1 + R_2 = 5\,\Omega$, is in parallel with the resistance, $R_3 + R_4 = 5\,\Omega$, making an equivalent resistance of 2.5 ohms as shown in the circuit of Fig. E-8.4.

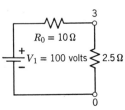

FIG. E-8.4

168

The use of voltage ratio in Fig. E-8.4 gives the voltage v_{30}.

$$v_{30} = \left(\frac{2.5}{12.5}\right)V_1 = \frac{2.5(100)}{12.5} = 20 \text{ volts.} \quad (1)$$

Then voltage ratio may be used in Fig. E-8.3 to obtain voltages v_{31} and v_{32}, from which we can then determine v_{12}.

$$v_{31} = \left(\frac{R_1}{R_1 + R_2}\right)v_{30} = \left(\frac{1}{1+4}\right)20 = 4 \text{ volts,}$$

$$v_{32} = \left(\frac{R_3}{R_3 + R_4}\right)v_{30} = \left(\frac{2}{2+3}\right)20 = 8 \text{ volts,}$$

and

$$v_{12} = v_{13} + v_{32} = -v_{31} + v_{32} = -4 + 8 = 4 \text{ volts.} \quad (2)$$

Now Thévenin's equivalent resistance will be determined by finding the driving point resistance between terminals 1 and 2 in Fig. E-8.3 with independent source, V_1, made zero. This means the resistance between terminals 1 and 2 in Fig. E-8.5 which is redrawn in Fig. E-8.6 to emphasize that the resistances of 10, 1, and 4 ohms are in a delta connection. This delta can then be replaced by an equivalent wye, using Equation 4.31, to obtain Fig. E-8.7, or its equivalent Fig. E-8.8.

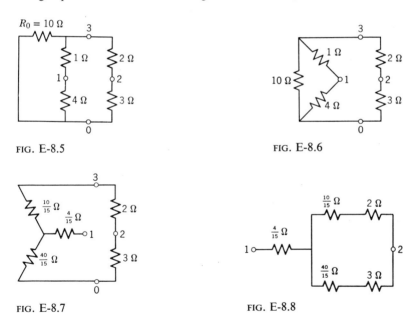

FIG. E-8.5

FIG. E-8.6

FIG. E-8.7

FIG. E-8.8

Thévenin's resistance can now be calculated from Fig. E-8.8 and is

$$R_T = R_{12} = 2.06 \, \Omega. \quad (3)$$

Refer to Fig. E-8.2 and insert values for V_T and R_T as shown in Fig. E-8.9.

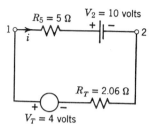

FIG. E-8.9

The current i may be calculated and is

$$i = \frac{V_T - V_2}{R_T + R_5} = \frac{4 - 10}{2.06 + 5} = -0.85 \text{ amp.} \tag{4}$$

Let us now determine Thévenin's resistance by an alternate method, as the ratio of the open circuit voltage across terminals 1–2 to the current from terminals 1 to 2 when these are short circuited. The open circuit voltage is already known, $v_{12} = V_T = 4$ volts.

With terminals 1 and 2 short-circuited the original circuit appears as shown in Fig. E-8.10. Since the short-circuit connection places the one- and two-ohm resistances in parallel and also the four- and three-ohm resistances in parallel, Fig. E-8.11 is an equivalent circuit. By voltage ratio

$$v_{32} = \frac{\frac{2}{3}(100)}{10 + \frac{2}{3} + \frac{12}{7}} = 5.39 \text{ volts,}$$

$$v_{20} = \frac{\frac{12}{7}(100)}{10 + \frac{2}{3} + \frac{12}{7}} = 13.8 \text{ volts.}$$

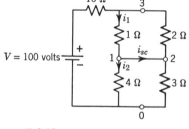

FIG. E-8.10

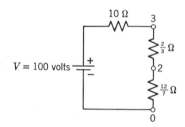

FIG. E-8.11

From Fig. E-8.10,

$$i_1 = \frac{v_{31}}{1} = \frac{v_{32}}{1} = 5.39 \text{ amp,}$$

$$i_2 = \frac{v_{10}}{4} = \frac{v_{21}}{4} = \frac{13.8}{4} = 3.45 \text{ amp,}$$

and

$i_{sc} = i_1 - i_2 = 1.94$ amp.

Finally Thévenin's resistance is

$$R_T = \frac{V_T}{i_{sc}} = \frac{4}{1.94} = 2.06 \, \Omega,$$

which checks with the previous results.

The latter method is useful in circuits with dependent sources.

4.3 GENERAL METHODS FOR CIRCUIT ANALYSIS (INDEPENDENT EQUATIONS)

The previous discussion has shown the usefulness of equivalent circuits in reducing the number of unknown quantities. In many circuits such reduction may be simply the preface to one of the general methods, the node voltage and loop current methods. These methods, which were introduced in Chapter 2, are applicable to any circuit and therefore should be studied in more detail.

For the general methods of Chapter 2 we discussed the need for writing a number of independent equations equal to the number of independent unknowns. With the relatively simple circuits we have studied so far there has been little need to question whether we really had independent equations or not. However, we are now able to systematize the writing of such equations and indicate the nature of their solution for any finite number of unknowns. We shall therefore study in more detail how to write independent equations equal in number to the number of unknowns.

4.3.1 Node Voltage Method

The node voltage method has the following steps: label the significant nodes, select a reference node,[1] determine the independent unknown voltages, write independent current equations about each node except the reference node in terms of the unknown voltages, and solve the equations for these voltages. We assume that the independent sources, including initial conditions, are known and that the passive elements, L, R, C and M are all known.

We shall illustrate the technique with the circuit of Fig. 4.13, which we shall also use for the branch-current method. The use of the transform circuit permits simplification since each branch may be considered as an immittance in series with a voltage source; furthermore

---

[1] It is not really necessary to select a reference node; it is possible to determine a group of independent unknown voltages by other methods. However, the use of the reference node is simple, systematic, and has no disadvantages.

if two immittances are in parallel, these branches may be replaced by a single immittance. We shall not replace voltage sources with equivalent current sources or vice versa, although this is possible. For the node-voltage method the immittances are labeled as admittances for convenience.

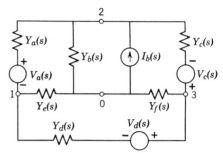

FIG. 4.13 Transform circuit for node-voltage method.

The nodes 0, 1, 2, and 3 are labeled as shown, and the node 0 is selected arbitrarily as the reference node. Then the independent unknown transform voltages are $V_{10}(s)$, $V_{20}(s)$, and $V_{30}(s)$. We need not label the nodes within the branches, such as the node between $V_a(s)$ and $Y_a(s)$, since the voltage $V_a(s)$ is known.

Since there are three independent unknown voltages, three independent equations involving these unknowns are needed. One such equation is obtained by applying Kirchhoff's current law at node 1; a second equation is obtained by applying Kirchhoff's current law at node 2, this equation being independent since elements are involved which did not appear in our first equation. For the same reason a current equation about node 3 is independent. If a current equation were **now** written about node 0 this equation would **not** be independent since all the terms have appeared in one of our previous equations. Hence for the node voltage method the proper number of independent equations is obtained by writing Kirchhoff's current equation about **every significant node except one node**, and this one node is customarily the reference node. The reason for omitting the equation about the reference node is that this will enable the immittance matrix to be symmetrical for a passive circuit.

In writing Kirchhoff's current law about each node it is helpful to consider the branch currents as positive leaving the node. Thus, for node 1 we indicate the transform currents as shown in Fig. 4.14.

For the sum of the transform currents leaving node 1 to be zero,

$$Y_a(s)[V_{12}(s) + V_a(s)] + Y_e(s)V_{10}(s) + Y_d[V_{13}(s) + V_d(s)] = 0. \quad (4.52)$$

The transform voltage $V_{12}(s)$ is replaced by its equivalence $[V_{10}(s) - V_{20}(s)]$ and $V_{13}(s)$ is replaced by its equivalence $[V_{10}(s) - V_{30}(s)]$, and the resulting equation is written in a systematic form. This systematic form consists in writing first the terms involving $V_{10}(s)$, then the terms involving $V_{20}(s)$, etc., and placing on the right-hand side of the equation such terms as do not involve the unknowns.

$$[Y_a(s) + Y_e(s) + Y_d(s)]V_{10}(s) - Y_a(s)V_{20}(s)$$
$$- Y_d(s)V_{30}(s) = -Y_a(s)V_a(s) - Y_d(s)V_d(s). \quad (4.53)$$

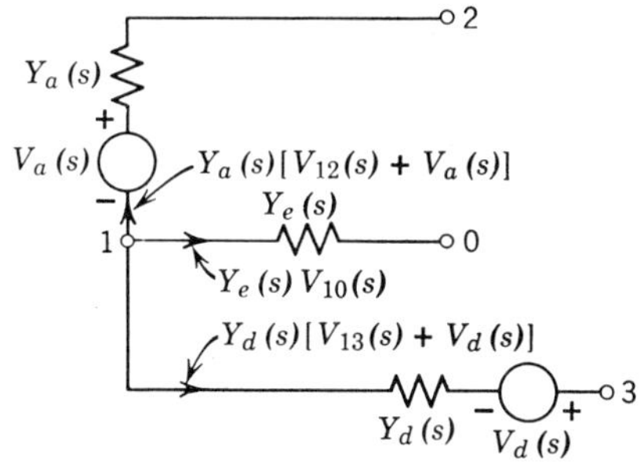

FIG. 4.14 Transform current about node 1 of Fig. 4.13.

In a similar way the following equations for nodes 2 and 3 are obtained.

$$-Y_a(s)V_{10}(s) + [Y_a(s) + Y_b(s) + Y_c(s)]V_{20}(s) - Y_c(s)V_{30}(s)$$
$$= Y_a(s)V_a(s) + I_b(s) - Y_c(s)V_c(s) \quad (4.54)$$

$$-Y_d(s)V_{10}(s) - Y_c(s)V_{20}(s) + [Y_c(s) + Y_d(s) + Y_f(s)]V_{30}(s)$$
$$= Y_c(s)V_c(s) + Y_d(s)V_d(s). \quad (4.55)$$

It is useful to generalize the above equations into the form of Equation 4.56.

$$Y_{11}(s)V_{10}(s) + Y_{12}(s)V_{20}(s) + Y_{13}(s)V_{30}(s) = I_1(s),$$
$$Y_{21}(s)V_{10}(s) + Y_{22}(s)V_{20}(s) + Y_{23}(s)V_{30}(s) = I_2(s), \quad (4.56)$$
$$Y_{31}(s)V_{10}(s) + Y_{32}(s)V_{20}(s) + Y_{33}(s)V_{30}(s) = I_3(s),$$

in which

$$Y_{11}(s) = Y_a(s) + Y_e(s) + Y_d(s), \quad Y_{12}(s) = -Y_a(s), Y_{13}(s) = -Y_d(s),$$

and

$$I_1(s) = -Y_a(s)V_a(s) - Y_d(s)V_d(s), \text{ etc.}$$

Note that $Y_{11}(s)$ represents the sum of all the branch transform admittances connected to node 1 and may be termed the self admittance of node 1. Also $Y_{12}(s) = Y_{21}(s)$ is the negative of the branch transform admittance connected between nodes 1 and 2 and may be termed the mutual admittance between these nodes. The transform current $I_1(s)$ is the sum of the equivalent current sources entering node 1, this may be demonstrated by replacing the branches containing $Y_a(s)$ and $Y_d(s)$ with equivalent circuits containing current sources.

It is important to practice writing these equations so that equations similar to Equations 4.53, 4.54, and 4.55 can be written directly from circuits.

Equation 4.56 may be solved for the unknown voltages by a systematic elimination of variables until an equation with a single unknown is obtained. Also the expression for any one of the unknown transform voltages can readily be written in determinant form. Thus

$$V_{20}(s) = \frac{\begin{vmatrix} Y_{11}(s) & I_1(s) & Y_{13}(s) \\ Y_{21}(s) & I_2(s) & Y_{23}(s) \\ Y_{31}(s) & I_3(s) & Y_{33}(s) \end{vmatrix}}{\begin{vmatrix} Y_{11}(s) & Y_{12}(s) & Y_{13}(s) \\ Y_{21}(s) & Y_{22}(s) & Y_{23}(s) \\ Y_{31}(s) & Y_{32}(s) & Y_{33}(s) \end{vmatrix}} \qquad (4.57)$$

In the denominator of Equation 4.57 is the determinant of a square matrix, the main diagonal consisting of the terms $Y_{11}(s)$, $Y_{22}(s)$, and $Y_{33}(s)$. The matrix is called symmetric if $Y_{12}(s) = Y_{21}(s)$, $Y_{13}(s) = Y_{31}(s)$, and $Y_{23}(s) = Y_{32}(s)$. For the elements we have defined this far **and for the systematic method of writing independent equations which we have used**, the admittance matrix will always be symmetric. We shall find later that systems with dependent sources may have admittance matrices that are not symmetric.

We are now in a position to write the transform current equations for a network having n unknown transform voltages. These are shown in Equation 4.58.

$$\begin{aligned} Y_{11}(s)V_{10}(s) + Y_{12}(s)V_{20}(s) + \ldots + Y_{1n}(s)V_{n0}(s) &= I_1(s) \\ Y_{21}(s)V_{10}(s) + Y_{22}(s)V_{20}(s) + \ldots + Y_{2n}(s)V_{n0}(s) &= I_2(s) \\ &\vdots \\ Y_{n1}(s)V_{10}(s) + Y_{n2}(s)V_{20}(s) + \ldots Y_{nn}(s)V_{n0}(s) &= I_n(s) \end{aligned} \qquad (4.58)$$

The expression for $V_{10}(s)$ may be written

$$V_{10}(s) = \frac{\Delta_{11}(s)I_1(s) + \Delta_{21}(s)I_2(s) + \Delta_{31}(s)I_3(s) + \ldots + \Delta_{n1}(s)I_n(s)}{\Delta(s)}$$

in which $\Delta(s)$ is the determinant of the transform admittance matrix and $\Delta_{jk}(s)$ is the cofactor which is equal to $(-1)^{j+k}$ times the minor determinant formed if the jth row and kth column are removed from the admittance matrix.

For any unknown $V_{k0}(s)$ the expression is

$$V_{k0}(s) = \sum_{j=1}^{n} \frac{\Delta_{jk}(s)I_j(s)}{\Delta(s)}. \qquad (4.59)$$

Matrices[2] provide a compact form in which to write the general current equations, Equation 4.58, as well as the solution, Equation 4.59. The student may already have been introduced to matrices in his mathematics course; if not, he undoubtedly will be. In this text there is not a good opportunity to use matrices effectively and we shall not elaborate on them. We feel obliged, however, to demonstrate the conciseness of the matrix notation as applied to the writing of general equations.

Equation 4.58 may be written,

$$\| Y(s) \| \| V(s) \| = \| I(s) \|, \tag{4.60}$$

in which the perpendiculars, $\| \ \|$, symbolize a matrix, $\| Y(s) \|$ is the square admittance matrix. $\| V(s) \|$ is a column matrix, $\| I(s) \|$ is a column matrix, and the rules for multiplying two matrices are such that Equation 4.60 is equivalent to Equation 4.58. Furthermore, the solution for voltages may be obtained by multiplying Equation 4.60 by the inverse of the admittance matrix. Thus

$$\| V(s) \| = \| Y(s) \|^{-1} \| I(s) \|, \tag{4.61}$$

in which $\| Y(s) \|^{-1}$ is the inverse of $\| Y(s) \|$ and is so defined that $\| Y(s) \|^{-1} \| Y(s) \| \| V(s) \| = \| V(s) \|$.

Example 4.9

GIVEN: The circuit of Fig. E-9.1 with resistive passive elements and d-c sources.

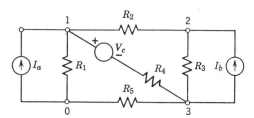

FIG. E-9.1

TO FIND: The node equations and then the evaluation of V_{10} for $R_1 = R_2 = R_3 = R_4 = R_5 = 1$ ohm, $I_a = 10$ amp, $I_b = 5$ amp, and $V_c = 20$ volts.

---

[2] A good reference is E. A. Guillemin, *The Mathematics of Circuit Analysis*, Wiley, New York, 1949.

SOLUTION: The node equations may be written directly as shown below.

$$\left(\frac{1}{R_1} + \frac{1}{R_2} + \frac{1}{R_4}\right)V_{10} - \left(\frac{1}{R_2}\right)V_{20} - \left(\frac{1}{R_4}\right)V_{30} = I_a + \frac{V_c}{R_4}, \qquad (1)$$

$$-\left(\frac{1}{R_2}\right)V_{10} + \left(\frac{1}{R_2} + \frac{1}{R_3}\right)V_{20} - \left(\frac{1}{R_3}\right)V_{30} = I_b, \qquad (2)$$

and

$$-\left(\frac{1}{R_4}\right)V_{10} - \left(\frac{1}{R_3}\right)V_{20} + \left(\frac{1}{R_3} + \frac{1}{R_4} + \frac{1}{R_5}\right)V_{30} = -I_b - \frac{V_c}{R_4} \qquad (3)$$

From these equations V_{10} may be determined by Cramer's rule.

$$V_{10} = \frac{\begin{vmatrix} I_a + \dfrac{V_c}{R_4} & -\dfrac{1}{R_2} & -\dfrac{1}{R_4} \\ I_b & \dfrac{1}{R_2}+\dfrac{1}{R_3} & -\dfrac{1}{R_3} \\ -I_b - \dfrac{V_c}{R_4} & -\dfrac{1}{R_3} & \dfrac{1}{R_3}+\dfrac{1}{R_4}+\dfrac{1}{R_5} \end{vmatrix}}{\begin{vmatrix} \dfrac{1}{R_1}+\dfrac{1}{R_2}+\dfrac{1}{R_4} & -\dfrac{1}{R_2} & -\dfrac{1}{R_4} \\ -\dfrac{1}{R_2} & \dfrac{1}{R_2}+\dfrac{1}{R_3} & -\dfrac{1}{R_3} \\ -\dfrac{1}{R_4} & -\dfrac{1}{R_3} & \dfrac{1}{R_3}+\dfrac{1}{R_4}+\dfrac{1}{R_5} \end{vmatrix}} \qquad (4)$$

In the denominator of Equation 4 is the determinant of the admittance matrix. By virtue of our systematic approach, it is symmetric, $[Y_{jk}(s) = Y_{kj}(s)]$; furthermore all terms that are not on the main diagonal have negative signs. V_{10} may readily be solved for by substituting numbers.

$$V_{10} = \frac{\begin{vmatrix} 30 & -1 & -1 \\ 5 & 2 & -1 \\ -25 & -1 & 3 \end{vmatrix}}{\begin{vmatrix} 3 & -1 & -1 \\ -1 & 2 & -1 \\ -1 & -1 & 3 \end{vmatrix}} = \frac{95}{8} = 11.9 \text{ volts} \qquad (5)$$

4.3.2 Branch-Current Method (Loop Method)

The branch-current method or loop method has the following steps: label each branch current in terms of independent unknown currents by applying Kirchhoff's current law at each node in succession until all branch currents are labeled, write Kirchhoff's voltage law around the "loops" of the independent unknown currents, and then solve these equations for the unknown currents.

We shall illustrate the technique with the same circuit as used for the node-voltage method except that the immittances will be labeled as impedances. This transform circuit is shown in Fig. 4.15. Although some simplification would be obtained by changing the current source $I_b(s)$ and the parallel impedance $Z_b(s)$ into an equivalent voltage source circuit, we deliberately retain the current source in order to show that such sources can readily be handled in this method. The source elements and the passive elements are again considered known.

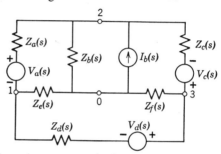

FIG. 4.15 Transform circuit for branch-current method.

There are a large number of possible selections for branch current labeling and thus for the loops of the independent currents. One choice is demonstrated in Fig. 4.16a, the labeling of some unknown currents being arbitrarily made at node 1 and then Kirchhoff's current law applied at the other nodes, a new unknown current being introduced when necessary. This is continued until all branch currents are labeled. Another choice of branch current labeling is shown in Fig. 4.17.

The unknown currents are independent because of the method used in labeling, no new unknown being introduced unless this is required by Kirchhoff's current law. In order to obtain a solution we need to write a number of equations equal to the number of unknowns and we need to assure ourselves that these equations are independent. One way of doing this is to adopt the mathematical concept of "loop" currents, in which all labeled currents are shown in a continuous path or loop. This concept of a loop is physically good since it satisfies Kirchhoff's current law for each current. If the independent equations we seek are obtained by writing Kirchhoff's voltage law around the loops of the unknown currents, we are assured of the proper number of independent equations. The equations are independent because our branch current labeling requires that each equation contain a term which is in no other equation.

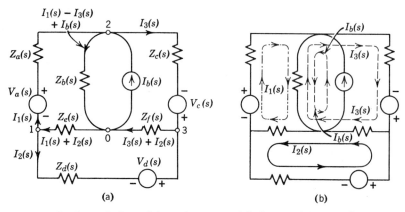

FIG. 4.16 One choice of branch-current labeling and loops of currents. (a) Labeling of branch currents. (b) Current loops.

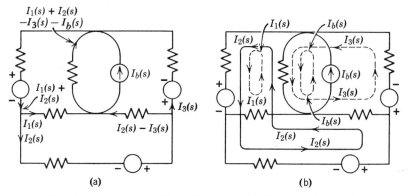

FIG. 4.17 Another choice of branch-current labeling and loops of currents. (a) Labeling of branch currents. (b) Current loops.

We write the voltage equations around the loop of $I_1(s)$ in Fig. 4.16.

$$-V_a(s) + Z_a(s)I_1(s) + Z_b(s)[I_1(s) - I_3(s) + I_b(s)]$$
$$+ Z_e(s)[I_1(s) + I_2(s)] = 0. \quad (4.62)$$

If the terms of Equation 4.62 are arranged into systematic form we obtain

$$[Z_a(s) + Z_b(s) + Z_e(s)]I_1(s) + Z_e(s)I_2(s) - Z_b(s)I_3(s)$$
$$= V_a(s) - Z_b(s)I_b(s). \quad (4.63)$$

Similar voltage equations around the loops of $I_2(s)$ and $I_3(s)$ give the following.

$$Z_e(s)I_1(s) + [Z_d(s) + Z_e(s) + Z_f(s)]I_2(s) + Z_f(s)I_3(s) = V_d(s) \quad (4.64)$$

$$-Z_b(s)I_1(s) + Z_f(s)I_2(s) + [Z_b(s) + Z_c(s) + Z_f(s)]I_3(s)$$
$$= V_c(s) + Z_b(s)I_b(s) \quad (4.65)$$

It is useful to generalize the above equations into the form of Equation 4.66.

$$\begin{aligned} Z_{11}(s)I_1(s) + Z_{12}(s)I_2(s) + Z_{13}(s)I_3(s) &= V_1(s) \\ Z_{21}(s)I_1(s) + Z_{22}(s)I_2(s) + Z_{23}(s)I_3(s) &= V_2(s) \\ Z_{31}(s)I_1(s) + Z_{32}(s)I_2(s) + Z_{33}(s)I_3(s) &= V_3(s) \end{aligned} \quad (4.66)$$

In these equations $Z_{11}(s) = Z_a(s) + Z_b(s) + Z_e(s)$, $Z_{12}(s) = Z_e(s)$, $Z_{13}(s) = -Z_b(s)$, etc. $Z_{12}(s)$ may be considered the mutual impedance between loops of $I_1(s)$ and $I_2(s)$, the sign being positive if the two currents have the same direction through the mutual impedance. The expression for any one of the transform currents can readily be written in determinant form. Thus

$$I_1(s) = \frac{\begin{vmatrix} V_1(s) & Z_{12}(s) & Z_{13}(s) \\ V_2(s) & Z_{22}(s) & Z_{23}(s) \\ V_3(s) & Z_{32}(s) & Z_{33}(s) \end{vmatrix}}{\begin{vmatrix} Z_{11}(s) & Z_{12}(s) & Z_{13}(s) \\ Z_{21}(s) & Z_{22}(s) & Z_{23}(s) \\ Z_{31}(s) & Z_{32}(s) & Z_{33}(s) \end{vmatrix}} \quad (4.67)$$

In the denominator of Equation 4.67 is the determinant of a square matrix. For the elements we have defined thus far and **for the systematic method we have used for writing independent equations**, this matrix will always be a symmetric matrix. We shall find later that systems with dependent sources may have impedance matrices that are not symmetric.

We can now write the voltage equations for a network having n unknown currents. This is shown in Equation 4.68.

$$\begin{aligned} Z_{11}(s)I_1(s) + Z_{12}(s)I_2(s) + \ldots + Z_{1n}(s)I_n(s) &= V_1(s) \\ Z_{21}(s)I_1(s) + Z_{22}(s)I_2(s) + \ldots + Z_{2n}(s)I_n(s) &= V_2(s) \\ &\vdots \\ Z_{n1}(s)I_1(s) + Z_{n2}(s)I_2(s) \ldots + Z_{nn}(s)I_n(s) &= V_n(s) \end{aligned} \quad (4.68)$$

The expression for $I_1(s)$ may be written

$$I_1(s) = \frac{\Delta_{11}(s)V_1(s) + \Delta_{21}(s)V_2(s) + \Delta_{31}(s)V_3(s)\ldots\Delta_{n1}(s)V_n(s)}{\Delta(s)}, \quad (4.69)$$

in which $\Delta(s)$ is the determinant of the transform impedance matrix and $\Delta_{jk}(s)$ is the cofactor which is equal to $(-1)^{j+k}$ times the minor determinant formed if the jth row and kth column are removed from the impedance matrix.

The expressions for any one of the currents may be written

$$I_k(s) = \sum_{j=1}^{n} \frac{\Delta_{jk}(s)V_j(s)}{\Delta(s)}. \tag{4.70}$$

Again we may write Equation 4.68 in the concise matrix form

$$\|Z(s)\|\,\|I(s)\| = \|V(s)\|, \tag{4.71}$$

in which $\|Z(s)\|$ is the square impedance matrix, and $\|I(s)\|$ and $\|V(s)\|$ are column matrices. Also Equation 4.71 may be multiplied by the inverse of the impedance matrix to give a concise solution which corresponds to Equation 4.70. Thus

$$\|I(s)\| = \|Z(s)\|^{-1}\|V(s)\|. \tag{4.72}$$

Example 4.10

GIVEN: The circuit of Fig. E-10.1 which is the same as for Example 4.9.

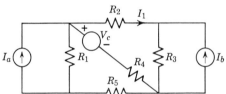

FIG. E-10.1

TO FIND: The loop equations and then the evaluation of I_1 for $R_1 = R_2 = R_3 = R_4 = R_5 = 1$ ohm, $I_a = 10$ amp, $I_b = 5$ amp, and $V_c = 20$ volts.

SOLUTION: The branch currents are labeled and the loop currents are shown in Fig. E-10.2.

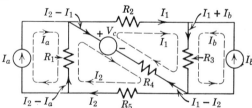

FIG. E-10.2

The voltage equations around the loops of the unknown currents I_1 and I_2 are

$$(R_2 + R_3 + R_4)I_1 - R_4 I_2 = V_c - R_3 I_b \tag{1}$$
$$-R_4 I_1 + (R_1 + R_4 + R_5)I_2 = -V_c + R_1 I_a \tag{2}$$

The expression for I_1 is

$$I_1 = \frac{\begin{vmatrix} V_c - R_3 I_b & -R_4 \\ -V_c + R_1 I_a & R_1 + R_4 + R_5 \end{vmatrix}}{\begin{vmatrix} R_2 + R_3 + R_4 & -R_4 \\ -R_4 & R_1 + R_4 + R_5 \end{vmatrix}}. \tag{3}$$

The impedance matrix is symmetric as we should expect for this circuit with our systematic method. The terms that are not on the main diagonal are negative because the currents I_1 and I_2 are in opposite directions in the common branch. Substitution of numerical values permits evaluation of I_1.

$$I_1 = \frac{\begin{vmatrix} 15 & -1 \\ -10 & 3 \end{vmatrix}}{\begin{vmatrix} 3 & -1 \\ -1 & 3 \end{vmatrix}} = \tfrac{35}{8} = 4.38 \text{ amp.} \tag{4}$$

Let us demonstrate in this example that the impedance matrix need not be symmetric. Suppose that in writing the voltage equations we wrote the first one around the loop of I_1 but for the second voltage equation we chose the loop that includes R_1, R_2, R_3, and R_5. The equations would be

$$(R_2 + R_3 + R_4)I_1 - R_4 I_2 = V_c - R_3 I_b, \tag{5}$$

and

$$(R_2 + R_3)I_1 + (R_1 + R_5)I_2 = R_1 I_a - R_3 I_b. \tag{6}$$

Equations 5 and 6 are independent and yield the correct solution; however, the impedance matrix of these equations is not symmetric.

4.3.2.1 *Mesh Current Labeling.* Is it possible to label loops immediately and not bother about branch-current labeling? As a general rule we are not able to ensure that the loops chosen will give independent equations of the proper number unless some scheme such as branch current labeling is used. For the special case of planar networks one can use a form of loop current labeling known as mesh current labeling.

A planar network is a network which can be sketched on a plane without crossovers. The open spaces of this network are meshes, or "windowpanes"; the loop currents that have a path around each open space are called mesh currents. Fig. 4.18a shows a nonplanar network; Fig. 4.18b shows a planar network on which the mesh currents are indicated, the number of independent currents being equal to the number of meshes. The mesh currents need not all be indicated as clockwise or counterclockwise. However, the advantage in choosing the directions all in one sense is that the sign of the mutual impedance in the impedance matrix would always be negative; a check on these signs thus serves to reduce the probability of error.

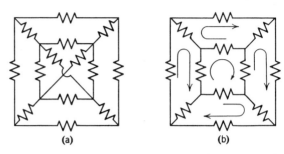

FIG. 4.18 A nonplanar and a planar network. (*a*) A nonplanar network. (*b*) Planar network with mesh currents.

Mesh current labeling has an advantage over branch current labeling for planar networks because it is easier to determine the number of independent unknown currents and the loops around which Kirchhoff's voltage equations are to be written. However, care must be used if current sources exist since these are known quantities. One way of handling this problem is to temporarily make the current sources zero while determining the independent unknown mesh currents. This is illustrated in Fig. 4.19. In Fig. 4.19*b* the current source is considered zero; there are two meshes, and two independent unknown currents are labeled as shown. In Fig. 4.19*c* the current source is replaced and sketched as a loop current which returns through one of the other branches. It does not matter which branch is selected, but a path must be selected and the voltage equations must be in accord with this path. For the circuit of Fig. 4.19*c* the loop voltage equations are written around the paths of the currents $I_1(s)$ and $I_2(s)$.

$$[Z_a(s) + Z_c(s)]I_1(s) - Z_c(s)I_2(s) = V_a(s) - Z_c(s)I_c(s)$$
$$-Z_c(s)I_1(s) + [Z_b(s) + Z_c(s)]I_2(s) = -V_b(s) + Z_c(s)I_c(s).$$
(4.73)

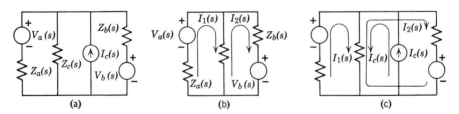

FIG. 4.19 Current labeling with current source. (*a*) Original circuit. (*b*) Mesh labeling with current source = 0. (*c*) Loop currents with current source.

The fact that **the** number of meshs is equal to the number of independent **unknown** currents for circuits without current sources can be determined as follows. For a one-mesh circuit there is one independent current. As an additional branch is added, thus changing the circuit to a two-mesh circuit, an additional independent current is required. This argument can be continued to show that a circuit of n meshes has n independent unknown currents.

In order to ensure independence of the voltage equations, these should be written around the paths of the mesh currents. This is apparent in such a circuit as that of Fig. 4.19c in which the loops of the currents $I_1(s)$ and $I_2(s)$ correspond to one possible choice of branch currents. For the circuit of Fig. 4.18b the central mesh current would not directly correspond to the usual choice of branch currents. This circuit is redrawn in Fig. 4.20. Fig. 4.20a shows the loop currents resulting from a branch-current labeling. The voltage equations to ensure independence would be those around the five loop currents shown. By a change of variable, $I_6(s) = I_5(s) + I_1(s)$, the mesh currents of Fig. 4.20b would be formed. It remains for us to show that a voltage equation around the path of $I_1(s)$ in Fig. 4.20b can be used to replace the voltage equation around the path of $I_1(s)$ in Fig. 4.20a. This is true because the former is equal to the latter minus the voltage equation around the path of $I_5(s)$ in Fig. 4.20a.

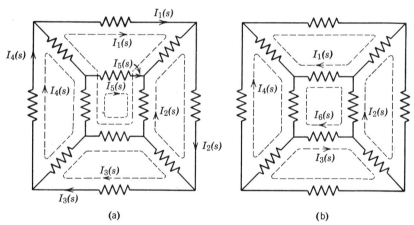

FIG. 4.20 Circuit to demonstrate independence of mesh-voltage equations. (a) Loops from branch labeling. (b) Mesh currents for $I_6(s) = I_5(s) + I_1(s)$.

4.3.2.2 Link Currents. Is the loop-current method the only general method for determining the number of independent unknown currents and the paths for Kirchhoff's voltage equations to ensure that these will be independent? There is another method which arises out of the branch of geometry known as topology and which is concerned with those properties of a figure which are unaffected by deformation without tearing or joining. In order for us to concentrate on the principles of this method we shall consider all our independent sources as zero. Thus, in the transform circuit each branch has a single element, a transform immittance. Then each branch is represented by a line and the network becomes a graph consisting of lines (branches) and nodes. This is demonstrated in Fig. 4.21 for the same circuit used in the branch current method.

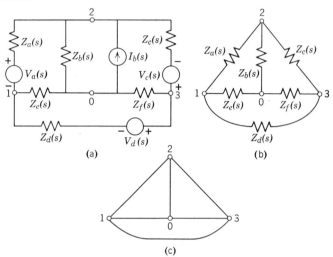

FIG. 4.21 Reduction of a circuit to a graph. (*a*) Original circuit. (*b*) Circuit with sources considered zero. (*c*) Graph of circuit.

A tree is defined as any set of branches in the graph which connects all nodes but forms no loops. The number of different trees that are possible increases with the complexity of the graph. For each tree that is chosen, the branches that are in the tree are called tree branches and the branches that are not in the tree are called links. Fig. 4.22 shows a number of trees for the graph of Fig. 4.21c, the tree branches being shown with solid lines, the links with dotted lines.

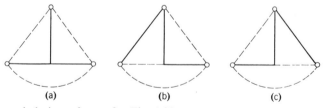

FIG. 4.22 Several choices of trees for Fig. 4.21c.

For a given tree the addition of each link to complete the graph will require the addition of one independent current in the circuit. Thus the

184

number of independent unknown currents is equal to the number of links. In order to ensure that the voltage equations will be independent we require that the link currents complete their paths through the tree branches and then write voltage equations around the loops of the link currents. For the trees chosen in Fig. 4.22, the loops of the link currents are shown in Fig. 4.23. The direction chosen as positive for these currents is arbitrary as usual.

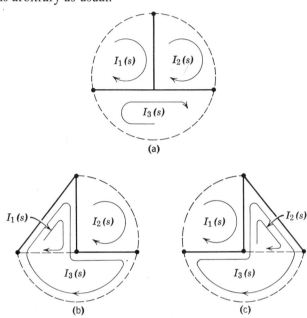

FIG. 4.23 Loops of link currents for trees of Fig. 4.22.

In order to write the voltage equations we now need to introduce the independent sources, use the link currents of a particular tree, and write the equations around the loops of the link currents. We demonstrate this for the tree of Fig. 4.23a. The circuit is shown in Fig. 4.24.

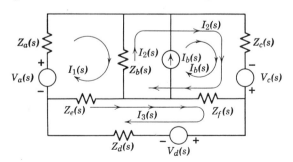

FIG. 4.24 Circuit with link currents of Fig. 4.23a.

The transform voltage equations are

$$[Z_a(s) + Z_b(s) + Z_e(s)]I_1(s) - Z_b(s)I_2(s) - Z_e(s)I_3(s) = V_a(s)$$

$$-Z_b(s)I_1(s) + [Z_b(s) + Z_c(s) + Z_f(s)]I_2(s)$$
$$- Z_f(s)I_3(s) = V_c(s) - [Z_c(s) + Z_f(s)]I_b(s)$$

$$-Z_e(s)I_1(s) - Z_f(s)I_2(s) + [Z_d(s) + Z_e(s) + Z_f(s)]I_3(s)$$
$$= -V_d(s) + Z_f(s)I_b(s). \quad (4.74)$$

Our choice of loop for the current source $I_b(s)$ means that the current in $Z_c(s)$ is actually $I_2(s) + I_b(s)$, not $I_2(s)$, and the current in $Z_f(s)$ is actually $I_3(s) - I_2(s) - I_b(s)$, not $I_3(s) - I_2(s)$.

The link-current method was demonstrated for a planar network for which the mesh-current method would have been equally satisfactory. However, the technique of the link-current method applies to nonplanar networks and is therefore a general method as is the branch or loop-current method.

For the circuits which have been illustrated the link-current method shows no advantage over the branch-current method. However, for more complex circuits the use of link currents with matrix methods and digital computers are useful for writing and solving Kirchhoff's voltage equations.

The number of tree branches correspond to the number of unknown independent voltages and thus one possible choice of independent voltages would be the voltages across the tree branches. One way to ensure independent current equations is to use, for each equation, a surface which cuts only one tree branch and which separates the network into two parts. This method has no advantages for simple circuits, but may have for complex circuits which warrant the use of matrix methods and digital computers in writing and solving the equations.

4.4 ADDITIONAL ITEMS IN CIRCUIT ANALYSIS

There are many items which are not properly categorized as equivalent circuits or general methods. The following items are selected ones that are considered appropriate at this time.

4.4.1 Resketching of Circuits

We are free to change the appearance of a circuit as long as we do not change any of the connections. Sometimes a change in the sketch of the circuit serves to clarify. In Fig. 4.25 are shown two sketches for wye and delta connections. In Fig. 4.26 are shown three sketches for a bridge network.

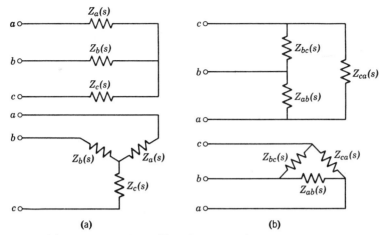

FIG. 4.25 (a) Wye connections. (b) Delta connections.

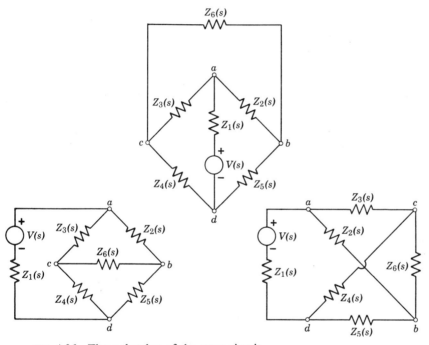

FIG. 4.26 Three sketches of the same circuit.

4.4.2 Symmetry in Networks

In certain networks symmetry permits a special form of simplification. This is usually possible because some branch currents are zero or because some voltage differences are zero. We will demonstrate the techniques with some illustrations.

In Fig. 4.27a we are seeking the driving point impedance, $Z(s) = [V(s)]/[I(s)]$. Symmetry requires that $I_1(s) = I_2(s)$, and therefore $I_0(s) = 0$. The branch with $I_0(s)$ may therefore be open-circuited, as shown in Fig. 4.27b and the driving point impedance calculated as $Z(s) = Z_1(s)$.

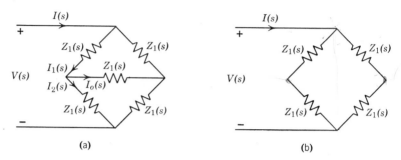

FIG. 4.27 A symmetrical network and its equivalent circuit. (a) Network. (b) Equivalent circuit.

In Fig. 4.28a we are seeking the driving point impedance $Z(s) = [V_{10}(s)]/[I(s)]$. Symmetry requires that the currents be equal in the branches connecting node 1 to nodes 2, 3, and 6. Therefore the voltage differences between nodes 2, 3, and 6 must be zero and these nodes may be connected together. For similar reasons the nodes 4, 5, and 7 may be connected together. The equivalent circuit is shown in Fig. 4.28b; the driving point impedance is equal to

$$Z(s) = \frac{Z_1(s)}{3} + \frac{Z_1(s)}{6} + \frac{Z_1(s)}{3} = \frac{5Z_1(s)}{6}.$$

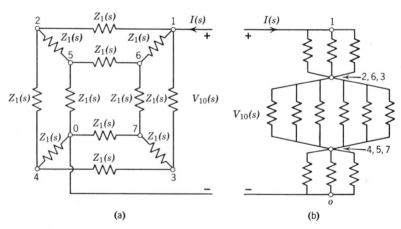

FIG. 4.28 A symmetrical network and its equivalent circuit. (a) Network. (b) Equivalent circuit.

Another example of symmetry is shown in Fig. 4.29a. If the impedance $Z_2(s)$ is replaced by an equivalent circuit consisting of two equal impedances in parallel, the circuit of Fig. 4.29b is obtained. Because of symmetry no current flows in the branches which cross the dashed lines, thus these branches may be open circuited to give the circuit of Fig. 4.29c.

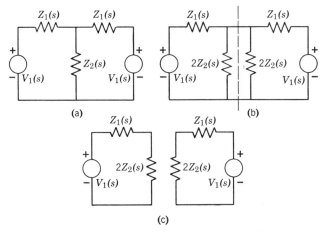

FIG. 4.29 Symmetrical network and equivalent. (a) Network. (b) Equivalent circuits. (c) Equivalent circuit.

If the voltage polarity of one of the sources in Fig. 4.29 is reversed, the circuit of Fig. 4.30 is obtained. Because of symmetry there is no current through $Z_2(s)$ and therefore the voltage difference between nodes a and b is zero. These terminals can therefore be shorted and the equivalent circuits sketched in Fig. 4.30b.

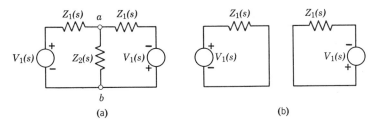

FIG. 4.30 Symmetrical network and equivalent circuit. (a) Symmetrical network. (b) Equivalent circuit.

4.4.3 Analysis of Ladder Network

A ladder network is one that can be sketched in the form of a ladder. If there is only one source, if this source is located at one end of the ladder, and if all currents and voltages are desired a special technique may be the easiest one to use. The technique is based on the fact that network functions are not functions of sources or responses, but only functions of the immittances of the network. We shall demonstrate with a simple example using a resistive network.

Example 4.11

GIVEN: Network of Fig. E-11.1 with resistances given in ohms.

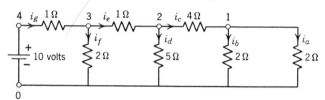

FIG. E-11.1

TO FIND: All currents and all independent voltages.

SOLUTION: Sketch the circuit of Fig. E-11.2 with the source $V = 10$ volts replaced with unknown source V'. Assume $i'_a = 1$ amp. Then calculate voltages and currents in succession until V' is calculated. Since the network functions are the same, the currents and voltages are proportional to the source function and can be calculated. For this circuit,

$v'_{10} = i'_a(2) = 2$ volts

$i'_b = \dfrac{v'_{10}}{2} = 1$ amp

$i'_c = i'_a + i'_b = 1 + 1 = 2$ amp

$v'_{20} = i'_c 4 + v'_{10} = 2(4) + 2 = 10$ volts

$i'_d = \dfrac{v'_{20}}{5} = \dfrac{10}{5} = 2$ amp

$i'_e = i'_c + i'_d = 2 + 2 = 4$ amp

$v'_{30} = i'_e(1) + v'_{20} = 4(1) + 10 = 14$ volts

$i'_f = \dfrac{v'_{30}}{2} = \dfrac{14}{2} = 7$ amp

$i'_g = i'_f + i'_e = 7 + 4 = 11$ amp

$V' = i'_g(1) + v'_{30} = 11 + 14 = 25$ volts.

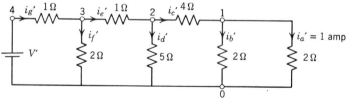

FIG. E-11.2

The actual currents and voltages in the original circuit are found by direct ratio.

$$v_{10} = v'_{10} \frac{V}{V'} = 2\left(\frac{10}{25}\right) = 0.8 \text{ volts}$$

$$i_b = i'_b \frac{V}{V'} = 1\left(\frac{10}{25}\right) = 0.4 \text{ amp}$$

$$i_c = i'_c \frac{V}{V'} = 2\left(\frac{10}{25}\right) = 0.8 \text{ amp}$$

$$v_{20} = v'_{20} \frac{V}{V'} = 4 \text{ volts}$$

$$i_d = \frac{2}{2.5} = 0.8 \text{ amp}$$

$$i_e = \frac{4}{2.5} = 1.6 \text{ amp}$$

$$v_{30} = \frac{14}{2.5} = 5.6 \text{ volts}$$

$$i_f = \frac{7}{2.5} = 2.8 \text{ amp}$$

and

$$i_g = \frac{11}{2.5} = 4.4 \text{ amp}$$

4.4.4 Reciprocity Theorem

For a network, whose immittance matrix may be written as a symmetric matrix, which has only one independent voltage source, and for which initial conditions are zero, the ratio of the independent transform voltage source to the transform current response is the same whether the source is inserted in series with branch c and the response determined in branch d or vice versa. The assumed positive direction of the current will have the same relation to the positive polarity of the voltage source in each of the two branches.

This theorem applies to networks for which the elements of the immittance matrix consists of some algebraic combination of the following terms: Ls, R, $1/Cs$, or Ms. For networks which have dependent sources (see Section 9.9), the immittance matrix can usually not be written in symmetric form.

A proof of this theorem can be made from Equation 4.70 which is repeated here,

$$I_k(s) = \sum_{j=1}^{n} \frac{\Delta_{jk}(s) V_j(s)}{\Delta(s)}. \tag{4.70}$$

If the single voltage source $V(s)$ is placed in branch c, the expression for the current response in branch d is

$$I_d(s) = \frac{\Delta_{cd}(s)V(s)}{\Delta(s)}. \tag{4.75}$$

If the voltage source is placed in branch d, the expression for the current response in branch c is

$$I_c(s) = \frac{\Delta_{dc}(s)V(s)}{\Delta(s)}. \tag{4.76}$$

Equations 4.75 and 4.76 are equal if $\Delta_{cd}(s) = \Delta_{dc}(s)$. These cofactors are always equal if the impedance matrix is symmetrical. A proof of this is left as an exercise for the student.

Figure 4.31 shows a simple circuit for which reciprocity may be demonstrated.

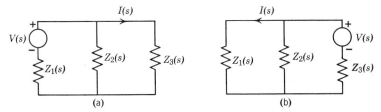

FIG. 4.31 Demonstration of reciprocity.

In Fig. 4.31a,

$$I(s) = \left[\frac{V(s)}{Z_1(s) + \dfrac{Z_2(s)Z_3(s)}{Z_2(s) + Z_3(s)}}\right]\left(\frac{Z_2(s)}{Z_2(s) + Z_3(s)}\right)$$

$$= \frac{V(s)Z_2(s)}{Z_1(s)Z_2(s) + Z_1(s)Z_3(s) + Z_2(s)Z_3(s)}. \tag{4.77}$$

In Fig. 4.31b,

$$I(s) = \left[\frac{V(s)}{Z_3(s) + \dfrac{Z_1(s)Z_2(s)}{Z_1(s) + Z_2(s)}}\right]\left(\frac{Z_2(s)}{Z_1(s) + Z_2(s)}\right)$$

$$= \frac{V(s)Z_2(s)}{Z_1(s)Z_3(s) + Z_2(s)Z_3(s) + Z_1(s)Z_2(s)}. \tag{4.78}$$

Thus the theorem of reciprocity is demonstrated for this simple circuit.

It is also possible to express the reciprocity theorem in terms of an independent current source and voltage response. Thus, with the same conditions as before, the ratio of an independent transform current source to the transform voltage response is the same whether the current source be placed across branch c and the voltage response determined across branch d or vice versa. A proof of this is possible from the general node voltage equations. However, it is easier to make this proof from the proof already made. Instead of applying a voltage source, $V(s)$, in series with branch c for Equation 4.75, we shall assume a current source, $I(s)$, placed **across** branch c; the equivalent voltage source would then be $I(s)Z_c(s)$. Then Equation 4.75 may be written

$$I_d(s) = \frac{\Delta_{cd}(s)I(s)Z_c(s)}{\Delta(s)}. \tag{4.79}$$

Now $I_d(s)$ may be written as $V_d(s)/Z_d(s)$, in which $V_d(s)$ represents the voltage response across branch d.

Thus

$$V_d(s) = \frac{\Delta_{cd}(s)I(s)Z_c(s)Z_d(s)}{\Delta(s)}. \tag{4.80}$$

In a similar way we manipulate Equation 4.76, considering $V(s)$ as equal to $I(s)Z_d(s)$, and finding the expression for $V_c(s)$ to be

$$V_c(s) = \frac{\Delta_{dc}(s)I(s)Z_d(s)Z_c(s)}{\Delta(s)}. \tag{4.81}$$

Since the ratio of $V_d(s)/I(s)$ of Equation 4.80 is the same as the ratio of $V_c(s)/I(s)$ of Equation 4.81, the reciprocity theorem is proved for the ratio of voltage response to current source.

4.4.5 Scaling and Normalizing

The student who has had some experience with electrical practice may wonder why we tend to use values of R, L, C, and ω with numbers usually ranging from 0.1 to 10. This is not a usual range for most of these quantities.

The main reason for our use of small integral values for most terms is to simplify the algebra, thus permitting us to concentrate on understanding the principles. Also the numbers we encounter in practice, if awkward to deal with, may be modified by suitable changes in the variables.

These changes in variable will be demonstrated for an equation which is a typical expression of Kirchhoff's laws.

$$L\frac{di}{dt} + Ri + \frac{1}{C}\int_0^t i\,dt + v_{ab}(0+) = V\cos(\omega t + \alpha). \tag{4.82}$$

The three changes in variable which might be considered are changes in the magnitude of time, current, and voltage. Let

$$t = k_1 t', \tag{4.83}$$

$$i = k_2 i', \tag{4.84}$$

and
$$v = k_3 v', \tag{4.85}$$

in which k_1, k_2, and k_3 are real constants.

Substitution of Equations 4.83, 4.84, and 4.85 into Equation 4.82 gives

$$L\frac{k_2}{k_1}\frac{di'}{dt'} + Rk_2 i' + \frac{k_2 k_1}{C}\int_0^{t'} i'\, dt' + k_3 v'_{ab}(0+) = k_3 V' \cos \omega k_1 t',$$

or

$$\frac{k_2 L}{k_1 k_3}\frac{di'}{dt'} + \frac{k_2 R i'}{k_3} + \frac{k_2 k_1}{k_3 C}\int_0^{t'} i'\, dt' + v'_{ab}(0+) = V' \cos \omega k_1 t',$$

or

$$L'\frac{di'}{dt'} + R'i' + \frac{1}{C'}\int_0^{t'} i'\, dt' + v'_{ab}(0+) = V' \cos \omega' t'. \tag{4.86}$$

Equation 4.86 is similar to Equation 4.82. However, all quantities have changed in numerical values as follows.

$$L' = \frac{k_2}{k_1 k_3} L, \tag{4.87}$$

$$R' = \frac{k_2}{k_3} R, \tag{4.88}$$

$$C' = \frac{k_3}{k_1 k_2} C, \tag{4.89}$$

$$v'_{ab}(0+) = \frac{v_{ab}(0+)}{k_3},$$

$$V' = \frac{V}{k_3},$$

and

$$\omega' = k_1 \omega. \tag{4.90}$$

Suppose we make a change only in the time scale, by making $k_1 \neq 1$. This has the effect of changing the numerical values of L, C, and ω, since $L' = L/k_1$, $C' = C/k_1$, and $\omega' = k_1 \omega$. However, $\omega' L' = \omega L$ and $\omega' C' = \omega C$; thus the phasor impedance of the circuit is unchanged. Such a change is often referred to as "frequency scaling" or "time scaling." When electronic analogue computers are used to simulate a physical system, time scaling is usually necessary to bring the frequency response within the capabilities of the computer.

A change in the voltage or current scale also affects the values of L, R, and C. The new transform impedance $Z'(s) = k_2/k_3\, Z(s)$, and thus these changes are referred to as magnitude scaling or impedance scaling.

It is possible to make the equations dimensionless by ascribing dimensions to the constants k_1, k_2, and k_3. If k_1 has the dimension of seconds, k_2 the dimension of amperes, and k_3 the dimension of volts, Equation 4.86 is dimensionless, and so are all the terms: L', R', C', $v'_{ab}(0+)$, V', and ω'. This is referred to as "normalizing," and has advantages in such matters as plotting of normalized values.

Example 4.12

GIVEN: The circuit of Fig. E-12.1 with the values given.

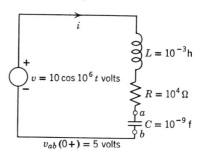

FIG. E-12.1

TO FIND: Changes in scales which will simplify the numerical work.

SOLUTION: The voltages are numbers of reasonable size so we choose $k_3 = 1$.

The angular velocity of $\omega = 10^6$ can be replaced by $\omega' = 1$ if k_1 is chosen as 10^{-6}. The resistance $R = 10^4$ can be replaced by $R' = 1$ if $k_2 = 10^{-4}$. In order to determine whether such choices result in simple values for L' and C' we calculate these from Equations 4.87 and 4.89.

$$L' = \frac{k_2}{k_1 k_3} L = \frac{10^{-4}}{10^{-6}}(10^{-3}) = 0.1 \text{ henry}$$

$$C' = \frac{k_3}{k_1 k_2} C = \frac{10^{-9}}{10^{-6} 10^{-4}} = 10 \text{ farads}$$

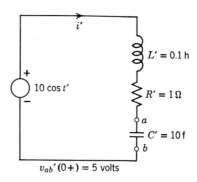

FIG. E-12.2

Network Functions, Equivalent Circuits, and General Network Methods 195

Our numbers are reasonable so we stay with the choices for k_1 and k_2. The effect of these scaling operations is to change the circuit to that shown in Fig. E-12.2.

After the current i' has been solved for (as a function of t'), the current i may be written as a function of t from Equations 4.83, 4.84, and 4.85.

4.4.6 Special Cases of Idealized Independent Sources

A physical power source can be represented either by a voltage source in series with an impedance or by a current source in parallel with an impedance. Sometimes we feel justified in assuming that the impedance in series with the voltage source may be considered zero or that the impedance in parallel with the current source may be considered infinite. Such assumptions cause unusual circuit situations which warrant brief consideration. We shall consider four cases.

A voltage source may be connected directly across some impedance as shown in Fig. 4.32a. Since the current in $Z_1(s)$ is known if $V(s)$ and $Z_1(s)$ are known, and since the current in $Z_1(s)$ does not affect the currents in $Z_2(s)$ and $Z_3(s)$, we are justified in removing $Z_1(s)$ as shown in Fig. 4.32b, in order to determine currents and voltages in the remainder of the circuit. However, if we wish the actual current in $V(s)$ we must include the current of $Z_1(s)$.

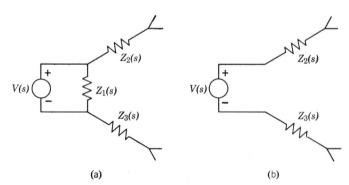

FIG. 4.32 Impedance in parallel with voltage source. (a) Portion of network. (b) Equivalent circuit.

A similar case occurs if a current source has a series impedance as shown in Fig. 4.33a. Since the voltage across $Z_1(s)$ is known if $Z_1(s)$ and $I(s)$ are known and since this voltage has no effect on the remainder of the circuit, we may short out this impedance as shown in Fig. 4.33b as far as the remainder of the network is concerned. Of course, if we wished eventually to determine the voltage across the current source in Fig. 4.33a, we would need to reintroduce the impedance $Z_1(s)$.

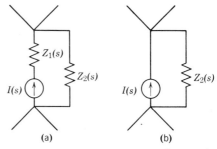

FIG. 4.33 Impedance in series with a current source. (*a*) Portion of network. (*b*) Equivalent circuit.

A third case arises if a voltage source is connected as shown in Fig. 4.34*a*. One can sketch the circuit of Fig. 4.34*b* as an equivalent circuit with the exception that now the actual current in the voltage source of Fig. 4.34*a* is the sum of the two currents in the voltage sources of Fig. 4.34*b*. The equivalence is seen by noting for Fig. 4.34*b* that the two terminals a and a' may be connected without disturbing the network.

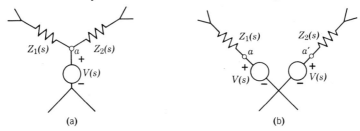

FIG. 4.34 Voltage source without simple series impedance. (*a*) Portion of network. (*b*) Equivalent circuit.

A fourth case arises if a current source is connected as shown in Fig. 4.35*a*. One can sketch an equivalent circuit as shown in Fig. 4.35*b*. Observe that the voltage across the current source of Fig. 4.35*a* is the sum of the two voltages across the current sources of Fig. 4.35*b*.

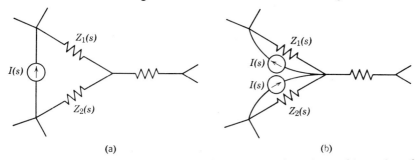

FIG. 4.35 Current source without simple parallel impedance. (*a*) Portion of network. (*b*) Equivalent network.

Network Functions, Equivalent Circuits, and General Network Methods   197

4.5 SUMMARY

The transform domain permits the definition of network functions. Network functions permit the development of network reduction methods and network theorems. The branch current (loop) method and the node voltage method are general methods that apply to both planar and nonplanar networks. The use of a network graph also provides general methods. The mesh current method does not apply to nonplanar networks.

Network reduction methods are valuable in simplifying circuits. The general methods are valuable for proving theorems and for solving circuits which have been simplified by reduction methods.

FURTHER READING

For further reading on the subject of this chapter, see M. E. VanValkenburg, *Network Analysis*, Chapter 9, Prentice-Hall, New York, 1955. For an excellent discussion of network graphs, see E. A. Guillemin, *Introductory Circuit Theory*, Chapter 1, Wiley, New York, 1953.

PROBLEMS

4.1 For each of the circuits shown in Fig. P-4.1, write the expression for the following.
a The transform driving point impedance.
b The transform driving point admittance.

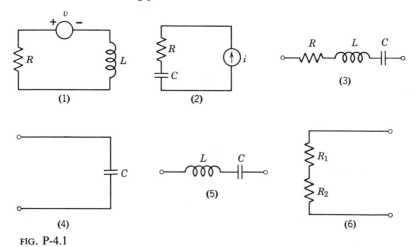

FIG. P-4.1

4.2 For the circuit shown in Fig. P-4.2 write, in terms of transform voltages and currents, as many transfer functions as you can. Tell whether the functions are immittance or gain functions.

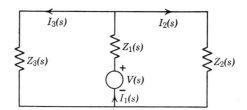

FIG. P-4.2

4.3 For each of the circuits shown in Fig. P-4.3 draw the simplest equivalent transform circuit and write the expression for the driving point impedance function of the equivalent circuit. In each case give the expression and/or numerical value of the equivalent R_0, L_0 and/or C_0 as appropriate.

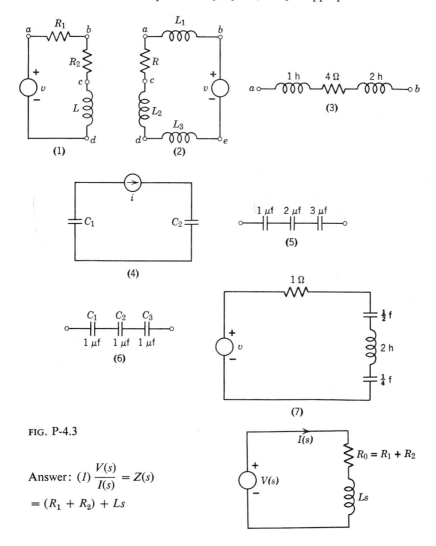

FIG. P-4.3

Answer: (1) $\dfrac{V(s)}{I(s)} = Z(s)$
$= (R_1 + R_2) + Ls$

Network Functions, Equivalent Circuits, and General Network Methods 199

4.4 For each of the circuits shown in Fig. P-4.4, assume all initial conditions are zero.

a Find the current i by means of the concept of transform driving point impedance.

b Find the voltage v_{ab} by means of the concept of transform transfer function (voltage ratio method).

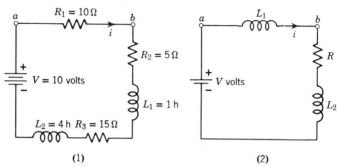

(1)

Partial answer:

$v_{ab} = \frac{1}{3}(10 - 10e^{-6t})$ volts

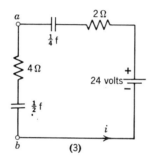

(3)

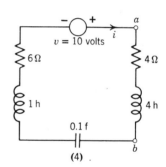

(4)

Partial answer:

$v_{ab} = 8e^{-t} \cos t$ volts

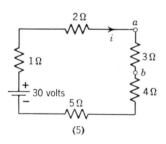

(5)

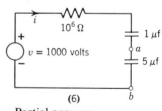

(6)

Partial answer:

$i = e^{-1.2t}$ milliamperes.

FIG. P-4.4

4.5 For each of the circuits shown in Fig. P-4.5 find the equivalent resistance, R_0, inductance, L_0, or capacitance, C_0, as appropriate.

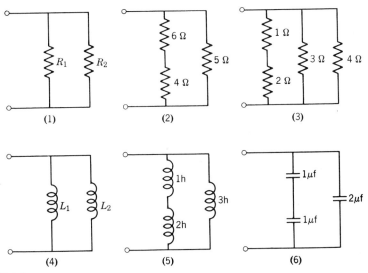

FIG. P-4.5

4.6 For each of the circuits shown in Fig. P-4.6 write the expression for the transform driving point admittance (all initial conditions are zero).

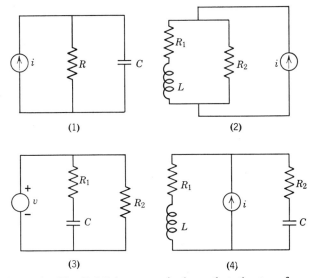

FIG. P-4.6

4.7 For the circuit shown in Fig. P-4.7 (next page) show that the transform current ratio $\dfrac{I_1(s)}{I_0(s)} = \dfrac{R_1}{L}\left[\dfrac{1}{s + \dfrac{R_1 + R_2}{L}}\right].$

Network Functions, Equivalent Circuits, and General Network Methods 201

FIG. P-4.7

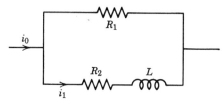

4.8 For the circuits shown in Fig. P-4.8 find the transform current ratio $[I_1(s)]/[I_0(s)]$.

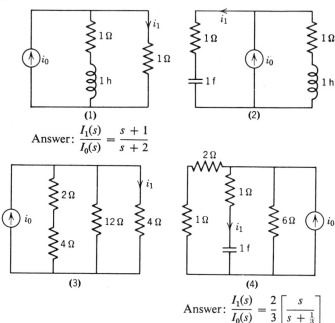

Answer: $\dfrac{I_1(s)}{I_0(s)} = \dfrac{s+1}{s+2}$

Answer: $\dfrac{I_1(s)}{I_0(s)} = \dfrac{2}{3}\left[\dfrac{s}{s+\frac{1}{3}}\right]$

FIG. P-4.8

4.9 For each of the circuits in Fig. P-4.8, let $i_0 = 10$ amperes direct current. Find i_1.

4.10 For the circuits shown in Fig. P-4.10 all initial conditions are zero.
  a Use network reduction methods on the transform network to find $I_1(s)$ and then find i_1.
  b Use current ratio to find $I_2(s)$ and then find i_2.
  c Use voltage ratio to find $V_{ab}(s)$ and then find v_{ab}.

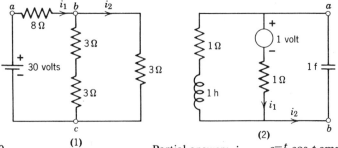

FIG. P-4.10

Partial answer: $i_2 = -e^{-t}\cos t$ amp.

202

Partial answer:

$i_1 = \frac{1}{2}[1 - e^{-t}(\cos t - \sin t)]$ amp

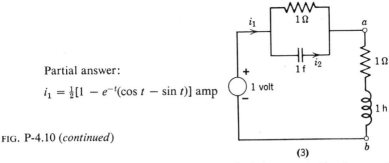

(3)

FIG. P-4.10 (continued)

4.11 For each of the circuits shown in Fig. P-4.11 find the current i by the use of superposition.

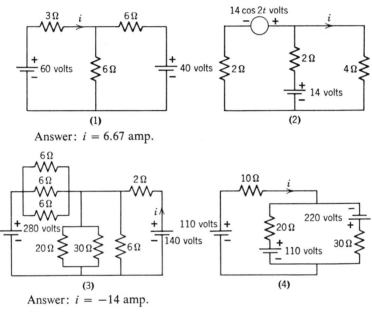

Answer: $i = 6.67$ amp.

Answer: $i = -14$ amp.

FIG. P-4.11

4.12 Given the circuit shown in Fig. P-4.12. Find the independent voltages by network reduction and voltage ratio techniques.

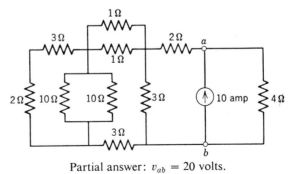

FIG. P-4.12   Partial answer: $v_{ab} = 20$ volts.

Network Functions, Equivalent Circuits, and General Network Methods   203

4.13[3] In Figs. P-4.13a and P-4.13b, connect a current source $I(s)$ to terminals a and c. For each circuit find the open circuit voltage $V_{bc}(s)$. Equate $V_{bc}(s)$ for Fig. P-4.13a to $V_{bc}(s)$ for Fig. P-4.13b showing that

$$Z_c(s) = \frac{Z_{ac}(s)Z_{bc}(s)}{Z_{ab}(s) + Z_{bc}(s) + Z_{ca}(s)}.$$

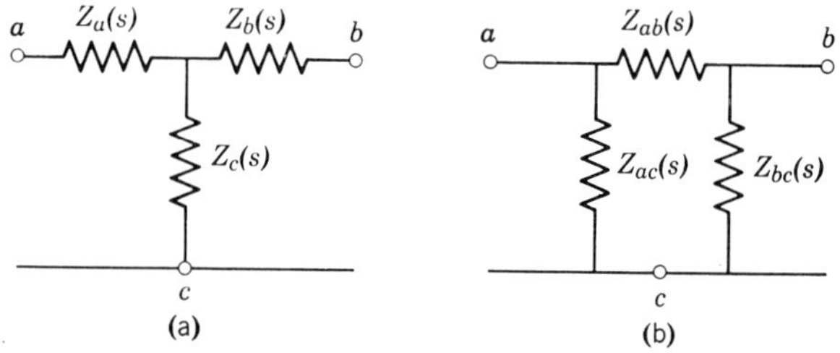

FIG. P-4.13

(a)   (b)

4.14[3] In Figs. P-4.13a and P-4.13b, connect a voltage source $V(s)$ to terminals a and c and short-circuit terminals b and c. Equate the short-circuit currents for each part of the figure to show that

$$Z_{ab}(s) = \frac{Z_a(s)Z_b(s) + Z_b(s)Z_c(s) + Z_c(s)Z_a(s)}{Z_c(s)}.$$

4.15 For each of the circuits shown in Fig. P-4.15, convert the given wye connection to an equivalent delta (or delta to wye) as appropriate.

Partial answer:

(a) $R_{ab} = 7.33\ \Omega$.

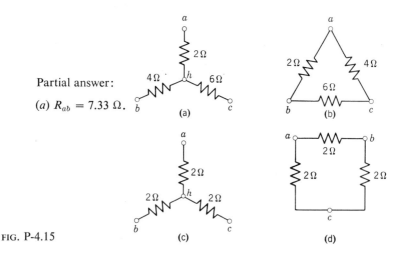

FIG. P-4.15

(a)   (b)   (c)   (d)

4.16 The network of Fig. P-4.16 is resistive, the values being shown in ohms. Determine by use of Δ-Y or Y-Δ conversions the following.

a The equivalent driving point impedance $Z_{ab}(s)$.
b The currents in each branch and the voltages v_{ab}, v_{ad}, and v_{ac}.

---

[3] J. E. Lindsay, "Wye and Delta Transformations," *Electrical Design News*, October, 1957, p. 48.

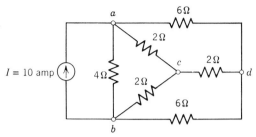

Partial answer:
The current in the 4 Ω resistance is 4.29 amp.

FIG. P-4.16

4.17 Obtain equivalent circuits for Fig. P-4.17, changing voltage sources to current sources and vice versa. Change to Laplace transform circuit if necessary. Assume initial conditions are zero unless stated otherwise.

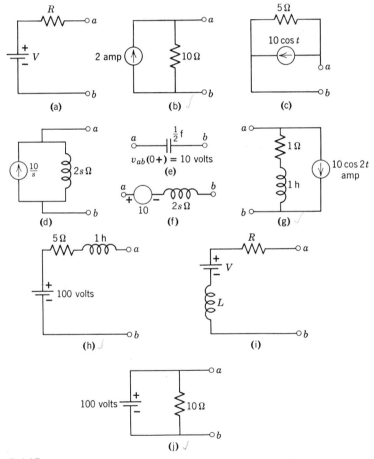

FIG. P-4.17

4.18 Devise some means to show that the equivalent source for Fig. P-4.17a is not equivalent internally.

4.19 Find the Thévenin's equivalent circuit at terminals *a-b* for each of the circuits shown in Fig. P-4.19. (Assume all initial conditions are zero where *L*'s or *C*'s are present.)

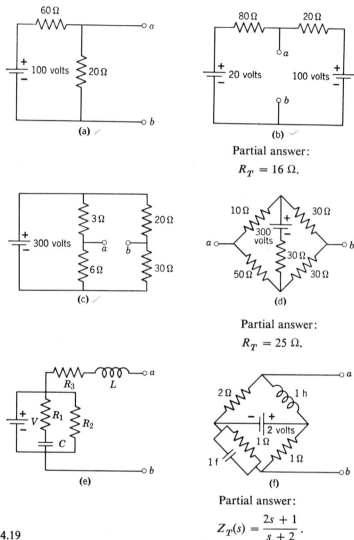

FIG. P-4.19

Partial answer (b): $R_T = 16\,\Omega$.

Partial answer (d): $R_T = 25\,\Omega$.

Partial answer (f): $Z_T(s) = \dfrac{2s + 1}{s + 2}$.

4.20 For the circuit shown in Fig. P-4.20, determine the number of independent voltages that would have to be solved for by the classical method (time domain), the number to be solved for by the Laplace transform method. Explain. (Assume initial conditions are all zero.)

206

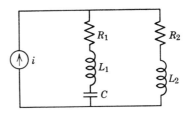

FIG. P-4.20

4.21 For the circuit shown in Fig. P-4.21, label currents leaving node 1. Assume node 0 as reference node and write expressions for each of the currents leaving node 1 in terms of the independent voltages and known values of sources and resistances. Show that the sum of the currents give the equation,

$$\left(\frac{1}{R_1} + \frac{1}{R_2}\right)V_{10} - \frac{1}{R_2}V_{20} = \frac{V}{R_2} - I_1.$$

Repeat the above and write the equations for nodes 2 and 3.

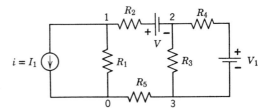

FIG. P-4.21

4.22 Find the complete solution for all the independent voltages in Fig. P-4.22, by the node voltage method.

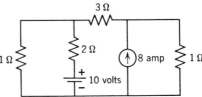

FIG. P-4.22

4.23 For the circuit shown in Fig. P-4.23, use node o as reference and find the complete solution for all of the independent voltages.

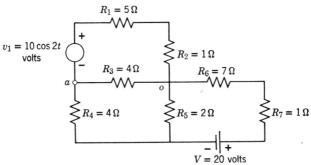

FIG. P-4.23   Partial answer: $v_{ao} = -1.2 - 2.8 \cos 2t$ volts.

Network Functions, Equivalent Circuits, and General Network Methods   207

4.24 For the circuit shown in Fig. P-4.24, use node 0 as reference and find the complete solution for all the independent voltages.

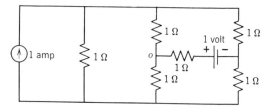

FIG. P-4.24

Partial answer: One of the independent voltages is 0.

4.25 For the circuit shown in Fig. P-4.25 assume that the initial conditions are not zero. Write the necessary equations to solve for the independent voltages. (DO NOT SOLVE.)

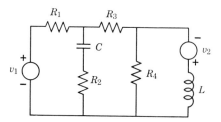

FIG. P-4.25

4.26 For each of the circuits shown in Fig. P-4.26, label branch currents. Clearly indicate the loops and write Kirchhoff's voltage equations for these loops. Solve the equations for the independent unknown currents.

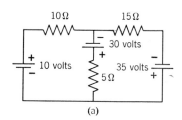

(a)

Partial answer: The current in the 5-Ω resistance is 2 amperes.

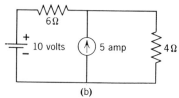

(b)

Partial answer: The magnitude of the current in the 6-Ω resistor is 1 amp.

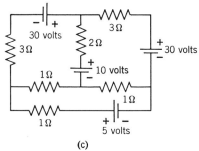

FIG. P-4.26

(c)

Partial answer: The magnitude of the current through the 5-volt source is 2 amp; through the 10-volt source is 5 amp.

4.27 Use the mesh-current method to solve for the independent currents in the circuits shown in Fig. P-4.26.

4.28 It has been suggested that loop currents be labeled directly, thereby removing the necessity for first labeling branch currents. Fig. P-4.28 shows a circuit on which loop currents were labeled directly.

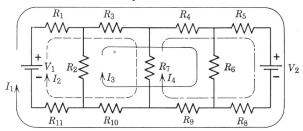

FIG. P-4.28

Will writing and solving the voltage equations about these loops give the correct solution? If not, explain why.

4.29 For the circuit shown in Fig. P-4.29, perform the following tasks.
  a Label branch currents and then show the loop currents.
  b Write the necessary Kirchhoff's voltage equations and solve for the independent currents.
  c Label mesh currents and write the necessary Kirchhoff's voltage equations and solve for the independent currents.

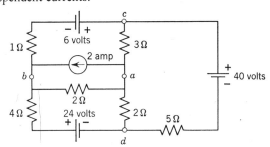

FIG. P-4.29

Partial answer: The 6-volt source is supplying 3.0 watts to the circuit. The current in the 3-Ω resistance is 4 amperes.

Network Functions, Equivalent Circuits, and General Network Methods   209

4.30 Repeat Problem 4.29 for the circuit shown in Fig. P-4.30.

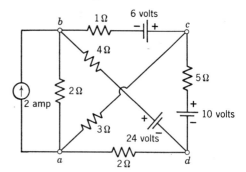

FIG. P-4.30.

Partial answer: The magnitude of the potential difference between points a and d is 5.62 volts.

4.31 For the circuits shown in Figs. P-4.26a and P-4.26c, construct a tree. Now determine the number of independent currents by adding the appropriate links.

4.32 Repeat Problem 4.31 for the circuits shown in Figures P-4.28, P-4.29, and P-4.30.

4.33 Figure P-4.33 shows a nonplanar circuit. For this circuit do the following things.
 a Determine the number of independent unknown currents.
 b Determine the number of independent unknown voltages.
 c Label the circuit clearly and write Kirchhoff's equations to solve for the unknown currents.
 d Write Kirchhoff's equations to solve for the unknown voltages.

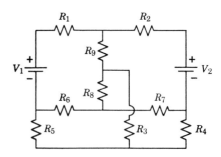

FIG. P-4.33

4.34 For the circuit of Fig. P-4.33, make these determinations:
 a Determine the number of unknown independent currents by the link-current method.
 b Determine the number of unknown independent voltages from a tree for this circuit.
 c Show several alternate trees.

210

4.35 If all the resistances in Fig. P-4.33 have a value of 1 ohm and $V_1 = 4$ volts and $V_2 = 2$ volts, use determinants to solve for the current in resistance R_1.

4.36 If all the resistances in Fig. P-4.33 have a value of 1 ohm and $V_1 = 20$ volts and $V_2 = 2$ volts, solve for the following.
  a Solve for all the currents by the loop current method.
  b Solve for all the independent voltages by the node-voltage method.
  Partial answer: The magnitude of some of the currents are:

  in R_1, 8.44 amperes R_3, 2.44 amperes
  R_2, 3.55 amperes R_5, 4.21 amperes

4.37 For the circuits shown in Fig. P-4.37, accomplish the following.
  a Use the node o as reference and find the voltage v_{mo} by the node-voltage method.
  b Find v_{mo} by the mesh-current method.
  c Find the current through resistance R_0 by Thévenin's theorem.

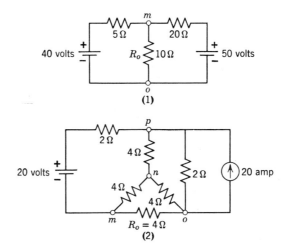

FIG. P-4.37

Partial answer: (2) $V_{mo} = 8$ volts.

4.38 For the circuits shown in P-4.37, retain the resistance R_0, and, by using equivalent circuit methods other than Thévenin's, reduce the rest of the circuit to a single source and resistance.

4.39 State the advantage, if any, of the link-current method over the mesh-current method in determining the number of independent currents to be solved for in a network.

4.40 For the circuit shown in Fig. P-4.40, use the principle of symmetry to find the current I.
  Answer: $I = 6$ amp.

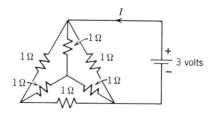

FIG. P-4.40

4.41 Figure P-4.41 shows twelve 1-Ω resistors connected to form a cube. Using the principle of symmetry, make the following determinations.
  a Find R_{10}.
  b Find R_{13}.
  c Find R_{14}.
  d Find R_{17}.

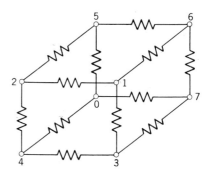

FIG. P-4.41

Partial answer:
$R_{10} = \frac{5}{6}$ Ω.
$R_{13} = \frac{7}{12}$ Ω.
$R_{14} = \frac{3}{4}$ Ω.

4.42 For the circuits shown in Fig. P-4.42, use the principle of symmetry to determine the current I. Check your answer by solving for I by some other method.

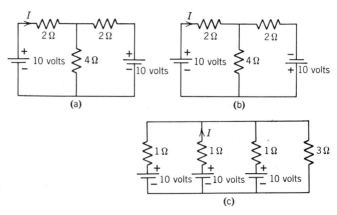

FIG. P-4.42

Answer: (a) $I = 1$ amp.
Answer: (b) $I = 5$ amp.
Answer: (c) $I = 1$ amp.

4.43 Use the ladder method to find the currents and voltages for the circuit shown in Fig. P-4.43.

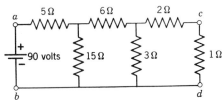

FIG. P-4.43

4.44 In the circuit shown in Fig. P-4.43 the 90-volt source is removed and then connected to terminals c-d and the 1-Ω resistance is removed and then connected to terminals a-b. What current will now flow in the 1-Ω resistance?

4.45 For the circuit shown in Fig. P-4.45, accomplish the following.
a Find the short-circuit current I.
b Verify the reciprocity theorem by inserting the voltage source in branch a-b, short-circuiting terminals c-d, and then calculating the current through the 6-Ω resistance.

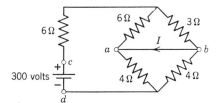

FIG. P-4.45

4.46 For each of the circuits shown in Fig. P-4.46, use scaling to give numerical values which will simplify the solutions.

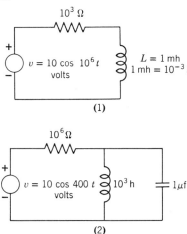

FIG. P-4.46

4.47 It is desired to demonstrate the response of the circuit shown in Fig. P-4.47 in the laboratory.

The following are found to be available. A signal generator with a voltage output variable from 1 to 100 volts and frequency output ranging from 100 to 10,000 cycles per second (628 to 62,800 radians per second). Resistors ranging from 1 to 10^6 ohms. Inductances ranging from 10^{-4} to 10 henries. Capacitors ranging from 10^{-12} to 10^{-3} farads. Draw a circuit showing values for the items available in the laboratory which will give the desired result.

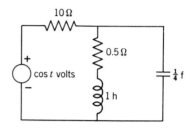

FIG. P-4.47

4.48 In the circuit shown in Fig. P-4.48 all initial conditions are zero.
  a Using the node voltage method and the Laplace transform, find the current i.
  b Use the principle of symmetry to find an equivalent transform circuit and then find i.
  c Use the result of part a to find the other branch currents.
  d Use the equivalent circuit of part b to find the other branch currents.
  e Use Thévenin's theorem to find i.

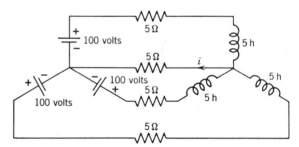

FIG. P-4.48

Partial answer: $i = 15(1 - e^{-4t})$ amp.

4.49 In the circuit shown in Fig. P-4.49, all initial conditions are zero. All values of R are equal, all values of L are equal, and all values of V are equal. Find the current i_0.

4.50 Use Thévenin's theorem to find the current i_0 in the circuit shown in Fig. P-4.50. Assume that all initial conditions are zero.

4.51 For the circuit shown in Fig. P-4.51 all initial conditions are zero. Find the voltage v_{ab}.

FIG. P-4.49

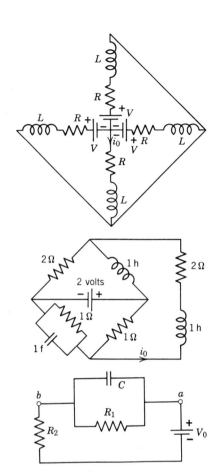

FIG. P-4.50

FIG. P-4.51

4.52 The circuit shown in Fig. P-4.52 is in the steady state with the switch open. At time $t = 0$ the switch is closed. Use Norton's theorem to find i for $t > 0$.

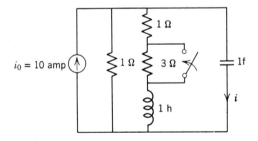

FIG. P-4.52

4.53 Using a general approach, similar to that used in Appendix E to prove Thévenin's theorem, use the node-voltage method to prove Norton's theorem.

4.54 Prove the reciprocity theorem using the general equations for the node-voltage method.

5 Steady-State Analyses with Sinusoidal Sources

The steady-state response to the sinusoidal driving function is of primary importance in the study of electrical circuits. The electrical energy for our homes and factories is supplied by generating and transmission systems in which voltages and currents vary sinusoidally at 60 cycles per second. Since our communications systems operate over limited ranges of frequencies, useful results are obtained from steady-state analyses of such systems. Also, any repetitive wave can be analyzed in terms of its sinusoidal frequency components. Even a single pulse can be represented by its frequency spectrum. We do not mean to infer that transient or free responses are not important. Rather, in complex systems the free response is extremely laborious to predict while the sinusoidal steady-state responses can be obtained readily and can be used to infer the free response.

Steady-state responses can be determined readily, not by the methods of Chapter 3, but by a "phasor" method which involves the algebra of complex numbers. We shall review the classical method of Chapter 3 and then gradually develop the phasor method. We shall find that all the methods and techniques for the transform circuits of Chapter 4 will apply to the "phasor" circuit. We shall also develop the phasor domain from the transform domain.

5.1 INTRODUCTION TO SINUSOIDAL WAVES

The term "sinusoidal functions of time" refers to either $A \cos(\omega t + \phi)$ or $A \sin(\omega t + \phi)$. This periodic function has a period of 2π radians or

T seconds with T being defined as

$$T = \Delta t = \frac{2\pi}{\omega}. \tag{5.1}$$

A is known as the amplitude of the wave; ω is the radian or angular frequency and is related to f, the frequency in cycles per second, by the expression,

$$\omega = 2\pi f. \tag{5.2}$$

The angle ϕ is known as the phase or phase angle of the wave with the dimensionless units of radians or degrees. The function $A \cos(\omega t + \phi)$ is said to "lead" the function $B \cos \omega t$ by the angle ϕ, or conversely the function $B \cos \omega t$ is said to "lag" the function $A \cos(\omega t + \phi)$ by the angle ϕ. Numerically, there is some arbitrariness about the angle ϕ since the sinusoidal function repeats in increments of

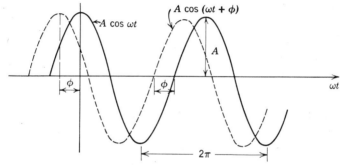

FIG. 5.1. Sinusoidal functions of time.

360°. Thus $A \cos(\omega t + 45°) = A \cos(\omega t - 315°) = A \cos(\omega t + 405°)$, etc. We shall normally use for ϕ that angle whose magnitude is less than 180°.

Figure 5.1 shows a sketch of the waves $A \cos \omega t$ and $A \cos(\omega t + \phi)$. From a physical standpoint we consider the minimum value of the wave to be zero, not $-A$. The reason is that for currents and voltages, negative values are simply opposite in sense to that which we arbitrarily call positive. A current of $10 \cos(\omega t + \phi)$ amperes has a maximum value (amplitude) of 10 amperes and a minimum value of zero.

5.2 The Addition of Sinusoidal Functions

This article reviews the classical method for determining steady-state solutions as applied to a simple circuit, points out the need for adding sinusoidal terms of the same frequency, and then develops a technique for using complex numbers in such additions.

Kirchhoff's voltage law for the circuit of Fig. 5.2 is

$$L\frac{di}{dt} + Ri = V\cos \omega t, \tag{5.3}$$

in which $i = i_{ss} + i_f$, and i_f satisfies the equation $L(di_f/dt) + Ri_f = 0$.

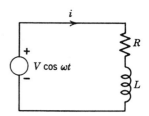

FIG. 5.2. Circuit with sinusoidal source.

The steady-state response, i_{ss}, must be of the form,

$$i_{ss} = K_1 \cos \omega t + K_2 \sin \omega t. \tag{5.4}$$

K_1 and K_2 may therefore be evaluated by the technique of Section 3.12 and become

$$K_1 = \frac{RV}{R^2 + \omega^2 L^2}, \tag{5.5}$$

and

$$K_2 = \frac{\omega LV}{R^2 + \omega^2 L^2}. \tag{5.6}$$

Until this time we have been willing to leave the steady-state expression for current in the form of Equation 5.4, but now we need to determine the amplitude of the steady-state current and its phase angle with respect to the applied voltage. In other words, we would like to determine C and ϕ in the following equation.

$$i_{ss} = K_1 \cos \omega t + K_2 \sin \omega t = C \cos(\omega t + \phi). \tag{5.7}$$

Expansion of the right-hand side gives

$$i_{ss} = K_1 \cos \omega t + K_2 \sin \omega t = C \cos \phi \cos \omega t - C \sin \phi \sin \omega t. \tag{5.8}$$

Equating coefficients of like terms, we obtain

$$\cos \phi = \frac{K_1}{C},$$

$$\sin \phi = -\frac{K_2}{C},$$

from which a triangle may be sketched as shown in Fig. 5.3. It is apparent that $C = \sqrt{K_1^2 + K_2^2}$ and $\phi = \tan^{-1}(-K_2/K_1)$, and

$$i_{ss} = K_1 \cos \omega t + K_2 \sin \omega t$$
$$= \sqrt{K_1^2 + K_2^2} \cos(\omega t - \tan^{-1} K_2/K_1). \tag{5.9}$$

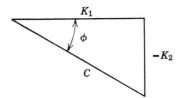

FIG. 5.3. Triangle for K_1, K_2, C, and ϕ.

Substitution of Equations 5.5 and 5.6 into Equation 5.9 gives

$$i_{ss} = \frac{V}{\sqrt{R^2 + \omega^2 L^2}} \cos\left[\omega t - \tan^{-1}\left(\frac{\omega L}{R}\right)\right]. \tag{5.10}$$

The amplitude of the current is $V/\sqrt{R^2 + \omega^2 L^2}$ and the steady-state current lags the voltage by an angle whose tangent is $\omega L/R$. The term ωL has the units of ohms; we shall call it inductive reactance. The angle ϕ would be zero if ωL were zero and approaches $-90°$ as R approaches zero. Thus for an $R - L$ circuit the steady-state current will lag the voltage by an angle between 0 and 90°.

This method for adding two sinusoidal functions of the same frequency is awkward and would be more difficult if the terms were of the form, $A \cos(\omega t + \alpha) + B \sin(\omega t + \beta)$, or if we were to add more than two terms. There is a more convenient method for performing the mathematics. This method is primarily based on the fact that every term has the same angular frequency, and thus the significant data is contained in amplitude and phase. It is also based on the algebra of complex numbers with which we have some familiarity. We recall Euler's formula (previously given in Chapter 3).

$$e^{j(\phi \pm n 2\pi)} = \cos \phi + j \sin \phi, \tag{5.11}$$

$n = 0, 1, 2, 3,$ etc.

We note again that

$$e^{j(\pi/2 \pm n 2\pi)} = j. \tag{5.12}$$

Equation 5.12 indicates that the imaginary unit, $j = \sqrt{-1}$, may be considered as an operator such that if a complex number is multiplied by j the effect is to rotate the complex number 90° in a counterclockwise sense. We may designate any number, $\hat{A}$, in the complex plane as follows.

$$\hat{A} = A e^{j\phi} = A \cos \phi + jA \sin \phi$$
$$= a + jb = \sqrt{a^2 + b^2} e^{j\phi}. \tag{5.13}$$

Figure 5.4 shows $\hat{A}$ as a point in the complex plane of numbers. Note that ϕ is measured counterclockwise from the positive axis of reals.

The term $a + jb$ is known as the rectangular form of the complex number $\hat{A}$. The term a is known as the real part of $\hat{A}$, or Re $\hat{A}$. The

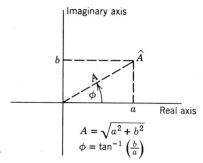

FIG. 5.4. Complex plane of number.

term b is the coefficient of the imaginary unit j, but this is usually worded as the imaginary part of $\hat{A}$, or Im $\hat{A}$. It is unfortunate that the word imaginary is used in this sense; it is also unfortunate that one of the axes is called the imaginary axis. We must live with this; let us try to adopt the philosophy that what is said to be imaginary may be real, or vice versa, and that what is real or imaginary in mathematics may have nothing to do with what is real or imaginary in physical phenomena.

The term $Ae^{j\phi}$ is known as the exponential form of the complex number $\hat{A}$, with A being the magnitude and ϕ the angle of $\hat{A}$. (The student will see the shorthand form $A/\underline{\phi}$ used in some texts and in some of the literature but this form lacks mathematical significance and should be avoided.) We need to distinguish between the complex number's magnitude, A, and the complex number, $\hat{A}$, and this is done with the caret (∧) placed over the letter to indicate that the quantity is complex. Certain complex quantities are never written in exponential form and these we shall consider as complex even though no caret mark is used. The complex frequency, s, is one of these and so may be the quantity, m, introduced in the classical method of Chapter 3.

We may summarize our discussion of the complex number $\hat{A}$ by saying that we shall use it in one of the following forms.

$$\hat{A} = Ae^{j\phi} = A\cos\phi + jA\sin\phi = a + jb,$$

in which $A = \sqrt{a^2 + b^2}$, $\phi = \tan^{-1}(b/a)$, $a = \text{Re } \hat{A}$, and $b = \text{Im } \hat{A}$.

Let us now consider the term $\hat{A}e^{j\omega t}$ in which $\hat{A} = Ae^{j\phi}$. Then we may expand $\hat{A}e^{j\omega t}$ as follows.

$$\hat{A}e^{j\omega t} = Ae^{j\phi}e^{j\omega t} = Ae^{j(\omega t + \phi)}$$
$$= A\cos(\omega t + \phi) + jA\sin(\omega t + \phi). \quad (5.14)$$

Thus

$$A\cos(\omega t + \phi) = \text{Re}(\hat{A}e^{j\omega t}),$$

and

$$A\sin(\omega t + \phi) = \text{Im}(\hat{A}e^{j\omega t}).$$

These results are interesting since both $\text{Re}(\hat{A}e^{j\omega t})$ and $\text{Im}(\hat{A}e^{j\omega t})$ are sinusoidal time functions whereas $\hat{A}e^{j\omega t}$ is a complex function of time. Although we realize the function $\hat{A}e^{j\omega t}$ does not exist physically, it is very useful to us in analysis.

Consider the locus of $\hat{A}e^{j\omega t}$ in the complex plane as t increases from zero. (See Fig. 5.5.) When $t = 0$, $\hat{A}e^{j\omega t} = \hat{A} = Ae^{j\phi}$. As t increases the locus of $\hat{A}e^{j\omega t}$ moves counterclockwise at a constant radius A and completes one revolution in a time $T = 2\pi/\omega$. The locus of $\hat{A}e^{j\omega t}$ is therefore a circle of radius A with center at the origin in the complex

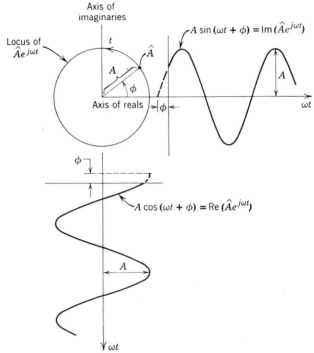

FIG. 5.5. Locus of $\hat{A}e^{j\omega t}$ and projections of $\hat{A}e^{j\omega t}$ on axes of reals and imaginaries.

plane. The projections on the axis of reals, $\text{Re}(\hat{A}e^{j\omega t})$, and on the axis of imaginaries, $\text{Im}(\hat{A}e^{j\omega t})$, are sinusoidal functions of time.

For certain mathematical operations it is possible to **replace** sinusoidal functions such as $A\cos(\omega t + \phi)$ or $\text{Re}(\hat{A}e^{j\omega t})$ with the corresponding complex time function, $\hat{A}e^{j\omega t}$. It then becomes possible to eliminate the time function and deal only with complex numbers, a distinct advantage in solving circuit problems.

We shall demonstrate this for the addition of two sinusoidal time functions of the same frequency.

Let
$$A_1 \cos(\omega t + \alpha_1) + A_2 \cos(\omega t + \alpha_2) = A_3 \cos(\omega t + \alpha_3) \qquad (5.15)$$
or
$$\text{Re}[A_1 e^{j(\omega t + \alpha_1)}] + \text{Re}[A_2 e^{j(\omega t + \alpha_2)}] = \text{Re}[A_3 e^{j(\omega t + \alpha_3)}],$$

in which A_1, A_2, α_1, and α_2 are known and it is desired to determine A_3 and α_3. Since Equation 5.15 holds for all values of time and since the terms are of the same frequency, a constant angle of any magnitude can be added to each angle without affecting the validity of the equation. If an angle of $-\pi/2$ radians be added, there results

$$A_1 \sin(\omega t + \alpha_1) + A_2 \sin(\omega t + \alpha_2) = A_3 \sin(\omega t + \alpha_3). \qquad (5.16)$$

Now multiply Equation 5.16 by j and add to Equation 5.15 to obtain

$$A_1[\cos(\omega t + \alpha_1) + j \sin(\omega t + \alpha_1)]$$
$$+ A_2[\cos(\omega t + \alpha_2) + j \sin(\omega t + \alpha_2)]$$
$$= A_3[\cos(\omega t + \alpha_3) + j \sin(\omega t + \alpha_3)],$$
or
$$A_1 e^{j(\omega t + \alpha_1)} + A_2 e^{j(\omega t + \alpha_2)} = A_3 e^{j(\omega t + \alpha_3)}. \qquad (5.17)$$

We have shown for equations of the form of Equation 5.15 that $\text{Re}[Ae^{j(\omega t + \alpha)}]$ may be **replaced** by $Ae^{j(\omega t + \alpha)}$.

The term $e^{j\omega t}$ may be factored out of Equation 5.17 to give

$$A_1 e^{j\alpha_1} + A_2 e^{j\alpha_2} = A_3 e^{j\alpha_3}. \qquad (5.18)$$

The quantities A_3 and α_3 may readily be obtained from Equation 5.18.

Example 5.1

GIVEN: The node of Fig. E-1.1 with

$i_1 = 10 \cos(\omega t + 60°)$ amp

$i_2 = 5 \sin \omega t$ amp

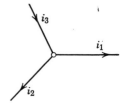

FIG. E-1.1

TO FIND: The expression for i_3.

SOLUTION: Kirchhoff's current law requires

$$i_3 = i_1 + i_2 = 10 \cos(\omega t + 60°) + 5 \sin \omega t = I_3 \cos(\omega t + \phi),$$

or

$$10\cos(\omega t + 60°) + 5\cos(\omega t - 90°) = I_3\cos(\omega t + \phi), \quad (1)$$

or

$$\text{Re}[10e^{j(\omega t+60°)}] + \text{Re}[5e^{j(\omega t-90°)}] = \text{Re}[I_3 e^{j(\omega t+\phi)}].$$

From the previous demonstration, we may replace Re [ ] by [ ], or
$$10e^{j(\omega t+60°)} + 5e^{j(\omega t-90°)} = I_3 e^{j(\omega t+\phi)},$$

or

$$10e^{j60°} + 5e^{-j90°} = I_3 e^{j\phi}. \quad (2)$$

In order to perform the addition we need to change the exponential form of complex numbers into rectangular form. The use of Euler's formula gives

$$10(\cos 60° + j\sin 60°) + 5[\cos(-90°) + j\sin(-90°)] = I_3 e^{j\phi}.$$
$$5 + j8.66 - j5 = I_3 e^{j\phi}$$
$$5 + j3.66 = \sqrt{5^2 + 3.66^2}\, e^{j\,\tan^{-1}(3.66/5)} = 6.2 e^{j36.2°} = I_3 e^{j\phi}$$

Finally,

$$i_3 = \text{Re}[I_3 e^{j(\omega t+\phi)}] = \text{Re}[6.2 e^{j(\omega t+36.2°)}]$$
$$= 6.2\cos(\omega t + 36.2°)\ \text{amp}. \quad (3)$$

In Fig. E-1.2 are shown the points in the complex plane, $10e^{j60°}$, $5e^{-j90°}$, and $6.2e^{j36.2°}$.

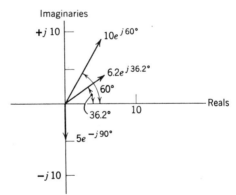

FIG. E-1.2

Figure E-1.3 shows the waveforms of our functions in the time domain as we might expect to see them on an oscilloscope in the laboratory.

In the previous example we note that the complex number which represents $\sin \omega t$ lies on the negative axis of imaginaries, and it is apparent that the complex number which represents $A \cos \omega t$ would lie on the positive axis of reals. Also the complex number which

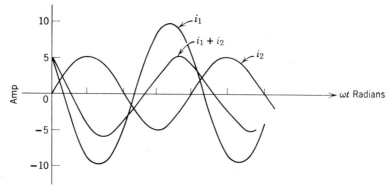

FIG. E-1.3

represents $-A \sin \omega t = A \cos(\omega t + \pi/2)$ would lie on the positive axis of imaginaries. These are the results of replacing $\text{Re}(\hat{A}e^{j\omega t})$ by $\hat{A}e^{j\omega t}$. For purposes of locating the complex numbers it is helpful to use the "crutch" of Fig. 5.6.

Figure 5.6 is also helpful in changing readily from one sinusoidal form to another.

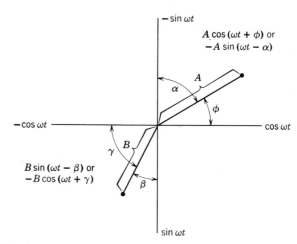

FIG. 5.6. "Crutch" for determining complex numbers to represent sinusoidal time functions.

Example 5.2

GIVEN: $v_{ab} = -10 \cos(\omega t + 60°)$ volts,
$v_{bc} = 8 \sin(\omega t + 120°)$ volts.

TO FIND: v_{ac}.

224

SOLUTION: The complex numbers $\hat{V}_{ab}$ and $\hat{V}_{bc}$ are located as shown in Fig. E-2.1.

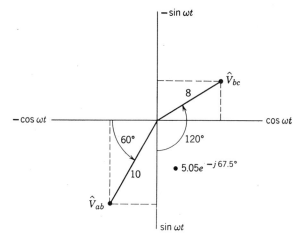

FIG. E-2.1

Then

$\hat{V}_{ab} = 10e^{-j120°} = -5 - j8.66.$

$\hat{V}_{bc} = 8e^{j30°} = 6.94 + j4.$

$\hat{V}_{ac} = \hat{V}_{ab} + \hat{V}_{bc} = 1.94 - j4.66 = \sqrt{1.94^2 + 4.66^2}\, e^{j\tan^{-1}(-4.66/1.94)}$

Then

$v_{ac} = \text{Re}[\hat{V}_{ac} e^{j\omega t}] = \text{Re}[5.05 e^{-j67.5°} e^{j\omega t}]$
$= 5.05 \cos(\omega t - 67.5°)$ volts.

5.3 COMPLEX QUANTITIES

The previous section has shown that complex quantities are useful for adding sinusoidal time functions. Such complex quantities are also the primary means for analyzing circuits in the sinusoidal steady state. Before we show that this is possible, it is appropriate to review the algebra of complex numbers and to introduce some additional items useful in circuit analysis.

A complex quantity, $\hat{A}$, may be expressed in either the rectangular (cartesian) form, $a + jb$, or in the exponential (polar) form, $Ae^{j\phi}$.

$\hat{A} = Ae^{j\phi} = A \cos \phi + jA \sin \phi = a + jb.$

The cartesian form may be obtained from the exponential form since $a = \text{Re } \hat{A} = A \cos \phi$ and $b = \text{Im } \hat{A} = A \sin \phi$.

The exponential form may be obtained from the cartesian since $A = \sqrt{a^2 + b^2}$ and $\phi = \tan^{-1}(b/a)$.

Addition or subtraction may be performed conveniently **only** in the rectangular form. For two quantities,

$\hat{A} = Ae^{j\phi} = a + jb,$

and

$\hat{D} = De^{j\beta} = g + jh$

$\hat{A} + \hat{D} = a + jb + (g + jh) = a + g + j(b + h),$

and

$\hat{A} - \hat{D} = a + jb - (g + jh) = a - g + j(b - h).$

Multiplication and division may be performed in both exponential and rectangular forms.

Thus

$\hat{A}\hat{D} = Ae^{j\phi}De^{j\beta} = ADe^{j(\phi+\beta)}$

or

$\hat{A}\hat{D} = (a + jb)(g + jh) = ag - bh + j(bg + ah).$

Also

$\dfrac{\hat{A}}{\hat{D}} = \dfrac{Ae^{j\phi}}{De^{j\beta}} = \dfrac{A}{D} e^{j(\phi-\beta)},$

and

$\dfrac{\hat{A}}{\hat{D}} = \dfrac{a + jb}{g + jh} = \dfrac{(a + jb)(g - jh)}{(g + jh)(g - jh)} = \dfrac{ag + bh + j(bg - ah)}{g^2 + h^2}.$

In the last operation the numerator and denominator were multiplied by the conjugate of the denominator in order to eliminate imaginary terms in the denominator. The conjugate of a complex number $\hat{A}$ is defined as having the same magnitude as $\hat{A}$ but having the negative angle of $\hat{A}$. The conjugate of $\hat{A}$ is written as $\hat{A}^*$. For our quantities

$\hat{A}^* = Ae^{-j\phi} = a - jb,$

and

$\hat{D}^* = De^{-j\beta} = g - jh.$

Roots of numbers may be obtained conveniently **only** in the exponential form. Let $\hat{A} = Ae^{j(\phi \pm n2\pi)}$, $n = 0, 1, 2, 3$, etc. The mth roots of $\hat{A}$ are obtained from

$(\hat{A})^{1/m} = [Ae^{j(\phi+n2\pi)}]^{1/m} = A^{1/m} e^{j[(\phi/m)+n(2\pi/m)]}.$

Example 5.3

GIVEN: The numbers $\hat{A} = 10e^{j53.1°} = 6 + j8$

$\hat{D} = 5e^{j150°} = -4.33 + j2.5$

TO FIND: The sum, difference, product, and ratio of these two numbers, also the 3rd roots of $\hat{D}$.

SOLUTION:

$\hat{A} + \hat{D} = 6 + j8 + (-4.33 + j2.5) = 1.67 + j10.5$

$\hat{A} - \hat{D} = 6 + j8 - (-4.33 + j2.5) = 10.33 + j5.5$

$\hat{A}\hat{D} = 10e^{j53.1°}5e^{j150°} = 50e^{j203.1°} = 50e^{-j156.9°}$

$\hat{A}\hat{D} = (6 + j8)(-4.33 + j2.5) = -26 - 20 - j34.6 + j15$

$\quad\quad = -46 - j19.6$

$\dfrac{\hat{A}}{\hat{D}} = \dfrac{10e^{j53.1°}}{5e^{j150°}} = 2e^{-j96.9°}$

$\dfrac{\hat{A}}{\hat{D}} = \dfrac{6 + j8}{-4.33 + j2.5} = \dfrac{(6 + j8)(-4.33 - j2.5)}{-4.33^2 + 2.5^2}$

$\quad\quad = \dfrac{-26.0 + 20 - j34.6 - j15}{18.8 + 6.25}$

$\quad\quad = \dfrac{-6 - j49.6}{25.05} = -0.24 - j1.97$

$(\hat{D})^{1/3} = [5e^{j(150° \pm n360°)}]^{1/3} = 5^{1/3}e^{j(50° + n120°)}$

$\quad\quad = 1.71e^{j(50° + n120°)}$

The three roots are $1.71e^{j50°}$, $1.71e^{j170°}$, and $1.71e^{-j70°}$.

It will be necessary to change a complex number from the cartesian form to the exponential form and vice versa. This should be practiced until good accuracy and speed have been developed. The following suggestions are made to assist in this development.

(1) Always make a sketch showing the approximate location of the number. This should help in avoiding major errors in angle.

(2) Keep in mind that for small angles $\sin \phi \simeq \tan \phi \simeq \phi$ in radians. 1 radian = $180°/\pi$ and is marked as R on many slide rules. For angles of 6° the error in using $\sin \phi \simeq \tan \phi$ and $\cos \phi \simeq 1$ is approximately 0.5%. However, one should not make the error of saying that $Ae^{j\phi°} \simeq Ae^{j0°}$ for small angles.

(3) There is a convenient method for converting complex numbers from cartesian to exponential form on some slide rules. We demonstrate for the K & E log log decitrig rule. To convert $A + jB$ to $Me^{j\phi}$ set the index of the C scale over the larger of the two numbers on the D scale and set the hairline over the smaller of the two numbers on the D scale. Now read the angle under the hairline on the T (tangent) scale. This is the angle ϕ. Now slide the slider until this angle on the S (sine-cosine scale) is under the hairline and read the magnitude of M on the D scale under the index of the C scale. To illustrate, suppose we wish to

Steady-State Analyses with Sinusoidal Sources

convert $3 + j4$ to exponential form. Set the right-hand index of the C scale over 4 and the hairline over 3 on the D scale. Read 53.1° on the T scale under the hairline. Now move the slider until 53.1° on the S scale appears under the hairline. Read 5 on the D scale under the index of the C scale. Therefore, $3 + j4 = 5e^{j53.1°}$. If we had had $4 + j3$ to convert we would have followed the same procedure but would have read the angle as 36.9° on the T scale. It is important to sketch the quantity to avoid errors in selecting the proper angle. To convert a complex number from exponential to cartesian form the reverse procedure is followed. The student should note that for complex numbers in cartesian form, if the ratio of A to B or B to A is greater than 10, or in exponential form where the tangent of the angle ϕ is greater than ten or less than $\frac{1}{10}$, this convenient method does not hold.

The complex number $\hat{A}$ may be represented by a directed line segment from the origin to the point $\hat{A}$. It is even more convenient to interpret Re $\hat{A}$ and Im $\hat{A}$ as the projections of the line segment on the real and imaginary axes respectively. Therefore, we are free to move the directed line segment anywhere in the complex plane, maintaining the length of the segment and the angle which the segment makes with the axis of reals. This is shown graphically in Fig. 5.7.

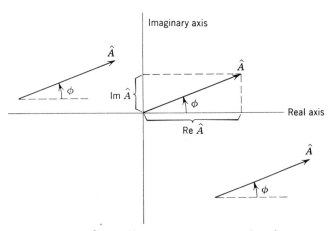

FIG. 5.7. Complex number, $\hat{A} = Ae^{j\phi}$, represented on complex plane.

The advantage of this method is that the addition and subtraction of complex numbers may be shown on the diagram. This is demonstrated in Fig. 5.8.

Note specifically that there is an easy way to indicate graphically the difference of two complex numbers, $\hat{A} - \hat{B}$. This is a directed line segment drawn from $\hat{B}$ to $\hat{A}$.

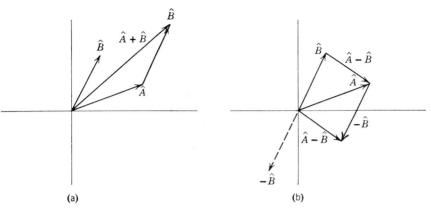

(a) (b)

FIG. 5.8. The graphical addition of complex numbers.

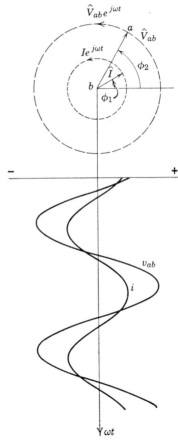

FIG. 5.9. Projection of $\hat{I}e^{j\omega t}$ and $\hat{V}_{ab}e^{j\omega t}$ on axis of reals.

Steady-State Analyses with Sinusoidal Sources 229

In circuit analysis we will use the identities $i = I\cos(\omega t + \phi_1) = \text{Re}(Ie^{j\phi_1}e^{j\omega t})$, and $v_{ab} = V_{ab}\cos(\omega t + \phi_2) = \text{Re}(V_{ab}e^{j\phi_2}e^{j\omega t})$, and then show the complex quantities $Ie^{j\phi_1}$ and $V_{ab}e^{j\phi_2}$ in the complex plane. What information is available from these line segments in the complex plane? Fig. 5.9 shows the complex quantities $\hat{I}$ and $\hat{V}_{ab}$, the complex time functions $\hat{I}e^{j\omega t}$ and $\hat{V}_{ab}e^{j\omega t}$ and their projections on the real axis. We observe that

$$\text{Re}(\hat{I}) = i\bigg|_{t=0} \quad \text{and} \quad \text{Re}(\hat{V}_{ab}) = v_{ab}\bigg|_{t=0}.$$

If the phase angle, ϕ_1, of $\hat{I}$ is in first or fourth quadrant, the current, $i(0)$, is positive in the sense indicated on some circuit diagram. If the phase of $\hat{I}$ is in the second or third quadrants, $i(0)$ is negative. If the phase of $\hat{V}_{ab}$ is in the first or fourth quadrant, $v_{ab}(0)$ is positive or a is plus with respect to b. If the phase of $\hat{V}_{ab}$ is in the second or third quadrant $v_{ab}(0)$ is negative or b is plus with respect to a. It is therefore possible to associate the first subscript for voltage with the head of the arrow or directed line segment and the second subscript with the tail of the arrow.

5.4 STEADY-STATE COMPLEX FORM (PHASORS) FROM THE TIME DOMAIN

We shall now show that sinusoidal steady-state responses may be obtained for linear systems by the direct use of complex numbers. This will be done by proving the following theorem.

Theorem. A linear differential equation with constant coefficients, with a single driving function, $f[t] = \text{Re}[\hat{D}e^{j\omega t}]$, and with a steady-state response, $x_{ss} = \text{Re}[\hat{A}e^{j\omega t}]$, is valid if $\text{Re}[\hat{D}e^{j\omega t}]$ is replaced by $\hat{D}e^{j\omega t}$ and $\text{Re}[\hat{A}e^{j\omega t}]$ is replaced by $\hat{A}e^{j\omega t}$.

Proof: Consider the equation,

$$\frac{d^n x}{dt^n} + a_{n-1}\frac{d^{n-1}x}{dt^{n-1}} \cdots a_1\frac{dx}{dt} + a_0 x = f(t), \tag{5.19}$$

for which the steady-state response, x_{ss}, is known to be $A\cos(\omega t + \phi)$ or $\text{Re}(\hat{A}e^{j\omega t})$ with $f(t) = D\cos(\omega t + \beta)$ or $\text{Re}(\hat{D}e^{j\omega t})$.
Thus,

$$\frac{d^n \text{Re}(\hat{A}e^{j\omega t})}{dt^n} + a_{n-1}\frac{d^{n-1}\text{Re}(\hat{A}e^{j\omega t})}{dt^{n-1}} \cdots$$
$$+ a_1 \frac{d\,\text{Re}(\hat{A}e^{j\omega t})}{dt} + a_0\text{Re}(\hat{A}e^{j\omega t}) = \text{Re}(\hat{D}e^{j\omega t}). \tag{5.20}$$

If the driving function were changed from $D\cos(\omega t + \beta)$ to $D\sin(\omega t + \beta) = \text{Im}(\hat{D}e^{j\omega t})$, a simple change in the phase reference of

$\pi/2$ radians, it is apparent that the steady-state response would also shift to $A \sin(\omega t + \phi) = \text{Im}(\hat{A}e^{j\omega t})$. Thus a second equation may be written:

$$\frac{d^n \text{Im}}{dt^n}(\hat{A}e^{j\omega t}) + a_{n-1}\frac{d^{n-1}\text{Im}(\hat{A}e^{j\omega t})}{dt^{n-1}} \cdots$$
$$+ a_1\frac{d\,\text{Im}(\hat{A}e^{j\omega t})}{dt} + a_0\,\text{Im}(\hat{A}e^{j\omega t}) = \text{Im}(\hat{D}e^{j\omega t}). \quad (5.21)$$

Now multiply Equation 5.21 by j, taking j inside the derivatives since j is not a function of time, and add the resulting equation to Equation 5.20. We obtain terms of the form, $\text{Re}(\hat{A}e^{j\omega t}) + j\,\text{Im}(\hat{A}e^{j\omega t})$, which are, of course, equal to $\hat{A}e^{j\omega t}$. The net result is the equation,

$$\frac{d^n}{dt^n}(\hat{A}e^{j\omega t}) + a_{n-1}\frac{d^{n-1}(\hat{A}e^{j\omega t})}{dt^{n-1}} \cdots a_1\frac{d(\hat{A}e^{j\omega t})}{dt}$$
$$+ a_0\hat{A}e^{j\omega t} = \hat{D}e^{j\omega t}. \quad (5.22)$$

This proves the theorem. The usefulness of this theorem arises from the fact that the differentiations can now readily be performed to yield,

$$(j\omega)^n\hat{A}e^{j\omega t} + a_{n-1}(j\omega)^{n-1}\hat{A}e^{j\omega t} \cdots a_1 j\omega \hat{A}e^{j\omega t} + a_0\hat{A}e^{j\omega t} = \hat{D}e^{j\omega t},$$

for which the term $e^{j\omega t}$ may be factored out to give

$$[(j\omega)^n + a_{n-1}(j\omega)^{n-1} \cdots a_1 j\omega + a_0]\hat{A} = \hat{D}. \quad (5.23)$$

The trigonometric time factors have been eliminated and we have some simple relations in complex quantities.

We shall apply the theorem to the electric circuit of Fig. 5.10a in which $v = V\cos(\omega t + \alpha)$ volts and we are seeking the steady-state response which we know to be of the form, $i_{ss} = I\cos(\omega t + \beta)$.

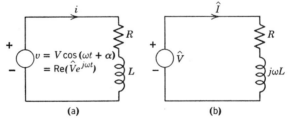

FIG. 5.10. An RL circuit and its steady-state complex equivalent.

The equation for the steady state response is

$$L\frac{di_{ss}}{dt} + Ri_{ss} = V\cos(\omega t + \alpha). \quad (5.24)$$

Using the theorem, we **replace** $V\cos(\omega t + \alpha)$ with $\hat{V}e^{j\omega t}$ and i_{ss} with $\hat{I}e^{j\omega t}$, in which $\hat{V} = Ve^{j\alpha}$ and $\hat{I} = Ie^{j\beta}$.

Then
$$L\frac{d}{dt}(\hat{I}e^{j\omega t}) + R\hat{I}e^{j\omega t} = \hat{V}e^{j\omega t},$$
or
$$(j\omega L + R)\hat{I}e^{j\omega t} = \hat{V}e^{j\omega t},$$
or
$$(R + j\omega L)\hat{I} = \hat{V}. \tag{5.25}$$

From Equation 5.25,
$$\hat{I} = \frac{\hat{V}}{R + j\omega L} = \frac{Ve^{j\alpha}}{\sqrt{R^2 + \omega^2 L^2}\, e^{j\theta}} = \frac{V}{\sqrt{R^2 + \omega^2 L^2}}\, e^{j(\alpha-\theta)},$$

with $\theta = \tan^{-1}\frac{\omega L}{R}$, and $\alpha - \theta = \beta$.

We can now write the solution for i_{ss} as
$$i_{ss} = \text{Re}(\hat{I}e^{j\omega t}) = \text{Re}\left[\frac{V}{\sqrt{R^2 + \omega^2 L^2}}\, e^{j(\alpha-\theta)} e^{j\omega t}\right],$$
$$= \frac{V}{\sqrt{R^2 + \omega^2 L^2}} \cos(\omega t + \alpha - \theta). \tag{5.26}$$

The current i_{ss} has the amplitude $V/\sqrt{R^2 + \omega^2 L^2}$ and has a phase angle equal to the phase angle of the voltage minus θ. We say that the current lags the voltage by the angle θ.

Figure 5.10b shows the circuit which may be sketched from the relations of Equation 5.25. Note that i_{ss} has become $\hat{I}$, v has become $\hat{V}$, L has become $j\omega L$, and R is still R. This reminds us of the Laplace transform in which i became $I(s)$, v became $V(s)$, L became Ls with an initial condition source, and R remained R. However, in this chapter we performed no transformation, we simply justified the replacement of sinusoidal time functions with complex time functions of the form $\hat{B}e^{j\omega t}$, and then eliminated $e^{j\omega t}$ as a common factor.

Let us apply our technique to a series RLC circuit shown in Fig. 5.11a. Kirchhoff's voltage equation for the complete response, i, is

$$L\frac{di}{dt} + Ri + \frac{1}{C}\int_0^t i\, dt + v_{cd}(0+) = v. \tag{5.27}$$

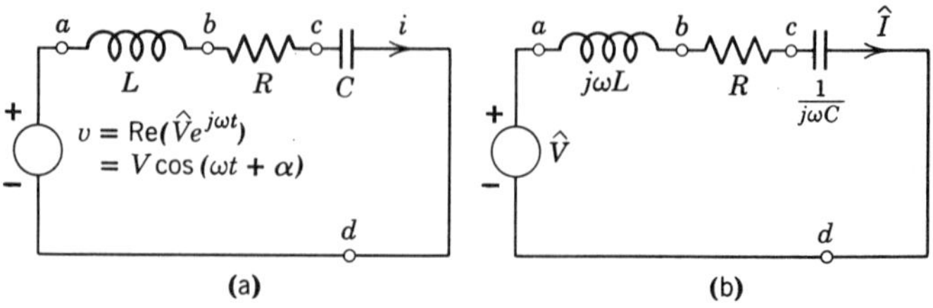

FIG. 5.11. An RLC circuit and its steady-state complex equivalent.

In order to apply the theorem, we differentiate this equation once to obtain

$$L\frac{d^2i}{dt^2} + R\frac{di}{dt} + \frac{i}{C} = \frac{dv}{dt},$$

for which the steady-state current i_{ss} must satisfy a similar equation,

$$L\frac{d^2i_{ss}}{dt^2} + \frac{Rdi_{ss}}{dt} + \frac{i_{ss}}{C} = \frac{dv}{dt}. \qquad (5.28)$$

Now we **replace** i_{ss} with $\hat{I}e^{j\omega t}$ and v with $\hat{V}e^{j\omega t}$ to obtain

$$\left[L(j\omega)^2 + R(j\omega) + \frac{1}{C}\right]\hat{I} = j\omega\hat{V},$$

or

$$\left(j\omega L + R + \frac{1}{j\omega C}\right)\hat{I} = \hat{V} \qquad (5.29)$$

The circuit for this equation is shown in Fig. 5.11b.

The phasor $\hat{I}$ may be written

$$\hat{I} = \frac{\hat{V}}{R + j\left(\omega L - \frac{1}{\omega C}\right)} = \frac{Ve^{j(\alpha-\theta)}}{\sqrt{R^2 + \left(\omega L - \frac{1}{\omega C}\right)^2}}$$

with

$$\theta = \tan^{-1}\left(\frac{\omega L - \frac{1}{\omega C}}{R}\right).$$

We really did not need to differentiate the original equation. Inspection of Equation 5.27 shows that i_{ss} must satisfy the expression,

$$L\frac{di_{ss}}{dt} + Ri_{ss} + \frac{1}{C}\int^t i_{ss}\,dt = v, \qquad (5.30)$$

the lower limit of the integral being omitted since this is a constant which must be part of the free response.

Replacement of i_{ss} with $\hat{I}e^{j\omega t}$ and v with $\hat{V}e^{j\omega t}$ in Equation 5.30 results in

$$\left(j\omega L + R + \frac{1}{j\omega C}\right)\hat{I} = \hat{V}, \qquad (5.31)$$

which is the same as Equation 5.29.

One might wonder whether the term $e^{-j\omega t} = \cos \omega t - j \sin \omega t$ could have been used as effectively as $e^{j\omega t}$. The answer is yes, this could have been done but there would be a change in the signs of the angles of all complex terms. If there are two possible systems of equal value, one of which has been well established, it is best not to "fight the system." We will use only $e^{j\omega t}$ and urge the student to do likewise.

Another question might arise. Suppose the source voltage were given as $V\sin(\omega t + \alpha)$. Should one change this to $V\cos(\omega t + \alpha - \pi/2)$ so one may use $\text{Re}[Ve^{j(\alpha-\pi/2)}e^{j\omega t}]$, or would it be all right to use $\text{Im}(Ve^{j\alpha}e^{j\omega t})$, and with this use $i_{ss} = \text{Im}(\hat{I}e^{j\omega t})$? Either method is all right as long as we can remember which one we started with. Those of us with short memories will prefer to use only $\text{Re}(\hat{V}e^{j\omega t})$ and $\text{Re}(\hat{I}e^{j\omega t})$.

It is readily shown that Kirchhoff's laws in the steady-state complex form are

$$\sum \hat{I} = 0. \tag{5.32}$$

and

$$\sum \hat{V} = 0. \tag{5.33}$$

$\hat{V}$ and $\hat{I}$ are complex quantities which are called **phasors**. The word phasor reminds us that these quantities arise from the steady-state sinusoidal solution of linear systems. Also we need to distinguish phasor voltage and current from transform voltage and current which are also complex, being functions of the complex frequency variable, s. At one time $\hat{V}$ and $\hat{I}$ were called vectors, since they may be thought of as a form of two-dimensional vectors; however, this usage has essentially stopped because of the confusion which occurs when complex vector quantities are encountered.

In the phasor domain we need also to define network functions, both the driving point immittances and the transfer functions. We shall call these **phasor network functions** to avoid confusing them with transform network functions although the words phasor or transform may be omitted when there is no chance of confusion. For phasor impedance, which is the ratio of phasor voltage to phasor current, with the units of ohms, we shall use the symbol $\hat{Z}$ or $Z(j\omega)$. Thus

$$\hat{Z} = Z(j\omega) = \frac{\hat{V}}{\hat{I}}. \tag{5.34}$$

For the circuit of Fig. 5.11b the phasor driving point impedance is $R + j\omega L + 1/j\omega C$.

The reciprocal of phasor impedance is called phasor admittance with the symbol $\hat{Y}$ or $Y(j\omega)$ and the units of mhos. Thus

$$\hat{Y} = Y(j\omega) = \frac{\hat{I}}{\hat{V}}. \tag{5.35}$$

Phasor immittance means phasor impedance and/or phasor admittance.

A phasor transfer function is the ratio of a phasor voltage or current to another phasor voltage or current with the exception that this ratio must not be a phasor driving point immittance. The general symbol used will be $\hat{T}$ or $T(j\omega)$.

A phasor diagram is the graphical representation of phasor quantities in the complex plane using a directed line segment to represent each phasor.

Example 5.4

GIVEN: The circuit of Fig. E-4.1.

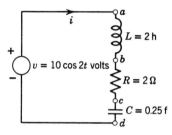

FIG. E-4.1

TO FIND: (a) i_{ss} by the use of phasors.
(b) A phasor diagram showing $\hat{V}$, $\hat{I}$, $\hat{V}_{ab}$, $\hat{V}_{bc}$, and $\hat{V}_{cd}$.

SOLUTION: Kirchhoff's voltage equation for steady-state conditions is

$$L\frac{di_{ss}}{dt} + Ri_{ss} + \frac{1}{C}\int^t i_{ss}\,dt = V\cos\omega t. \tag{1}$$

Since $i_{ss} = \text{Re}(\hat{I}e^{j\omega t})$ and $v = \text{Re}(\hat{V}e^{j\omega t})$, the theorem permits replacement of i_{ss} by $\hat{I}e^{j\omega t}$ and v by $\hat{V}e^{j\omega t}$. This results in

$$\left(j\omega L\hat{I} + R\hat{I} + \frac{\hat{I}}{j\omega C}\right)e^{j\omega t} = \hat{V}e^{j\omega t},$$

or the phasor equation

$$\left(j\omega L + R + \frac{1}{j\omega C}\right)\hat{I} = \hat{V}. \tag{2}$$

The circuit for this equation is shown in Fig. E-4.2. The phasor driving point impedance is

$$\hat{Z} = j\omega L + R + \frac{1}{j\omega C}$$
$$= j4 + 2 - j2 = 2 + j2$$
$$= 2.83e^{j45°}\text{ ohms}.$$

From Equation 2

$$\hat{I} = \frac{\hat{V}}{\hat{Z}} = \frac{10e^{j0}}{2.83e^{j45°}} = 3.53e^{-j45°}\text{ amp}$$

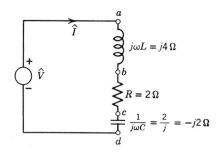

FIG. E-4.2

and

$$i_{ss} = \text{Re}(\hat{I}e^{j\omega t})$$
$$= \text{Re}(3.53e^{-j45°}e^{j2t})$$
$$= 3.53\cos(2t - 45°) \text{ amp.} \quad (3)$$

The phasor voltages may be calculated as follows.

$$\hat{V}_{ab} = j\omega L\hat{I} = j4(3.53e^{-j45°}) = 14.1e^{j45°} \text{ volts}$$
$$\hat{V}_{bc} = R\hat{I} = 2(3.53e^{-j45°}) = 7.06e^{-j45°} \text{ volts}$$
$$\hat{V}_{cd} = \frac{\hat{I}}{j\omega C} = -j2(3.53e^{-j45°}) = 7.06e^{-j135°} \text{ volts.}$$

The phasors $\hat{V}$, $\hat{I}$, $\hat{V}_{ab}$, $\hat{V}_{bc}$, and $\hat{V}_{cd}$ are shown in the phasor diagram of Fig. E-4.3.

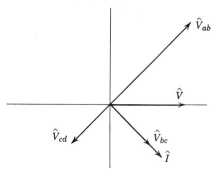

FIG. E-4.3

The phasor voltages are sketched to the same scale so that $\hat{V}_{ab} + \hat{V}_{bc} + \hat{V}_{cd} = \hat{V}$. Note that for the frequency given $V_{ab} > V$, which means that the phasor voltage across one element of a series circuit may have a greater amplitude than the phasor voltage across the entire series circuit. The current phasor is sketched to a different scale than the voltage phasors.

We shall sometimes not know the time function in detail, only knowing that all sources and responses are varying sinusoidally with time. We are then free to choose one of these quantities as having a zero phase angle with cos ωt and would call its phasor the **reference** phasor.

Example 5.5

GIVEN: The circuit of Fig. E-5.1.

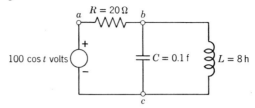

FIG. E-5.1

TO FIND: The steady-state currents and voltages. Also show a phasor diagram of these quantities.

SOLUTION: The node voltage method is selected with c as the reference node. Kirchhoff's steady-state current equation about node b is

$$C \frac{dv_{bc_{ss}}}{dt} + \frac{v_{bc_{ss}}}{R} + \frac{1}{L} \int^t v_{bc_{ss}} \, dt = \frac{V \cos \omega t}{R} \tag{1}$$

In the phasor domain Equation 1 becomes

$$\left[j\omega C + \frac{1}{R} + \frac{1}{j\omega L} \right] \hat{V}_{bc} = \frac{\hat{V}}{R} . \tag{2}$$

The bracketed term which is the coefficient of $\hat{V}_{bc}$ is a phasor admittance. The circuit in the phasor domain, in which each passive element is labeled as an admittance, is shown in Fig. E-5.2.

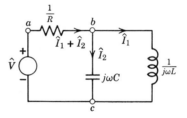

FIG. E-5.2

Substitution of numerical values into Equation 2 gives

$$(j.1 + 0.05 - j.125)\hat{V}_{bc} = \frac{100}{20}$$

or

$$\hat{V}_{bc} = \frac{5}{0.05 - j.025} = \frac{500}{5.58e^{-j26.5}} = 89.5 e^{j26.5°} \text{ volts.}$$

$$\hat{I}_1 = \frac{\hat{V}_{bc}}{j\omega L} = \frac{89.5 e^{j26.5°}}{j8} = 11.2 e^{-j63.5°} \text{ amp.}$$

$$\hat{I}_2 = j\omega C \hat{V}_{bc} = j.1(89.5 e^{j26.5°}) = 8.95 e^{j116.5°} \text{ amp.}$$

$$\hat{I}_1 + \hat{I}_2 = 11.2 e^{-j63.5°} + 8.95 e^{j116.5°}$$
$$= 5 - j10 + (-4 + j8)$$
$$= 1 - j2 = 2.23 e^{-j63.5°} \text{ amp.}$$

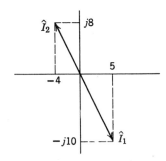

$\hat{V}_{ab}$ may be calculated from $R(\hat{I}_1 + \hat{I}_2)$ or from $\hat{V}_{ab} = \hat{V} - \hat{V}_{bc}$. Using the former we obtain

$$\hat{V}_{ab} = R(\hat{I}_1 + \hat{I}_2) = 20(2.23e^{-j63.5°}) = 44.6e^{-j63.5°} \text{ volts.}$$

Then

$v_{bc(ss)} = \text{Re}(\hat{V}_{bc}e^{j\omega t}) = \text{Re}(89.5e^{j26.5°}e^{j\omega t}) = 89.5\cos(t + 26.5°)$ volts.
$i_{1ss} = \text{Re}(\hat{I}_1 e^{j\omega t}) = 11.2\cos(t - 63.5°)$ amp
$i_{2ss} = \text{Re}(\hat{I}_2 e^{j\omega t}) = 8.95\cos(t + 116.5°)$ amp
$(i_1 + i_2)_{ss} = \text{Re}[(\hat{I}_1 + \hat{I}_2)e^{j\omega t}] = 2.23\cos(t - 63.5°)$ amp
and
$V_{ab(ss)} = \text{Re}(\hat{V}_{ab}e^{j\omega t}) = 44.6\cos(t - 63.5°).$

A phasor diagram is shown in Fig. E-5.3. For this circuit the sum of two currents has less amplitude than the amplitude of either current. Note that the current $\hat{I}_1$ through the inductive reactance lags the voltages $\hat{V}_{bc}$ by 90°, while the current $\hat{I}_2$ through the capacitive reactance leads the voltage $\hat{V}_{bc}$ by 90°.

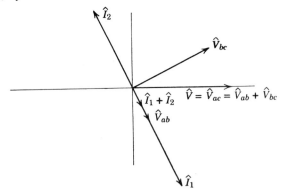

FIG. E-5.3

5.5 ANALYSIS IN THE PHASOR DOMAIN

For a circuit with sinusoidal driving functions we now have three separate forms of Kirchhoff's laws and three separate circuits. These are in the time domain, the Laplace transform domain and the phasor domain. Fig. 5.12 shows a simple circuit sketched in these three domains. For the time domain circuit Kirchhoff's equations are differential equations to be solved by the classical method. Initial conditions may or may not be shown in the sketch. For the transform circuit Kirchhoff's equations are algebraic and initial conditions are always shown on the sketch. Either the classical or the transform methods will yield complete solutions for the response functions. The phasor domain circuit has sources of only one frequency and will yield only the steady state sinusoidal response at this frequency.

For the time domain circuit the passive elements are labeled as R, L, or C. For the transform or phasor circuits the passive elements are labeled as either impedances or admittances; in Fig. 5.12 they are labeled as impedances.

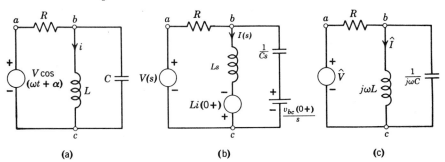

FIG. 5.12. Circuit in three domains. (*a*) Time domain. (*b*) Transform domain. (*c*) Phasor domain.

Network functions are not defined physically in the time domain. Thus although one may obtain complete solutions for voltage and current in the time domain by taking the inverse Laplace transform of transform voltages and currents, the inverse Laplace transform of such functions as $Z(s)$, $Y(s)$, or $T(s)$ is not defined physically. Correspondingly, one may obtain steady state solutions for voltage and current in the time domain from phasor voltages and current but one should not try to change the phasor network functions $\hat{Z}$, $\hat{Y}$, or $\hat{T}$, into functions of time. This change is simply not defined.

It is very important to realize that for the phasor domain the equations are algebraic, as they are for the transform domain. Therefore, all the techniques developed in Chapter 4 for the transform circuits apply to the phasor circuits. This applies to the node voltage and branch current methods and to all the network reduction methods.

5.5.1 Phasor Driving Point Immittances

Phasor driving point immittances may be expressed as complex numbers whereas transform immittances may be expressed only in algebraic form. For this reason and also because the uses of phasor driving point immittances are very great, some special nomenclature has developed and needs to be learned.

The definition of phasor driving point immittance is similar to that for transform driving point immittance. Phasor driving point impedance applies to a phasor circuit in which there is only one independent source and is the ratio of the phasor voltage across the source to the phasor current through the source, the direction of the current being from − to + through the source. The words phasor and driving point may be omitted if this leads to no confusion. When we speak of the phasor impedance of a branch of a network we refer to the value of impedance which would be obtained if a single source were applied to the branch while isolated from the network. A phasor immittance is represented in a circuit diagram by a jagged line, ⌐∧∧∧∧⌐, the same symbol used for transform immittance or the element resistance.

Phasor driving point impedance is given the phase angle θ. We shall try not to use θ for other quantities. Also the real part of $\hat{Z}$ is labeled R and called resistance; the imaginary part of $\hat{Z}$ is labeled X and called reactance. Thus

$$\hat{Z} = Ze^{j\theta} = R + jX. \tag{5.36}$$

For passive networks R cannot be negative, so θ cannot be less than $-90°$ nor greater than $+90°$. (θ is called the power factor angle and $\cos \theta$ is called the power factor. The reason θ is associated with power will be taken up in the next section.) If $-90° < \theta < 0°$ the impedance is said to be capacitive and the power factor is said to be a leading power factor, implying that the phasor current leads the phasor voltage. If $0 < \theta < 90°$, the impedance is said to be inductive and the power factor is said to be a lagging power factor, implying that the phasor current lags the phasor voltage. If $\theta = 0$ the impedance is said to be resistive.

The phasor impedance, $\hat{Z}$, of a network is an algebraic combination of the impedances of the various branches. Therefore, we should not expect $R = \text{Re }\hat{Z}$ to be a function only of the resistances of the network nor should we expect $X = \text{Im }\hat{Z}$ to be a function only of the reactances of the network. If X is positive the impedance is said to be inductive; if X is negative, the impedance is said to be capacitive.

The element, inductance, has a phasor impedance, $\hat{Z} = R + jX = 0 + j\omega L$. Thus, the reactance of an inductance is written

$$X_L = \omega L. \tag{5.37}$$

The element, capacitance, has a phasor impedance, $\hat{Z} = R + jX = 0 + 1/j\omega C = 0 - j/\omega C$. Thus the reactance of a capacitance is written

$$X_C = -\frac{1}{\omega C}. \tag{5.38}$$

Phasor driving point admittance is the reciprocal of the phasor driving point impedance, and is written

$$\hat{Y} = \frac{1}{\hat{Z}} = \frac{1}{Ze^{j\theta}} = Ye^{-j\theta} = G + jB. \qquad (5.39)$$

G, Re $\hat{Y}$, is termed conductance; B, Im $\hat{Y}$, is termed susceptance. Since the angle of the admittance is the negative of the angle of the impedance, X and B will be of opposite sign. Also $B_L = -1/\omega L$, and $B_C = \omega C$.

Example 5.6

GIVEN: Two impedances $R_1 + j0$ and $0 + jX_1$ in parallel as shown in Fig. E-6.1a.

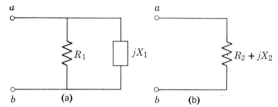

FIG. E-6.1

TO FIND: (a) The equivalent series impedance as shown in Fig. E-6.1b. (b) R_2 and X_2 if $R_1 = 10\ \Omega$ and $X_1 = 5\ \Omega$.

SOLUTION: (a) For transform impedances we showed that two impedances in parallel could be replaced by an impedance equal to the product of the impedances divided by the sum of the impedances. This relationship applies equally well to phasor impedances. So

$$\hat{Z}_{ab} = R_2 + jX_2 = \frac{(R_1)(jX_1)}{R_1 + jX_1} = \frac{jR_1X_1(R_1 - jX_1)}{R_1^2 + X_1^2}$$

$$= \frac{R_1X_1^2}{R_1^2 + X_1^2} + \frac{jR_1^2X_1}{R_1^2 + X_1^2}.$$

Thus

$$R_2 = \frac{R_1 X_1^2}{R_1^2 + X_1^2}, \qquad (1)$$

and

$$X_2 = \frac{R_1^2 X_1}{R_1^2 + X_1^2}. \qquad (2)$$

Note that R_2 and X_2 are each functions of both R_1 and X_1. Also note that the sign of the reactance, X_2, is the same as that of the reactance X_1.

(b) For the numerical values of $R_1 = 10\ \Omega$ and $X_1 = 5\ \Omega$, corresponding to $G_1 = 0.1$ mhos and $B_1 = -0.2$ mhos, it is easier to solve directly than to use the literal formulas. Thus,

$$\hat{Z}_{ab} = R_2 + jX_2 = \frac{1}{G_1 + jB_1} = \frac{1}{0.1 - j0.2}$$

$$= \frac{0.1 + j0.2}{0.05} = 2 + j4 \text{ ohms,}$$

or, $R_2 = 2 \, \Omega$ \hfill (3)

and $X_2 = 4 \, \Omega$. \hfill (4)

The following additional examples demonstrate some of the techniques useful in the phasor domain.

Example 5.7

GIVEN: The circuit of Fig. E-7.1 with $\hat{V}_{ab} = 100 + j0$ volts.

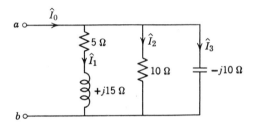

FIG. E-7.1

TO FIND: The currents $\hat{I}_1$, $\hat{I}_2$, $\hat{I}_3$, and $\hat{I}_0$ using a phasor diagram.

SOLUTION: This is a type of problem in which a phasor diagram should be sketched first to reduce the chances for numerical errors that constantly beset us. A phasor diagram is shown in Fig. E-7.2a, with an alternate polygon form in Fig. E-7.2b. $\hat{I}_2$ is in phase with $\hat{V}_{ab}$, whereas $\hat{I}_3$ must lead $\hat{V}_{ab}$ by 90° because its impedance is a capacitive reactance. $\hat{I}_2$ and $\hat{I}_3$ have the same magnitudes while $\hat{I}_1$ has a magnitude which is less by a factor of 10 to about 16. Furthermore, $\hat{I}_1$ lags $\hat{V}_{ab}$ by an angle

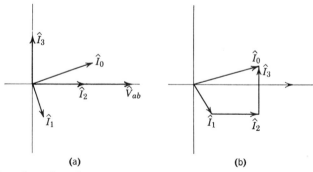

FIG. E-7.2. $\hat{I}_0 = \hat{I}_1 + \hat{I}_2 + \hat{I}_3$.

242

whose tangent is 3. Thus these phasors and their sum $\hat{I}_0$ can be located in the diagram of Fig. E-7.2. If our numerical answers differ markedly from the sketch, we should expect an error in our numerical work.

We now perform the calculations:

$$\hat{I}_1 = \frac{100 + j0}{5 + j15} = \frac{100 + j0}{5(1 + j3)} \cdot \frac{1 - j3}{1 - j3} = \frac{20 - j60}{10} = 2 - j6 \text{ amp.}$$

$$\hat{I}_2 = \frac{100 + j0}{10 + j0} = 10 + j0 \text{ amp.}$$

$$\hat{I}_3 = \frac{100 + j0}{0 - j10} = 0 + j10 \text{ amp.}$$

$$\hat{I}_0 = \hat{I}_1 + \hat{I}_2 + \hat{I}_3 = 2 - j6 + 10 + j10 = 12 + j4 \text{ amp.}$$

The calculations were performed in rectangular form since the final operation is an addition.

Example 5.8

GIVEN: The circuit of Fig. E-8.1 with $\hat{V}_{ac} = 100 + j0$ volts.

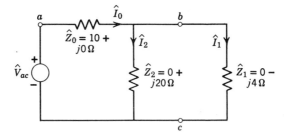

FIG. E-8.1

TO FIND: Several methods for determining phasor currents and voltages, performing the calculations for one of these.

SOLUTION: The branch- or mesh-current methods would involve solving two equations with two unknowns, whereas the node voltage method would involve only one unknown. Thus we reject the branch- or mesh-current methods. Furthermore we have worked a similar problem by the node voltage method (see Example 5.5) so we consider network reduction methods. The impedances $\hat{Z}_1$ and $\hat{Z}_2$ are in parallel and may thus be replaced by an equivalent impedance, $\hat{Z}_3$, as shown in Fig. E-8.2.

$$\hat{Z}_3 = \frac{\hat{Z}_1 \hat{Z}_2}{\hat{Z}_1 + \hat{Z}_2}$$

$$= \frac{(0 - j4)(0 + j20)}{-j4 + j20}$$

$$= \frac{80}{j16} = 0 - j5 \, \Omega.$$

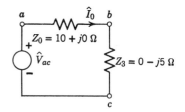

FIG. E-8.2

We could now find

$$\hat{I}_0 = \frac{\hat{V}_{ac}}{\hat{Z}_0 + \hat{Z}_3},$$

and then determine $\hat{I}_1$ and $\hat{I}_2$ by current ratio. Or we could determine $\hat{V}_{ab}$ and $\hat{V}_{bc}$ by voltage ratio. We arbitrarily select the latter method.

$$\hat{V}_{ab} = \frac{\hat{Z}_0 \hat{V}_{ac}}{\hat{Z}_0 + \hat{Z}_3} = \frac{10(100)}{10 - j5} = \frac{200}{2 - j1} = \frac{200(2 + j1)}{4 + 1}$$
$$= 80 + j40 \text{ volts}$$

$\hat{V}_{bc} = \hat{V}_{ac} - \hat{V}_{ab} = 100 - (80 + j40) = 20 - j40$ volts.

The currents $\hat{I}_0$, $\hat{I}_1$, and $\hat{I}_2$ are

$$\hat{I}_0 = \frac{\hat{V}_{ab}}{\hat{Z}_0} = \frac{80 + j40}{10} = 8 + j4 \text{ amp}$$

$$\hat{I}_1 = \frac{\hat{V}_{bc}}{\hat{Z}_1} = \frac{20 - j40}{0 - j4} = \frac{5 - j10}{-j1} = 10 + j5 \text{ amp}$$

$$\hat{I}_2 = \frac{\hat{V}_{bc}}{\hat{Z}_2} = \frac{20 - j40}{0 + j20} = \frac{1 - j2}{j1} = -2 - j1 \text{ amp}.$$

$\hat{I}_0$ does equal $\hat{I}_1 + \hat{I}_2$, as it should, and the phasor diagram shown in Fig. E-8.3 serves to show that $\hat{I}_1$ and $\hat{I}_2$ do differ from $\hat{V}_{bc}$ by 90° as they should.

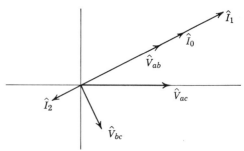

FIG. E-8.3

Example 5.9

GIVEN: The circuit of Fig. E-9.1.
TO FIND: Several methods for determining $\hat{I}_5$.

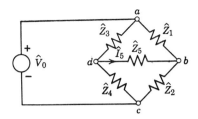

FIG. E-9.1

SOLUTION: Since only one quantity, $\hat{I}_5$, is desired it is logical to replace the rest of the circuit, other than $\hat{Z}_5$, by an equivalent circuit using either Thévenin's theorem or some other reduction methods. Before showing these, a preliminary check shows that the mesh current method requires three equations in three unknowns and the node voltage method requires two equations in two unknowns. Furthermore the node voltage method would require an additional operation to determine $\hat{I}_5$.

(a) We shall first outline the method using Thévenin's theorem. Thévenin's voltage may be considered as the voltage $\hat{V}_{db}$ in Fig. E-9.2.

$$\hat{V}_T = \hat{V}_{db} = \hat{V}_{dc} - \hat{V}_{bc}, \qquad (1)$$

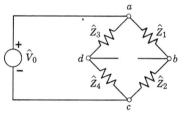

FIG. E-9.2

and the voltages $\hat{V}_{dc}$ and $\hat{V}_{bc}$ may be found by voltage ratio.

$$\hat{V}_{dc} = \frac{\hat{Z}_4 \hat{V}_0}{\hat{Z}_3 + \hat{Z}_4} \quad \text{and} \quad \hat{V}_{bc} = \frac{\hat{Z}_2 \hat{V}_0}{\hat{Z}_1 + \hat{Z}_2}.$$

Thévenin's impedance, $\hat{Z}_T$, may be found by determining the driving point impedance $\hat{Z}_{db}$ in Fig. E-9.3a in which the source voltage $\hat{V}_0$ is made zero.

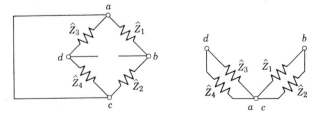

FIG. E-9.3

Since points a and c are at the same voltage the nodes may be connected as shown in Fig. E-9.3b and it is obvious now that impedances $\hat{Z}_3$ and $\hat{Z}_4$ are in parallel. Also impedances $\hat{Z}_1$ and $\hat{Z}_2$ are in parallel.

Steady-State Analyses with Sinusoidal Sources 245

Thus

$$\hat{Z}_T = \frac{\hat{Z}_3 \hat{Z}_4}{\hat{Z}_3 + \hat{Z}_4} + \frac{\hat{Z}_1 \hat{Z}_2}{\hat{Z}_1 + \hat{Z}_2}. \qquad (2)$$

Finally $\hat{I}_5$ may be calculated from the circuit of Fig. E-9.4, in which $\hat{V}_T$ and $\hat{Z}_T$ are known from Equations 1 and 2.

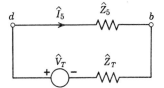

FIG. E-9.4

(b) Let us now consider using some other network reduction methods. Wye-delta or delta-wye conversions are possible. However, the algebra of such conversions are formidable unless the three impedances have the same phase. Furthermore, in this case such conversion would involve $\hat{Z}_5$ which we did not wish to disturb. Let us resketch as shown in Fig. E-9.5. This is permissible since there is still no voltage

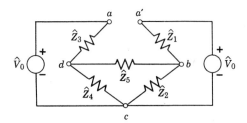

FIG. E-9.5

difference between terminals a and a'. We can now replace the voltage sources by equivalent current sources as shown in Fig. E-9.6.

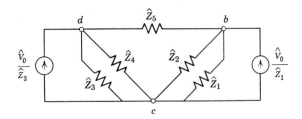

FIG. E-9.6

The parallel impedances may be replaced by their equivalence, as shown in Fig. E-9.7 and then the current sources replaced by their equivalent voltage sources as shown in Fig. E-9.8.

$$\hat{Z}_7 = \frac{\hat{Z}_3 \hat{Z}_4}{\hat{Z}_3 + \hat{Z}_4}, \qquad \hat{Z}_6 = \frac{\hat{Z}_1 \hat{Z}_2}{\hat{Z}_1 + \hat{Z}_2}$$

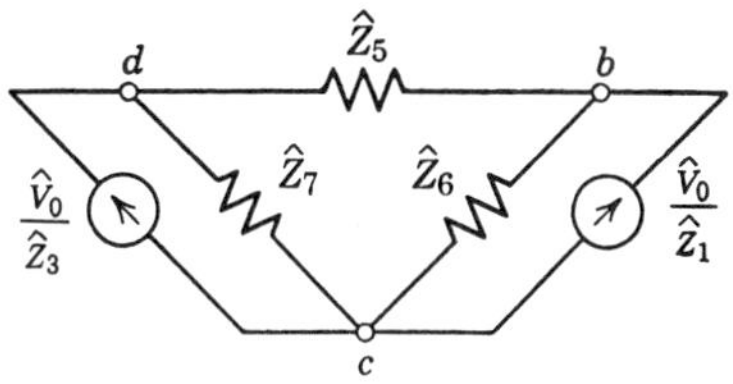

FIG. E-9.7

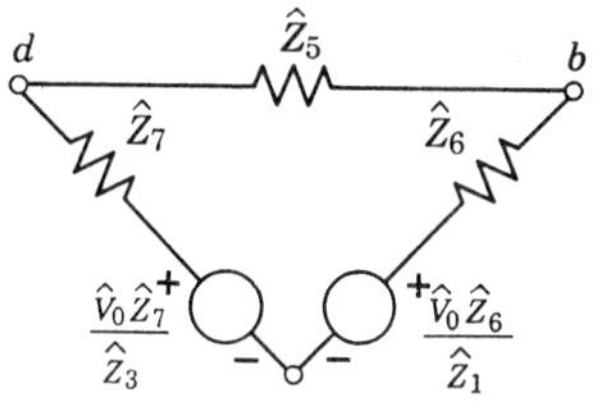

FIG. E-9.8

Example 5.10

GIVEN: The delta impedances of Fig. E-10.1a.

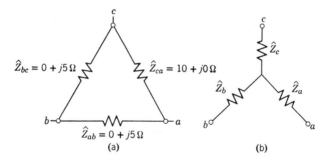

FIG. E-10.1

TO FIND: The equivalent wye impedances.

SOLUTION: We may use the delta-wye impedance equivalence determined in Chapter 4. Thus

$$\hat{Z}_b = \frac{\hat{Z}_{ab}\hat{Z}_{bc}}{\hat{Z}_{ab} + \hat{Z}_{bc} + \hat{Z}_{ca}} = \frac{j5(j5)}{10 + j10} = \frac{-25}{10 + j10}$$

$$= \frac{-2.5(1 - j1)}{1 + 1} = -1.25 + j1.25 \, \Omega$$

$$\hat{Z}_a = \frac{\hat{Z}_{ac}\hat{Z}_{ab}}{\hat{Z}_{ab} + \hat{Z}_{bc} + \hat{Z}_{ca}} = \frac{j5(10)}{10 + j10} = \frac{j5(1 - j1)}{1 + 1}$$

$$= 2.5 + j2.5 \, \Omega$$

$$\hat{Z}_c = \frac{\hat{Z}_{ca}\hat{Z}_{bc}}{\hat{Z}_{ab} + \hat{Z}_{bc} + \hat{Z}_{ca}} = \frac{j5(10)}{10 + j10} = 2.5 + j2.5 \, \Omega$$

If each impedance had nonzero values for both real and imaginary parts the algebra would have been considerable. (The exception occurs if the phase of the impedances are equal.) The purpose of this example was to demonstrate the possibility of having a mathematical equivalence which is not physically realizable with passive elements. $\hat{Z}_b$ has a negative real part of -1.25 ohms, a negative resistance. The wye circuit is still equivalent to the delta circuit with regard to external effects.

Example 5.11

GIVEN: The circuit of Fig. E-11.1.

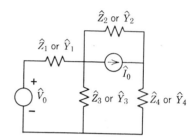

FIG. E-11.1

TO FIND: (a) Branch current equations and solutions for this circuit using $\hat{V}_0 = 100 + j0$ volts, $\hat{I}_0 = 10 - j10$ amp, $\hat{Z}_1 = 10 + j0$ ohms, $\hat{Z}_2 = 0 - j10$ ohms, $\hat{Z}_3 = 5 + j0$ ohms, and $\hat{Z}_4 = 0 + j5$ ohms.
(b) Node voltage equations and solutions.

SOLUTION: (a) For the branch current method we use impedances and select arbitrarily the currents of Fig. E-11.2.

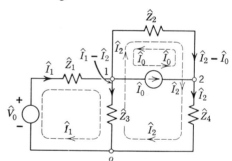

FIG. E-11.2

Kirchhoff's voltage equations around the loops of the independent currents $\hat{I}_1$ and $\hat{I}_2$ are

$$(\hat{Z}_1 + \hat{Z}_3)\hat{I}_1 - \hat{Z}_3\hat{I}_2 = \hat{V}_0 \tag{1}$$
$$-\hat{Z}_3\hat{I}_1 + (\hat{Z}_2 + \hat{Z}_3 + \hat{Z}_4)\hat{I}_2 = \hat{Z}_2\hat{I}_0. \tag{2}$$

The insertion of numerical values gives

$$15\hat{I}_1 - 5\hat{I}_2 = 100 + j0,$$

and

$$-5\hat{I}_1 + (5 - j5)\hat{I}_2 = -100 - j100.$$

The use of Cramer's rule gives the solution for the branch currents,

$$\hat{I}_1 = \frac{\begin{vmatrix} 100 & -5 \\ -100-j100 & 5-j5 \end{vmatrix}}{\begin{vmatrix} 15 & -5 \\ -5 & 5-j5 \end{vmatrix}} = \frac{-j1000}{50-j75} = \frac{-j40}{2-j3}$$

$$= \frac{40e^{-j90}}{3.61e^{-j56.3}} = 11.1e^{-j33.7°} \text{ amp.} \tag{3}$$

$$\hat{I}_2 = \frac{\begin{vmatrix} 15 & 100 \\ -5 & -100-j100 \end{vmatrix}}{50-j75} = \frac{-1000-j1500}{50-j75} = \frac{-40-j60}{2-j3}$$

$$= \frac{72.2e^{-j123.7°}}{3.61e^{-j56.3}} = 20e^{-j67.4°} \text{ amp.} \tag{4}$$

(b) For the node-voltage method we use admittances and arbitrarily select 0 as the reference node. The circuit is shown in Fig. E-11.3.

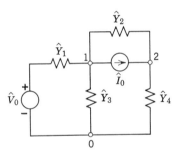

FIG. E-11.3

Kirchhoff's current equations about the nodes 1 and 2 are

$$(\hat{Y}_1 + \hat{Y}_2 + \hat{Y}_3)\hat{V}_{10} - \hat{Y}_2\hat{V}_{20} = \hat{Y}_1\hat{V}_0 - \hat{I}_0, \tag{5}$$

and

$$-\hat{Y}_2\hat{V}_{10} + (\hat{Y}_2 + \hat{Y}_4)\hat{V}_{20} = \hat{I}_0. \tag{6}$$

The numerical values for the admittances can readily be obtained by taking the reciprocal of the impedances. Thus $\hat{Y}_1 = 1/\hat{Z}_1 = 1/(10+j0) = 0.1 + j0$ mhos, $\hat{Y}_2 = 0 + j0.1$ mhos, $\hat{Y}_3 = 0.2 + j0$ mhos, and $\hat{Y}_4 = 0 - j0.2$ mhos. Insertion of these numerical values into Equations 5 and 6 gives,

$$(0.3 + j0.1)\hat{V}_{10} - j0.1\,\hat{V}_{20} = j10,$$

and

$$-j0.1\,\hat{V}_{10} + (0 - j0.1)\hat{V}_{20} = 10 - j10.$$

Steady-State Analyses with Sinusoidal Sources

The use of Cramer's rule gives the solutions for $\hat{V}_{10}$ and $\hat{V}_{20}$.

$$\hat{V}_{10} = \frac{\begin{vmatrix} j10 & -j0.1 \\ 10-j10 & -j0.1 \\ 0.3+j0.1 & -j0.1 \\ -j0.1 & -j0.1 \end{vmatrix}}{} = \frac{2+j1}{0.02+j0.03}$$

$$= \frac{2.23e^{j26.5}}{0.0361e^{-j56.3}} = 61.8e^{j82.8°} \text{ volts.} \tag{7}$$

$$\hat{V}_{20} = \frac{\begin{vmatrix} 0.3+j.1 & j10 \\ -j0.1 & 10-j10 \end{vmatrix}}{0.02-j0.03} = \frac{3-j2}{0.0361e^{-j56.3°}}$$

$$= \frac{3.61e^{-j33.7°}}{0.0361e^{-j56.3°}} = 100e^{j22.6°} \text{ volts.} \tag{8}$$

A partial check on the answers is afforded by the fact that $\hat{V}_{20} = \hat{Z}_4 \hat{I}_2 = j5\hat{I}_2$. Substitution of $\hat{I}_2$ from Equation 4 shows a check with Equation 8.

5.6 POWER UNDER SINUSOIDAL STEADY-STATE CONDITIONS (WATTMETER)

For voltages and currents varying sinusoidally with time the rate of energy transfer or power is **generally not sinusoidal**. We should expect this because if the power varied sinusoidally with time the average power over integral periods of the wave would be zero and there would be no useful transfer of electrical energy. Although analysis is easier to perform in terms of voltage and current, our end objective is with energy transfer or the time rate of energy transfer, power.

The sketch in Fig. 5.13 shows two portions of a network connected by two leads for which the current is

$$i = I\cos(\omega t + \phi_i) \tag{5.40}$$

and the voltage difference is

$$v = V\cos(\omega t + \phi_v). \tag{5.41}$$

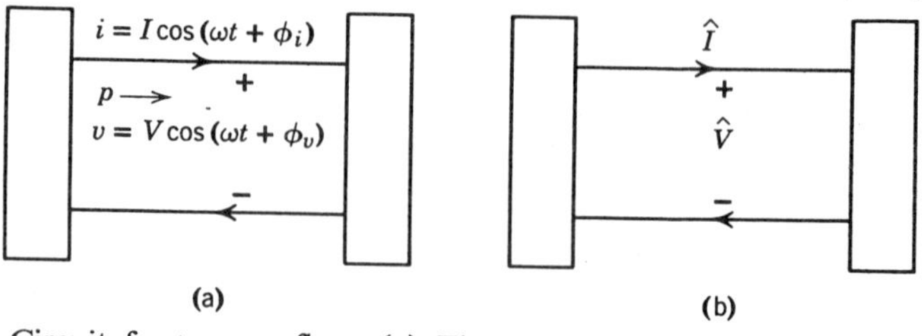

FIG. 5.13. Circuit for energy flow. (*a*) Time domain circuit. (*b*) Phasor circuit.

There may be sources in either part of the network but all independent sources must vary sinusoidally at the same radian frequency.

For our conventions the power to the right is

$$p = vi = VI \cos(\omega t + \phi_v) \cos(\omega t + \phi_i) \tag{5.42}$$

The product of two cosine functions can readily be changed to more significant terms from the trigonometric identity,

$$\cos(\alpha + \beta) = \cos \alpha \cos \beta - \sin \alpha \sin \beta,$$

from which,

$$\cos(\alpha - \beta) = \cos \alpha \cos \beta + \sin \alpha \sin \beta,$$

and thus

$$\cos \alpha \cos \beta = \tfrac{1}{2}[\cos(\alpha + \beta) + \cos(\alpha - \beta)]$$

Equation 5.42 becomes

$$p = \frac{VI}{2} [\cos(2\omega t + \phi_v + \phi_i) + \cos(\phi_v - \phi_i)] \tag{5.43}$$

The capital letter P is used to designate average power, whereas the lower case letter p is a function of time.

Figure 5.14 shows voltage, current, and power as a function of time. At those instants of time at which power is negative, energy is flowing to the left in Fig. 5.13.

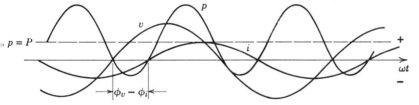

FIG. 5.14. Voltage, current, and power as functions of time.

For all finite frequencies except $\omega = 0$, the first term of the power expression, $\dfrac{VI}{2} \cos(2\omega t + \phi_v + \phi_i)$, has a frequency double that of the voltage or current and has an average value of zero over integral half periods of the voltage or current. The second term, $\dfrac{VI}{2} \cos(\phi_v - \phi_i)$ is a constant, so the average power, P, to the right in Fig. 5.13, is

$$P = \frac{VI}{2} \cos(\phi_v - \phi_i). \quad (\omega \neq 0) \tag{5.44}$$

The most common application occurs with a network which has no sources and can therefore be represented by an impedance $\hat{Z}$, as shown in Fig. 5.15a or by an admittance $\hat{Y}$ as in Fig. 5.15b.

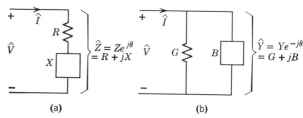

(a)   (b)

FIG. 5.15. Passive circuits for energy flow.

In this case the average power is

$$P = \frac{VI}{2}\cos(\phi_v - \phi_i) = \frac{VI}{2}\cos\theta. \quad (\omega \neq 0) \tag{5.45}$$

For the circuit of Fig. 5.15a, this becomes

$$P = \frac{I^2 Z \cos\theta}{2} = \frac{I^2 R}{2}, \quad (\omega \neq 0) \tag{5.46}$$

since $V = ZI$ and $Z\cos\theta = R$. $\cos\theta$ is defined as the power factor and θ is termed the power factor angle.

For the circuit of Fig. 5.15b, the average power is

$$P = \frac{V^2 Y \cos\theta}{2} = \frac{V^2 G}{2}, \quad (\omega \neq 0) \tag{5.47}$$

since $I = YV$ and $Y\cos\theta = G$.

Be sure to note that in using terms such as $I^2R/2$ to obtain average power, I must be the amplitude of the sinusoidal current through R. Similarly, in using $V^2G/2$, V must be the amplitude of the sinusoidal voltage across G.

Although instantaneous power is obtained for the circuit of Fig. 5.13a from the product of instantaneous voltage and current, $p = vi$, no significant result is obtained by multiplying $\hat{V}$ by $\hat{I}$, since

$$\hat{V}\hat{I} = Ve^{j\phi_v}Ie^{j\phi_i} = VIe^{j(\phi_v+\phi_i)} = VI\cos(\phi_v + \phi_i) + jVI\sin(\phi_v + \phi_i).$$

However, if either $\hat{V}$ or $\hat{I}$ is conjugated, the real part of the product is twice the average power. It is customary to conjugate the current. Then

$$\hat{V}\hat{I}^* = Ve^{j\phi_v}Ie^{-j\phi_i} = VI\cos(\phi_v - \phi_i) + jVI\sin(\phi_v - \phi_i).$$

Thus an alternate method of determining average power is

$$P = \tfrac{1}{2}\operatorname{Re}(\hat{V}\hat{I}^*). \tag{5.48}$$

Extensions of this conjugate method will be taken up in Chapter 8.

One of the instruments capable of measuring power is of the electrodynamometer type (see Appendix D). This instrument is constructed so that the torque developed is proportional to the product of the currents in two coils. If this instrument is connected as shown in Fig. 5.16a the torque will be proportional to the product vi. One coil is in series with the current i; the other coil is placed in series with a high resistance R across the voltage v so that its current is essentially proportional to v. If the frequency is low enough and the inertia of the moving parts made small enough, this instrument will indicate instantaneous values of vi. For most practical applications the inertia is so large that the deflection is proportional to the average value of vi. If the voltage and current are sinusoidal functions of time and the instrument is properly calibrated, the reading will be $\dfrac{VI}{2}\cos(\phi_v - \phi_i)$. If the instrument is connected as shown in Fig. 5.16, this reading will be the average power.

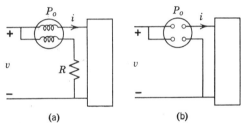

FIG. 5.16. Connection of wattmeter for measuring power. (a) Wattmeter, P_0, showing coils. (b) Wattmeter, P_0, as we shall show it.

The instrument is not perfect. The current coil has an impedance and thus there is a voltage across this coil. The voltage coil has a finite impedance and requires some current and power. We shall usually assume that the current coil has zero impedance and that the voltage coil has infinite impedance and show the wattmeter, P_o, as indicated in Fig. 5.16b. The student will find in the laboratory that such idealizations are not always realistic.

Example 5.12

GIVEN: The circuit of Fig. E-12.1.
TO FIND: The reading of the wattmeter, P_o.
SOLUTION: The current $\hat{I}$ may be solved for since

$$\hat{I} = \frac{\hat{V}}{\hat{Z}} = \frac{100e^{j30°}}{6 + j8} = \frac{100e^{j30°}}{10e^{j53.1°}} = 10e^{-j23.1°} \text{ amp.}$$

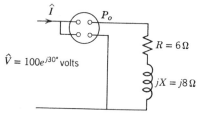

FIG. E-12.1

The wattmeter reading may now be calculated in several ways:

$$P_o = \frac{VI}{2} \cos(\phi_v - \phi_i) = \frac{100(10)}{2} \cos[30° - (-23.1°)]$$

$$= \frac{100(10)}{2} \cos 53.1° = 300 \text{ watts,}$$

$$P_o = \frac{I^2 R}{2} = \frac{10^2(6)}{2} = 300 \text{ watts,}$$

or

$$P_o = \tfrac{1}{2} \operatorname{Re}(\hat{V}\hat{I}^*) = \tfrac{1}{2} \operatorname{Re}[100 e^{j30°} \, 10 e^{j23.1°}]$$

$$= 500 \operatorname{Re}(e^{j53.1°}) = 300 \text{ watts.}$$

5.7 ROOT-MEAN-SQUARE VALUES FOR CURRENT AND VOLTAGE (INSTRUMENTS)

It is convenient to relate the magnitude of a periodic current to the average power dissipated as heat in the circuit of the current. This relationship is developed for periodic currents in general and then to the sinusoidal current as a special case.

Consider the circuit of Fig. 5.17 in which i is a periodic function with period T and for which steady state exists. The "average value of a periodic function" is defined as the average value during one period or an integral number of periods. Thus the average power of a periodic function is

$$P = \operatorname{avg} p = \frac{1}{T} \int_{K}^{K+T} p \, dt, \qquad (5.49)$$

in which K is a real constant.

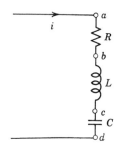

FIG. 5.17. Periodic current i in RLC branch.

The elements of inductance, L, and capacitance, C, have been so defined that they are capable of storing energy but not of dissipating it. Since the energy stored in the inductance, $\frac{1}{2}Li^2$, is the same at the beginning and end of a period, there has been no change in stored energy and thus no average power is supplied to the inductance. Similarly the energy stored in the capacitance, $\frac{1}{2}Cv_{cd}^2$, is the same at the beginning and end of a period and thus no average power is supplied to the capacitance. For the resistance, the instantaneous power is

$$p = v_{ab}i = Ri^2, \tag{5.50}$$

and the average power is

$$P = \frac{1}{T}\int_{K}^{K+T} Ri^2\, dt. \tag{5.51}$$

Equation 5.51 may be written

$$P = [\text{avg } i^2]R = I_{\text{rms}}^2 R, \tag{5.52}$$

thus defining

$$I_{\text{rms}} = \sqrt{\text{avg } i^2}. \tag{5.53}$$

The subscripts rms stand for "root mean square" or the "square root of the mean of the squared function." This tells us that in order to find the rms value of a periodic function we first square the function, then obtain the average value during one period, then take the square root of this average value. The word "effective" is also used in much of the literature to mean rms. We prefer rms because the word effective is given other meanings in other situations; there is only one meaning to rms.

Example 5.13

GIVEN: The current waves shown in Fig. E-13.1.

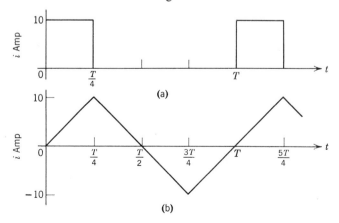

FIG. E-13.1

TO FIND: I_{rms}.

SOLUTION: (a) We first square the function, thus obtaining a sketch shown in Fig. E-13.2.

FIG. E-13.2

The average of this squared function is $100/4 = 25$ by inspection and then

$$I_{\text{rms}} = \sqrt{\text{avg } i^2} = \sqrt{25} = 5 \text{ amp.}$$

If this current wave were in a resistance of 10 ohms, the average power would be $I_{\text{rms}}^2 R = 5^2(10) = 250$ watts.

(b) Again we square the function, obtaining a group of parabolic curves as shown in Fig. E-13.3.

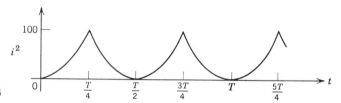

FIG. E-13.3

The function has a symmetrical shape such that the average during the first quarter period is the same as the average during an entire period. Also it is easy to show for this simple parabolic curve that the average ordinate is one third of the maximum ordinate. Thus

$\text{avg } i^2 = \frac{100}{3} = 33.3$,

and

$I_{\text{rms}} = \sqrt{33.3} = 5.77$ amp.

We have introduced the concept of rms value by using a periodic current; we could in Fig. 5.17 equally well have used the voltage v_{ab}. The instantaneous power in the resistor R is v_{ab}^2/R and the average power is

$$\frac{1}{T} \int_{K}^{K+T} \frac{v_{ab}^2}{R} \, dt.$$

Thus we could define

$$V_{ab(\text{rms})} = \sqrt{\text{avg } v_{ab}^2}. \tag{5.54}$$

As a matter of fact we are not of course, limited to currents or voltages; the term rms can be applied to any periodic function, $f(t)$.

$$F_{\text{rms}} = \sqrt{\frac{1}{T} \int_{K}^{K+T} [f(t)]^2 \, dt}. \tag{5.55}$$

For the special case of sinusoidal functions of time, the last article has shown that, for a current $I \cos(\omega t + \phi)$, an average power of $I^2 R/2$ was dissipated in a resistance R. From this we could conclude immediately that $I_{\text{rms}} = I/\sqrt{2}$ for a sinusoidal wave. However, let us prove this from our definition of rms values.

$$\begin{aligned} I_{\text{rms}} &= \sqrt{\frac{1}{T} \int_{K}^{K+T} i^2 \, dt} \\ &= \sqrt{\frac{1}{2\pi} \int_{K_1}^{K_1+2\pi} I^2 \cos^2(\omega t + \phi) \, d(\omega t)} \\ &= I \sqrt{\frac{1}{2\pi} \int_{K_1}^{K_1+2\pi} \cos^2(\omega t + \phi) \, d(\omega t)}, \end{aligned} \tag{5.56}$$

the change in variable being made for convenience in integrating. The integral may be evaluated mathematically or one may deduce from the trigonometric identity, $\cos^2 \alpha = (\cos 2\alpha + 1)/2$, that the average value of $\cos^2(\omega t + \phi)$ is $\frac{1}{2}$, and thus

$$I_{\text{rms}} = \frac{I}{\sqrt{2}}, \tag{5.57}$$

for a sinusoidal wave.

The rms value of a sinusoidal voltage wave is related to its amplitude in the same way; that is, $V_{\text{rms}} = V/\sqrt{2}$.

Many texts use V_m and I_m to indicate the amplitudes of sinusoidal voltages and currents and then use V and I to indicate their rms values. This is convenient if only phasor quantities are being used.

Instruments of the electrodynamometer type are suitable for reading rms values (see Appendix D). The deflection of the instrument is proportional to the product of the currents in the two coils and thus if the coils are connected in series the deflection will be proportional to the square of the current. Because of inertia the actual deflection will be the average of this current squared. A suitable scale plus proper calibration results in a fairly good instrument. We shall indicate ammeters and voltmeters by circles with associated letters. Usually we will assume the ammeters to have zero impedance and the voltmeters to have infinite impedance.

Example 5.14

GIVEN: The circuit of Fig. E-14.1 in which R_1 and $\hat{Z}_2$ are unknown, A_0 reads 5 amperes, V_1 reads 50 volts, V_2 reads 80 volts, and V_0 reads 113 volts. Readings are rms values. Waves are sinusoids.

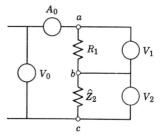

FIG. E-14.1

TO FIND: Magnitudes of R_1 and $\hat{Z}_2$ and phase angle of $\hat{Z}_2$ if possible.

SOLUTION: Since all readings are amplitudes divided by $\sqrt{2}$,

$$R_1 = \frac{V_{ab}}{I} = \frac{V_1}{A_0} = \frac{50}{5} = 10 \text{ ohms}$$

$$Z_2 = \frac{V_{bc}}{I} = \frac{V_2}{A_0} = \frac{80}{5} = 16 \text{ ohms}$$

$$Z_{ac} = \frac{V_{ac}}{I} = \frac{V_0}{A_0} = \frac{113}{5} = 22.6 \text{ ohms}$$

We do not know the phase angles of $\hat{Z}_2$ or $\hat{Z}_{ac}$ but we do know that $\hat{Z}_{ac} = R_1 + \hat{Z}_2$. Therefore a diagram may give us some possible phase angles. Fig. E-14.2 shows R_1 as a directed line segment.

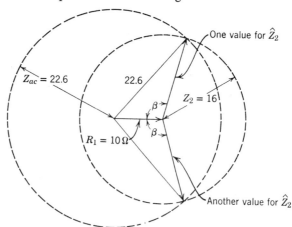

FIG. E-14.2

Using the head of the directed line segment as its center a circle is sketched with a magnitude of 16 ohms; using the tail of the line segment as its center a circle is sketched with a magnitude of 22.6 ohms. The intersection of these circles represent solutions to the equation $\hat{Z}_{ac} = R + \hat{Z}_2$. The angle β may be calculated from the law of cosines.

$$\cos \beta = \frac{10^2 + 16^2 - 22.6^2}{2(10)(16)} = -0.482.$$

Then

$$\beta = 118.8°,$$

and angle of $\hat{Z}_2 = \pm 61.2°$.

The sign of this angle cannot be determined from the data.

One could have used a voltage diagram rather than an impedance diagram.

5.8 MAXIMUM POWER TRANSFER

In communications systems it is often desirable to design the load impedance so that maximum average power is supplied to the load. For purposes of analysis we may replace the rest of the network by its Thévenin's equivalent circuit which is represented by $\hat{V}$ and $\hat{Z}_g$ in Fig. 5.18.

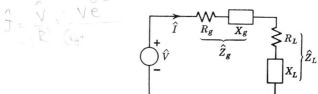

FIG. 5.18

The expression for the average power supplied to the load is

$$P = \frac{I^2 R_L}{2} = \frac{V^2 R_L}{2[(R_g + R_L)^2 + (X_g + X_L)^2]} \tag{5.58}$$

Assume that we are free to vary both X_L and R_L in order to maximize the power. Let us first hold R_L constant and consider the effect of varying X_L. Equation 5.58 shows that if $X_L = -X_g$, the average power P would be a maximum. Then let us consider the effect of varying R, assuming X_L is held constant. The variation of P with R_L is not directly evident. If R_L is either zero or infinitely large, P will be zero. Let us differentiate P with respect to R_L and set the derivative equal to zero; the maximum value should occur when $dP/dR_L = 0$.

$$\frac{dP}{dR_L} = \frac{V^2}{2} \left[\frac{(R_g + R_L)^2 + (X_g + X_L)^2 - 2R_L(R_g + R_L)}{[(R_g + R_L)^2 + (X_g + X_L)^2]^2} \right]. \tag{5.59}$$

The numerator of Equation 5.59 is zero if

$$R_L = \sqrt{R_g^2 + (X_g + X_L)^2}. \tag{5.60}$$

If $X_L = -X_g$,

$$R_L = R_g. \tag{5.61}$$

The conclusion is that if both R_L and X_L may be varied, maximum average power is transferred to R_L if

$$\hat{Z}_L = \hat{Z}_g{}^*. \tag{5.62}$$

If only R_L may be varied the criterion for R_L is given by Equation 5.60.

5.9 PHASOR DOMAIN FROM THE LAPLACE TRANSFORM DOMAIN

In Section 5.4 the phasor domain was obtained from the time domain. It should be possible to obtain the phasor domain from the transform domain. The following represents such a derivation. The technique used permits the complete response to be obtained, and can best be shown by applying it to a specific case. Consider the circuit of Fig. 5.19.

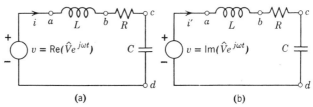

FIG. 5.19. An *RLC* circuit with sinusoidal driving functions.

Kirchhoff's voltage equation for Fig. 5.19a is

$$L\frac{di}{dt} + Ri + \frac{1}{C}\int_0^t i\,dt + v_{cd}(0+) = \text{Re}(\hat{V}e^{j\omega t}), \tag{5.63}$$

in which i is the **complete response**, which includes the free response, i_f, and the steady-state response, $i_{ss} = \text{Re}(\hat{I}e^{j\omega t})$.

Suppose that for this same circuit we impress the voltage,

$$V\sin(\omega t + \alpha) = \text{Im}(\hat{V}e^{j\omega t}),$$

as shown in Fig. 5.19b. We use i' to indicate the *complete* current response for this case, and write for this circuit

$$L\frac{di'}{dt} + Ri' + \frac{1}{C}\int_0^t i'\,dt + v'_{cd}(0+) = \text{Im}(\hat{V}e^{j\omega t}). \tag{5.64}$$

We do not need to have the same initial conditions in the two circuits.

We multiply Equation 5.64 by j and add the resulting equation to Equation 5.63 to obtain

$$L\frac{d}{dt}(i + ji') + R(i + ji') + \frac{1}{C}\int_0^t (i + ji')\,dt + v_{cd}(0+)$$
$$+ jv'_{cd}(0+) = \hat{V}e^{j\omega t}. \quad (5.65)$$

Now let us define

$$\hat{\imath} = i + ji', \quad (5.66)$$

in which $\hat{\imath}$ is then a complex function of time, such that $i = \text{Re}(\hat{\imath})$. Also we let

$$\hat{\imath}(0+) = i(0+) + ji'(0+),$$

and

$$\hat{v}_{cd}(0+) = v_{cd}(0+) + jv'_{cd}(0+).$$

Then Equation 5.65 becomes

$$L\frac{d\hat{\imath}}{dt} + R\hat{\imath} + \frac{1}{C}\int_0^t \hat{\imath}\,dt + \hat{v}_{cd}(0+) = \hat{V}e^{j\omega t}. \quad (5.67)$$

It is important to observe that in changing Equation 5.63 to Equation 5.67 we have justified replacing a real function of time by a complex function of time. This is similar to the theorem of Section 5.4 except that now we have extended the theorem to apply to the complete response rather than limiting it to the steady state response. We could therefore use the classical method to obtain the complete solution, $\hat{\imath}$, for Equation 5.67. Then $i = \text{Re}(\hat{\imath})$ would be the **complete** response to the original circuit problem. However, our main purpose is to relate the Laplace transform domain and the phasor domain so we proceed to take the Laplace transform of Equation 5.67. This becomes

$$\left(Ls + R + \frac{1}{Cs}\right)\hat{I}(s) - L\hat{\imath}(0+) + \frac{\hat{v}_{cd}(0+)}{s} = \frac{\hat{V}}{s - j\omega}, \quad (5.68)$$

in which $\hat{I}(s) = \mathscr{L}(\hat{\imath})$, and $\hat{I}(s)$ should therefore be interpreted as the Laplace transform of a complex function of time. $\hat{I}(s)$ may be solved for in Equation 5.68.

$$\hat{I}(s) = \frac{s\hat{V} + [Ls\hat{\imath}(0+) - \hat{v}_{cd}(0+)](s - j\omega)}{L\left(s^2 + \frac{R}{L}s + \frac{1}{LC}\right)(s - j\omega)}, \quad (5.69)$$

or

$$\hat{I}(s) = \frac{\hat{K}_1}{s - j\omega} + \frac{\hat{A}_1 s + \hat{A}_2}{s^2 + \frac{R}{L}s + \frac{1}{LC}}, \quad (5.70)$$

Steady-State Analyses with Sinusoidal Sources

the constants $\hat{K}_1$, $\hat{A}_1$, and $\hat{A}_2$ being indicated as complex since $\hat{I}(s)$ is complex. Also $\hat{K}_1$ is the coefficient associated with the steady-state response, whereas $\hat{A}_1$ and $\hat{A}_2$ are coefficients associated with the free response. The complex steady-state time response is

$$\hat{\imath}_{ss} = \mathscr{L}^{-1}\left(\frac{\hat{K}_1}{s - j\omega}\right) = \hat{K}_1 e^{j\omega t}. \tag{5.71}$$

Recall that $i = \mathrm{Re}(\hat{\imath})$ and that $i_{ss} = \mathrm{Re}(\hat{I}e^{j\omega t})$
Then

$$i_{ss} = \mathrm{Re}(\hat{I}e^{j\omega t}) = \mathrm{Re}(\hat{\imath}_{ss}) = \mathrm{Re}(\hat{K}_1 e^{j\omega t}), \tag{5.72}$$

from which

$$\hat{I} = \hat{K}_1. \tag{5.73}$$

$\hat{K}_1$ can be evaluated by Heaviside's expansion method. For Equation 5.70,

$$\hat{K}_1 = \hat{I}(s)[s - j\omega]\big|_{s=j\omega}, \tag{5.74}$$

and thus the phasor domain response and the Laplace transform of the complex time response may be related. Since $\hat{K}_1 = \hat{I}$, Equation 5.74 may be written

$$\hat{I} = \hat{I}(s)[s - j\omega]\big|_{s=j\omega}. \tag{5.75}$$

Similar results would have been obtained if we had sought phasor voltage responses. For $\hat{V}_{cd}$ we would have found

$$\hat{V}_{cd} = \hat{V}_{cd}(s)[s - j\omega]\big|_{s=j\omega}. \tag{5.76}$$

These results were obtained for a specific circuit; however, if we have followed the argument we can extend the results to a general circuit. If $I(s)$ is the transform response to a driving function $V(s) = \mathscr{L}[V\cos(\omega t + \alpha)]$ then $\hat{I}(s)$ is the transform response to a driving function $\hat{V}(s) = \mathscr{L}(\hat{V}e^{j\omega t})$ and $\hat{I} = \hat{I}(s)[s - j\omega]\big|_{s=j\omega}$. The argument is also not restricted to the case of a current response to a voltage driving function. It applies to either a current or voltage response and either a current or voltage driving function. The only restriction is that the driving function be varying sinusoidally with time. **If $R(s)$ is the transform response to a driving function, $D(s) = \mathscr{L}[D\cos(\omega t + \alpha)]$, then $\hat{R}(s)$ is the transform response to the driving function, $\hat{D}(s) = \mathscr{L}(\hat{D}e^{j\omega t})$, and $\hat{R} = \hat{R}(s)[s - j\omega]\big|_{s=j\omega}$.**

What about the relationship between phasor network functions and transform network functions? We have probably noticed that these were amazingly similar, the s in the transform function being replaced by $j\omega$ in the phasor function. In order to prove this we shall use the above results with the restriction that is inherent in the definition of network functions, that initial conditions be zero. For a driving-point immittance,

$$\hat{Z} = \frac{\hat{V}}{\hat{I}} = \frac{\hat{V}(s)[s-j\omega]|_{s=j\omega}}{\hat{I}(s)[s-j\omega]|_{s=j\omega}} = \frac{\hat{V}(s)}{\hat{I}(s)}\bigg|_{s=j\omega} = \frac{V(s)}{I(s)}\bigg|_{s=j\omega} = Z(s)\bigg|_{s=j\omega} \quad (5.77)$$

or

$$\hat{Y} = \frac{\hat{I}}{\hat{V}} = \frac{\hat{I}(s)[s-j\omega]|_{s=j\omega}}{\hat{V}(s)[s-j\omega]|_{s=j\omega}} = \frac{\hat{I}(s)}{\hat{V}(s)}\bigg|_{s=j\omega} = \frac{I(s)}{V(s)}\bigg|_{s=j\omega} = Y(s)\bigg|_{s=j\omega}. \quad (5.78)$$

The ratio, $\hat{V}(s)/\hat{I}(s)$, is equal to the ratio, $V(s)/I(s)$, because network functions are constants in linear systems. For the transfer function,

$$\hat{T} = \frac{\hat{R}}{\hat{D}} = \frac{\hat{R}(s)[s-j\omega]|_{s=j\omega}}{\hat{D}(s)[s-j\omega]|_{s=j\omega}} = \frac{\hat{R}(s)}{\hat{D}(s)}\bigg|_{s=j\omega} = T(s)\bigg|_{s=j\omega}. \quad (5.79)$$

Thus the phasor domain is related to the transform domain. We should keep in mind that $s = \sigma + j\omega_s$ and that therefore setting $s = j\omega$ means $\sigma = 0$ and $\omega_s = \omega$. For sinusoidal driving functions we may therefore interpret the positive imaginary axis of the s domain as equal to the angular frequency of the driving function.

Example 5.15

GIVEN: The circuit of Fig. E-15.1 with $i(0+) = 2$ amperes.

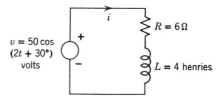

FIG. E-15.1

TO FIND: The complete response i demonstrating the use of the Laplace transform of a complex time function.

SOLUTION: Kirchhoff's voltage equation is

$$L\frac{di}{dt} + Ri = \text{Re}(\hat{V}e^{j\omega t}). \quad (1)$$

Using the technique of Section 5.9, we may write the equation

$$L\frac{d\hat{i}}{dt} + R\hat{i} = \hat{V}e^{j\omega t}, \quad (2)$$

in which $\hat{i} = i + ji'$, and $i'(0+)$ may have any finite real value.

The Laplace transform of Equation 2 is

$$(Ls + R)\hat{I}(s) - L\hat{i}(0+) = \frac{\hat{V}}{s - j\omega}. \quad (3)$$

If we now arbitrarily select $i'(0+) = 0$, $\hat{\imath}(0+) = 2 + j0$. The substitution of numerical values into Equation 3 results in

$$(4s + 6)\hat{I}(s) - 4(2) = \frac{50e^{j30°}}{s - j2},$$

or

$$\hat{I}(s) = \frac{50e^{j30°} + 8(s - j2)}{(s - j2)(4s + 6)}$$

$$= \frac{12.5e^{j30°} + 2(s - j2)}{(s - j2)(s + 1.5)}$$

$$= \frac{\hat{K}_1}{s - j2} + \frac{\hat{A}_1}{s + 1.5}. \tag{4}$$

The first term in the partial fraction expansion is the complex steady-state response, the second term is the complex free response. The solution for $\hat{\imath}$ is

$$\hat{\imath} = \hat{K}_1 e^{j2t} + \hat{A}_1 e^{-1.5t},$$

and the solution for i is

$$i = \text{Re}(\hat{\imath}) = \underbrace{\text{Re}(\hat{K}_1 e^{j2t})}_{i_{ss}} + \underbrace{\text{Re}(\hat{A}_1 e^{-1.5t})}_{i_f}.$$

If we wish only i_{ss}, we need to solve only for $\hat{K}_1$. From Equation 4,

$$\hat{K}_1 = \frac{12.5e^{j30°} + 2(s - j2)}{s + 1.5}\bigg|_{s=j2} = \frac{12.5e^{j30°}}{j2 + 1.5}$$

$$\frac{12.5e^{j30°}}{2.5e^{j53.1°}} = 5e^{-j23.1°},$$

and

$$i_{ss} = \text{Re}(\hat{K}_1 e^{j2t}) = \text{Re}(5e^{-j23.1°} e^{j2t})$$
$$= 5\cos(2t - 23.1°) \text{ amperes.}$$

The free response can also be found. From Equation 4,

$$\hat{A}_1 = \frac{12.5e^{j30°} + 2(s - j2)}{s - j2}\bigg|_{s=-1.5}$$

$$= \frac{12.5e^{j30°} + 2(-1.5 - j2)}{-1.5 - j2} = \frac{12.5e^{j30°}}{-2.5e^{j53.1°}} + 2$$

$$= -5e^{-j23.1°} + 2 = -5(0.918 - j.392) + 2 = -2.59 + j1.96.$$

Then

$$i_f = \text{Re}(\hat{A}_1 e^{-1.5t}) = -2.59 e^{-1.5t},$$

and the complete response is

$$i = i_{ss} + i_f = 5\cos(2t - 23.1°) - 2.59 e^{-1.5t} \text{ amp.}$$

5.10 SUMMARY

The use of the phasor domain is preferred to the time domain for determining steady state sinusoidal responses because it is easier to deal with complex numbers than sinusoidal time functions. The phasor domain was obtained from both the time domain and the Laplace transform domain by developing a complex time function, $\hat{\imath}$, such that $\hat{\imath}_{ss} = \hat{I}e^{j\omega t}$, and $i_{ss} = \text{Re}(\hat{I}e^{j\omega t})$. The phasor domain resembles the transform domain in being algebraic; however, phasor voltage and current have the dimensions of volts and amperes respectively while transform voltage and current have the dimensions of volt seconds and ampere seconds respectively. A phasor network function may be obtained directly from a transform network function by substituting $j\omega$ for s. However, a phasor voltage or current cannot be obtained directly; it must be obtained from the transform of a complex time function. As an example,

$$\hat{V} = \hat{V}(s)[s - j\omega]\big|_{s=j\omega}.$$

Root-mean-square values are defined for periodic functions in general and then applied to sinusoidal functions. They are particularly related to average power and the reading of instruments which measure periodic voltages or currents.

FURTHER READING

All introductory circuit texts discuss the phasor domain and relate this domain to the time domain. Good discussions are in H. A. Thompson's *Alternating-Current and Transient Circuit Analysis*, Chapter 5, McGraw-Hill, New York, 1955, and in P. R. Clement and W. C. Johnson's *Electrical Engineering Science*, Chapters 10 and 11, McGraw-Hill, New York, 1960. In some of the literature steady-state response is obtained to an exponential driving function of the form Ae^{st}, with s being a complex quantity. It is easy to confuse this with the s of the Laplace transform, particularly since the network functions appear to be identical. However, this is not true since voltages and currents are not in the transform domain. One reference which relates the Laplace transform and phasor domains is D. Cheng's *Analysis of Linear Systems*, Art. 7–6, Addison-Wesley, Reading, Mass., 1959.

PROBLEMS

5.1 A sinusoidal, steady-state voltage is defined by the expression,

$v = 100 \cos(377t + 60°)$ volts.

   a Plot two complete cycles of this wave. Show both a scale of ωt, radians, and a scale of t, seconds.

b Give the amplitude; the angular frequency; the frequency in cycles per second; the period in seconds; the magnitude of the voltage at times: $t = 0$ sec, $t = 1/180$ sec, $t = 1/240$ sec; the minimum value; and the phase.
c Write the expression for v as a sine function.

5.2 A steady-state sinusoidal current is expressed as $i_{ss} = 3 \cos 100t + 4 \sin 100t$ amperes.
  a Make a plot of i_{ss} versus ωt for two complete cycles of each component of i_{ss}, ($3 \cos 100t$, $4 \sin 100t$).
  b Express i_{ss} as a single cosine function. Plot two cycles of i_{ss} on the same axis as part a.
  c For the expression of i_{ss} in part b give the amplitude, radian frequency, frequency in cycles per second (cps), the phase, and the period in seconds.

5.3 For the voltage waveform shown in Fig. P-5.3:
  a Determine T, f, ω, and the amplitude.
  b Write an expression for v as a cosine function with appropriate phase angle.
  c Write an expression for v as a sine function with appropriate phase angle.
  d Write an expression for v as the sum of a cosine and sine function where neither has a phase angle.

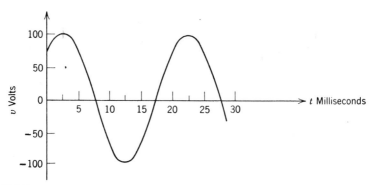

FIG. P-5.3

Partial answer: c, $v = 100 \sin(100\pi t + 45°)$ volts.

5.4 For the circuit shown in Fig. P-5.4.

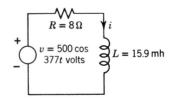

FIG. P-5.4

  a Determine the steady-state current i_{ss}.
  b Plot the voltage v and the current i_{ss} for two cycles of i_{ss}.

c What is the amplitude, angular frequency, and period of i_{ss}?
d What is the phase of i_{ss}? Does i_{ss} lead or lag v?
Partial answer: a, $i_{ss} = 50 \cos(377t - 36.9°)$ amp.

5.5 The waveforms shown in Fig. P-5.5a are for the steady-state voltages v_{ab} and v_{bc} in the circuit shown in Fig. P-5.5b.

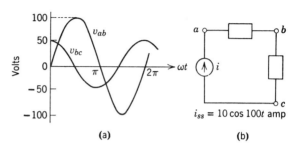

FIG. P-5.5   (a)   (b)

a Determine the steady-state voltage v_{ac}.
b Show a sketch of v_{ac}, v_{ab}, and v_{bc} versus ωt on one set of axis.
c On another set of axis, sketch v_{ac} and i_{ss} versus ωt.
d On the sketch of part c, show the phase angle. Does the current lead or lag the voltage?
e What is the phase angle between v_{ab} and i_{ss}; between v_{bc} and i_{ss}? (Specify which quantity leads or lags the other quantity.)

5.6 Use Euler's formula to show that:

a $10e^{j53.1°} = 6 + j8$.
b $100e^{-j120°} = -50 - j86.6$.
c $\sqrt{2}e^{j225°} = -1 - j1$.
d $-3e^{-j26.6°} = -2.68 + j1.34$.
e $5e^{j(2t+30°)} = 5\cos(2t + 30°) + j5\sin(2t + 30°)$.
f $\operatorname{Re}(\check{V}e^{j\omega t}) = V\cos(\omega t + \phi)$.
g $\operatorname{Im}(10e^{-j\omega t}e^{-j\phi}) = -10\sin(\omega t + \phi)$.

5.7 Why is it sometimes useful to replace $\operatorname{Re}[Ve^{j(\omega t+\phi)}]$ with $Ve^{j(\omega t+\phi)}$?

5.8 Two steady-state currents are expressed as

$i_1 = 10\cos(2t + 30°)$,

and

$i_2 = 5\cos(2t - 60°)$.

a Find $i = i_1 + i_2$ by replacing $\operatorname{Re}[\hat{I}_1 e^{j2t}]$ and $\operatorname{Re}[\hat{I}_2 e^{j2t}]$ with $\hat{I}_1 e^{j2t}$ and $\hat{I}_2 e^{j2t}$.
b Show the locus of $\hat{I}_1 e^{j2t}$, $\hat{I}_2 e^{j2t}$ and $\hat{I}e^{j2t}$ in the complex plane as t increases from zero.

c Sketch the projections of $\hat{I}e^{j2t}$, $\hat{I}_1e^{j2t}$ and $\hat{I}_2e^{j2t}$ on the axis of reals.
d Show that $\text{Re}(\hat{I}_1) = i_1(0)$, $\text{Re}(\hat{I}_2) = i_2(0)$ and $\text{Re}(\hat{I}) = i(0)$.

5.9 Express each of the following as positive cosine functions with appropriate angle.

a $10 \sin 10t$.
b $-10 \cos(2t + 60°)$.
c $-5 \sin(10t - 30°)$.
d $-10 \cos(10t + 90°)$.
e $100 \sin(2t - 120°)$.
f $-86.6 \sin(2t + 150°)$.

5.10 Repeat problem 5.9 but express each function given as a positive sine function with appropriate angle.

5.11 Convert the following complex numbers to exponential form.
a $3 + j4$
b $8 - j6$
c $-3 + j6$
d $-5 - j5$
e $147 + j47$
f $2.1 - j0.2$

5.12 Convert the following complex numbers to rectangular form.
a $10e^{-j36.9°}$
b $5e^{-j53.1°}$
c $-5e^{+j60°}$
d $-3.02e^{-j76.2°}$
e $100e^{-j120°}$
f $173.2e^{-j90°}$
g $2e^{j(\pi/6)}$
h $710e^{-j(3\pi/6)}$

5.13 Perform the indicated operations and give answers in both rectangular and exponential form.
a $10e^{j60°} + 3e^{-j30°}$
b $-5e^{+j60°} - 5e^{-j60°}$
c $5e^{j53.1°} + (3 + j4)$
d $10 + j0 + 10e^{+j120°}$

5.14 Perform the indicated operations and express answers in both rectangular and exponential form.
a $(10e^{j60°})(5e^{j(\pi/2)})$
b $(170e^{-j150°})(0.32e^{+j28°})$
c $(-3 + j4)(8 - j2)$
d $\dfrac{10e^{j53.1°}}{0.1 + j12}$
e $\dfrac{4 + j3.1}{5e^{-j53.1°}}$
f $(-6 + j2)^2$
g $\dfrac{3e^{j14°}}{2 + j6} - (1 + j1)$
h $120[e^{j(\pi/3)} + e^{j\pi} - e^{j(\pi/2)}]$
i $120(e^{j0°} + e^{j120°} + e^{j240°})$

5.15 Given two complex numbers $\hat{A}$ and $\hat{B}$, prove that $\text{Re}\,\hat{A} + \text{Re}\,\hat{B} = \text{Re}\,[\hat{A} + \hat{B}]$.

5.16 Find the real part of the following.

a $10e^{j(\pi/3)}$

b $10e^{(1+j2)}$

c $100e^{j(\omega t+\alpha)}$

d $10e^{j(2t+30°)} + 5e^{j(2t+60°)}$

e $-j10e^{j30°}$

f $\dfrac{3+j4}{2+j2}$

5.17 a Find the cube roots of 64 as complex quantities.

b Find the sum of the three roots.

5.18 Show by expansion of $e^{j\alpha}e^{j\beta} = e^{j(\alpha+\beta)}$, that $\cos(\alpha+\beta) = \cos\alpha\cos\beta - \sin\alpha\sin\beta$, and $\sin(\alpha+\beta) = \sin\alpha\cos\beta + \cos\alpha\sin\beta$.

5.19 a Prove that the theorem of Section 5.4 also applies if $\text{Re}(\hat{D}e^{-j\omega t})$ and $\text{Re}(\hat{A}e^{-j\omega t})$ are replaced by $\hat{D}e^{-j\omega t}$ and $\hat{A}e^{-j\omega t}$ respectively.

b Show that the effect of this for the circuit of Fig. 5.10 is to change the sign of the angles of $\hat{V}$ and $\hat{I}$, but that there is no change in the time domain equation for i_{ss} (Equation 5.26).

5.20 a Show that Equation 5.26 may be written

$$i_{ss} = \dfrac{VR}{R^2+\omega^2L^2}\cos(\omega t + \alpha) + \dfrac{V\omega L}{R^2+\omega^2L^2}\sin(\omega t + \alpha).$$

b Using the classical technique of Chapter 3 to determine i_{ss} for Equation 5.24, assume

$$i_{ss} = K_1\cos(\omega t + \alpha) + K_2\sin(\omega t + \alpha)$$

and substitute this into Equation 5.24 to show that

$$K_1 = \dfrac{VR}{R^2+\omega^2L^2} \quad \text{and} \quad K_2 = \dfrac{V\omega L}{R^2+\omega^2L^2}.$$

5.21 Given the circuit shown in Fig. P-5.21.

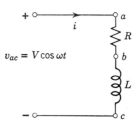

FIG. P-5.21

a Write the expressions for i_{ss}, $v_{ab(ss)}$, and $v_{bc(ss)}$.

b Show that the amplitude of either $v_{ab(ss)}$ or $v_{bc(ss)}$ can not exceed V.

c Assume $\omega = 0$ and show that the steady-state values of current and voltages are in accord with the direct current case.

d For $V = 100$ volts, $R = 10$ ohms, $L = 2$ henries, $\omega = 5$ radians per second, sketch the steady-state values of v_{ac}, i, v_{ab}, and v_{bc} for at least one cycle.

e On a phasor diagram show $\hat{V}_{ac}$, $\hat{V}_{ab}$, $\hat{V}_{bc}$, and $\hat{I}$.

Steady-State Analyses with Sinusoidal Sources 269

5.22 For the circuit of Fig. P-5.22 note that $\hat{V}_{ab} = j\omega L \hat{I}$, $\hat{V}_{bc} = R\hat{I}$ and $\hat{V}_{cd} = \hat{I}/j\omega C$.
  a Show that the magnitude of V_{ab} or V_{cd}, may be larger than V, and that this is specifically true if $\omega L = 1/\omega C$, and $\omega L > R$.
  b Show that V_{bc} is never larger than V.

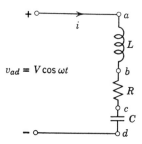

FIG. P-5.22

5.23 Use phasors to find the sum of the two currents, $i_1 = 10 \cos 10t$ amp and $i_2 = 10 \cos(10t + 60°)$ amp.

5.24 The sum of two voltages, $v_1 = 100 \cos(377t + 30°)$ and $v_2 = 100 \sin(377t + 60°)$ was to be found by using phasors. The following were submitted showing part of the solution and the answers found.

$\hat{V}_1 = 86.6 + j50$
$\hat{V}_2 = 86.6 - j50$
$\hat{V}_1 + \hat{V}_2 = 173.2 + j0$
$v_1 + v_2$
$= 173.2 \cos 377t$ volts
(1)

$\hat{V}_1 = 86.6 + j50$
$\hat{V}_2 = 50.0 + j86.6$
$\hat{V}_1 + \hat{V}_2 = 136.6 + j136.6$
$v_1 + v_2$
$= 193.4 \cos(377t + 45°)$ volts
(2)

$\hat{V}_1 = -50.0 + j86.6$
$\hat{V}_2 = +50.0 + j86.6$
$\hat{V}_1 + \hat{V}_2 = 0 + j173.2$
$v_1 + v_2$
$= 173.2 \sin(377t + 90°)$ volts
(3)

$\hat{V}_1 = -50 + j86.6$
$\hat{V}_2 = +86.6 - j50.0$
$\hat{V}_1 + \hat{V}_2 = 36.6 + j36.6$
$v_1 + v_2$
$= 51.7 \sin(377t + 45°)$ volts
(4)

a Which, if any, of the answers is correct?
b For any incorrect answers, explain what error was made.

5.25 For the circuit shown in Fig. P-5.25:
For each of the values of v and i given below.
  a Find the phasor impedance $\hat{Z}$.
  b Find the phasor admittance $\hat{Y}$.
  [Give answers in both rectangular and exponential form.]
  1. $v = 200 \cos 377t$ volts,
     $i = 10 \cos 377t$ amp.
  2. $v = 10 \cos(10t + 45°)$ volts,
     $i = 2 \cos(10t + 35°)$ amp.
  3. $v = 100 \cos(2t + 30°)$
     $i = 5 \cos(2t - 60°)$ amp.
  4. $v = 40 \cos(100t + 17°)$ volts,
     $i = 8 \cos 100t$ amp.
  5. $v = 100 \cos(\pi t - 15°)$ volts,
     $i = \sin(\pi t + 45°)$ amp.

270

FIG. P-5.25
Partial answer:
1. $\hat{Z} = 20e^{j0°}$
2. $\hat{Z} = 5e^{j10°}$
3. $\hat{Y} = 0 - j0.05$ mhos.
4. $\hat{Z} = 4.78 + j1.46$ Ω.

5.26 For each of the circuits shown in Fig. P-5.26 determine the simplest series combination of elements in the black box.

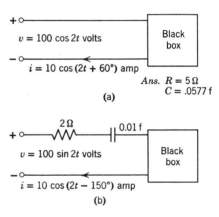

(a)
$v = 100 \cos 2t$ volts
$i = 10 \cos(2t + 60°)$ amp
Ans. $R = 5$ Ω
$C = .0577$ f

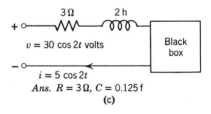

(b)
2 Ω, 0.01 f
$v = 100 \sin 2t$ volts
$i = 10 \cos(2t - 150°)$ amp

(c)
3 Ω, 2 h
$v = 30 \cos 2t$ volts
$i = 5 \cos 2t$
Ans. $R = 3$ Ω, $C = 0.125$ f

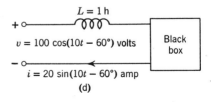

(d)
$L = 1$ h
$v = 100 \cos(10t - 60°)$ volts
$i = 20 \sin(10t - 60°)$ amp

FIG. P-5.26

Steady-State Analyses with Sinusoidal Sources 271

5.27 Given the circuit shown in Fig. P-5.27:
  a Use phasors to determine $\hat{I}$, $\hat{V}_{ab}$, $\hat{V}_{bc}$, and $\hat{V}_{cd}$.
  b Construct a phasor diagram showing $\hat{I}$, $\hat{V}_{ab}$, V_{bc} and $\hat{V}_{cd}$.
  c What is the phase angle between $\hat{I}$ and $\hat{V}_{ad}$? Is $\hat{I}$ leading or lagging $\hat{V}_{ad}$?
  d Write the expressions for i, v_{ab}, v_{bc}, and v_{cd}.
  e Show that $\hat{Z}_{ad} = R + j\omega L + \dfrac{1}{j\omega C} = \hat{V}_{ad}/\hat{I}$.

  Partial answer: $\hat{I} = 7.07 e^{-j45°}$ amp.

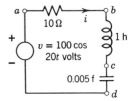

FIG. P-5.27

5.28 For each of the circuits shown in Fig. P-5.28, use phasors to find the voltages v_{ab}, v_{bc}, and v_{ac}. Show a phasor diagram with $\hat{I}$, $\hat{V}_{ab}$, $\hat{V}_{bc}$, and $\hat{V}_{ac}$ for each. What is $\hat{Z}_{ac}$ for each circuit?

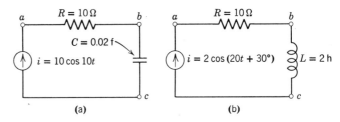

FIG. P-5.28

5.29 In what ways would you say the Laplace domain and phasor domain are similar in facilitating solutions of network problems? In what principal way do they differ?

5.30 For each of the statements below, fill in the blank.
  a Steady-state solutions to circuits with sinusoidal driving functions may be found by solving Kirchhoff's laws in any of three forms. They are the ―――――, ――――――――――― and ――――――――.
  b Phasor solutions give sinusoidal steady state responses that hold for only ――― ――――――.
  c For phasor circuits passive elements are labeled as ―――――――― or ――――――――.
  d Equations in the time domain are ――――――― whereas they are ――――――――― in the phasor domain.

e The inverse Laplace transform of ——— ——— is not defined.

f Driving point immittances are defined in the ——— or ——— but not in the ——— domain.

5.31 Draw the phasor circuit for each of the circuits shown in Fig. P-5.31. Find the driving point phasor impedance for each. Find the driving point phasor admittance of each.

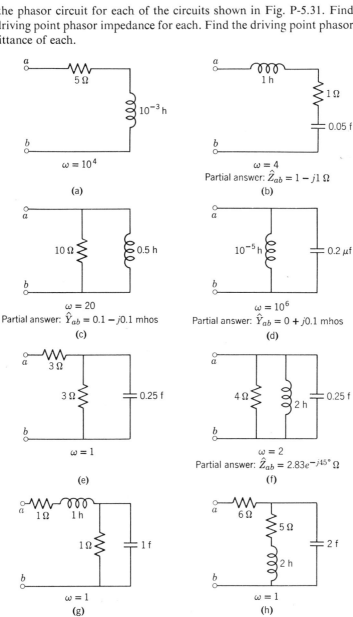

FIG. P-5.31

Steady-State Analyses with Sinusoidal Sources   273

5.32 Given $\hat{Z} = R + jX_L$; prove that

$$\hat{Y} = \frac{R}{R^2 + X_L^2} - j\frac{X_L}{R^2 + X_L^2}.$$

If $R = 15$ ohms and $X_L = 5$ ohms, what is the magnitude of the conductance and susceptance? Express $\hat{Y}$ in exponential form for the values given.

5.33 The driving point impedance of the circuit shown in Fig. P-5.33 is $\hat{Z} = 5 + j10$ ohms. Show a circuit consisting of a resistance in parallel with a reactance that will have the same driving point impedance. Give the values of R and X in the parallel circuit.

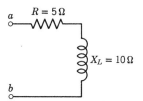

FIG. P-5.33

Partial answer: $R = 25\ \Omega$.

5.34 Given the phasor circuits shown in Fig. P-5.34, find the current in each branch. Use a phasor diagram to aid in the solution.

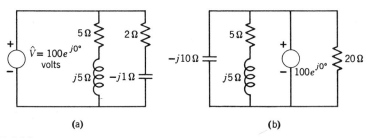

FIG. P-5.34

5.35 For each of the phasor circuits shown in Fig. P-5.35, use voltage ratio to find $\hat{V}_{ab}$ and $\hat{V}_{bc}$. Show a closed polygon phasor diagram of the voltages.

5.36 For each of the circuits shown in Fig. P-5.36 use current ratio to find the currents in each branch. Show a closed polygon phasor diagram of the currents and voltages.

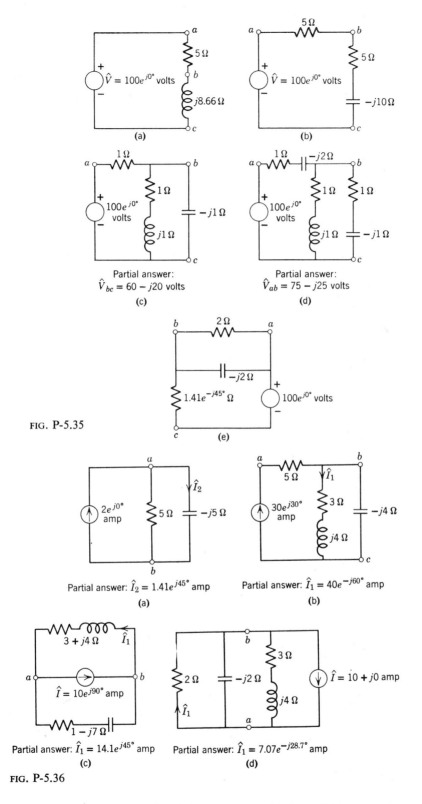

FIG. P-5.35

FIG. P-5.36

5.37 For the circuit shown in Fig. P-5.37, find the current $\hat{I}_0$ by:
a The mesh-current method.
b The node-voltage method.
c Thévenin's theorem.

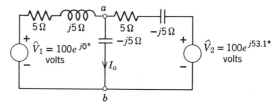

FIG. P-5.37

Answer: $\hat{I}_0 = 6.32e^{+j71.5°}$.

5.38 a In Fig. P-5.38, use the node voltage method to determine the independent voltages in complex form; then calculate the current through the capacitance.
b Find the current in the capacitance by using superposition.

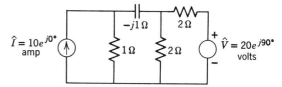

FIG. P-5.38

Partial answer: The magnitude of the current in the capacitance is 6.32 amp.

5.39 Use Thévenin's theorem to find the current $\hat{I}_0$ in each of the circuits shown in Fig. P-5.39.

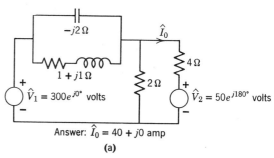

Answer: $\hat{I}_0 = 40 + j0$ amp
(a)

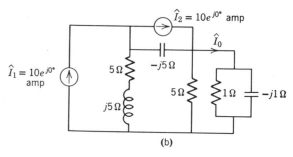

(b)

FIG. P-5.39

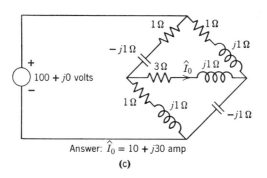

Answer: $\hat{I}_0 = 10 + j30$ amp

(c)

5.40 For the circuit shown in Fig. P-5.40, find the current $\hat{I}_0$ by:
 a The mesh-current method.
 b The node-voltage method.
 c Thévenin's theorem.

Answer: $\hat{I}_0 = 12 - j16$ amp.

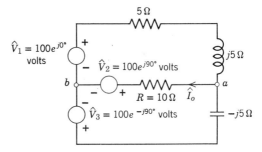

FIG. P-5.40

5.41 In Fig. P-5.41 convert the deltas to wyes and the wyes to deltas.

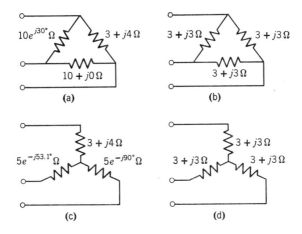

FIG. P-5.41

Steady-State Analyses with Sinusoidal Sources 277

5.42 For the circuit shown in Fig. P-5.42:
  a Find the current i.
  b Find the expression for the instantaneous power p.
  c Make a sketch of a few cycles of v, i, and p.
  d Determine the average power, P.

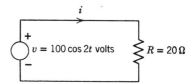

FIG. P-5.42

5.43 Repeat problem 5.42, but:
  a Replace R with a 0.01-f capacitance.
  b Replace R with a series combination of $R = 5\,\Omega$ and $L = 2.5$ henries.
  c Replace v with $v = 100\cos(2t + 45°)$ and R with a series combination of $R = 5\,\Omega$, $C = 0.2$ f and $L = 2.5$ henries.

5.44 For the circuit shown in Fig. P-5.44:
  a Solve for the current $\hat{I}$.
  b Determine the total average power, P.
  c Expand $\hat{V}\hat{I}^*$ in rectangular form and show that $\text{Re}(\hat{V}\hat{I}^*/2) = P$.
  d Show that the total average power is equal to the sum of each $I^2R/2$
  e What is the power factor of the circuit as seen from the source?

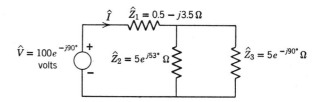

FIG. P-5.44

Partial answer: $P_{\text{total}} = 400$ watts

5.45 For a certain circuit the driving point impedance is $\hat{Z} = 20e^{j60°}$ ohms and the voltage applied is $\hat{V} = 100e^{-j30°}$ volts.
  a Determine the total power by three different ways.
  b What is the power factor?
  Answer: $P = 125$ watts, power factor $= 0.5$.

5.46 In the circuit shown in Fig. P-5.46 determine the reading of each of the wattmeters. What is the total average power?
  Partial answer: $P_2 = 7500$ watts, $P_3 = 2700$ watts, and $P_1 = 10{,}200$ watts.

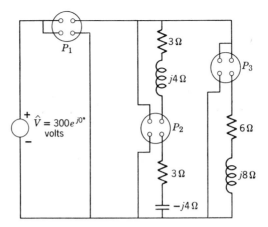

FIG. P-5.46

5.47 Determine the reading of the wattmeter P_1 in Fig. P-5.47. Draw a phasor diagram showing the voltage $\hat{V}_{ac}$ and the current $\hat{I}$. Is the reading of the wattmeter the actual average power of the circuit?

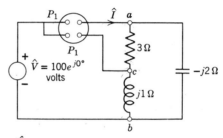

FIG. P-5.47

Partial answer: $P_1 = (V_{ac}/2)I \cos\angle\dfrac{\hat{V}_{ac}}{\hat{I}} = 750$ watts.

5.48 Two voltage sources are connected in series as indicated in Fig. P-5.48. $v_{ab} = V \cos \omega t$, $v_{cb} = V \cos(\omega t + \alpha)$.

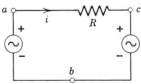

FIG. P-5.48

a For $\alpha = 60°$ sketch the phasors $\hat{V}_{ab}$, $\hat{V}_{cb}$, $\hat{V}_{ac}$, and $\hat{I}$. Determine the expression for the average power dissipated in R and the expressions for the average power into or out of the voltage sources.
b Repeat part *a* for $\alpha = 0°$ and $180°$.
c It is suggested that superposition be used in the following way. With the voltage source v_{ab} considered zero, the average power dissipated in R is calculated; then with the voltage source v_{cb} considered zero, the power dissipated in R is calculated. The total power dissipated in R is claimed to

be the sum of these powers. Why is this method in error? Would the method be all right if the two sources were of different frequencies?

5.49 For the circuit shown in Fig. P-5.49, determine the power for each source. Specify whether the power is into or out of each source. What is the power factor of each source?

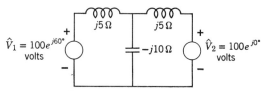

FIG. P-5.49

Partial answer: Power is into source $\hat{V}_2$.

5.50 For the waveform shown in Fig. P-5.50:
  a Show that $I_{rms} = A\sqrt{t_1/T}$.
  b Determine the average value of the current and show that I_{rms} will be larger than I_{avg} except for direct current under which conditions $I_{rms} = I_{avg}$. Are these conclusions valid for any waveshape?

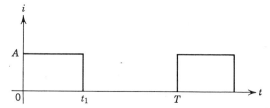

FIG. P-5.50

5.51 In Fig. P-5.51, A_1 is a D'Arsonval instrument which reads average current, A_2 is an electrodynamometer instrument which reads rms current, and P is an electrodynamometer type of wattmeter which reads average vi or, in this case, average $i^2 R$. For each current, determine the readings of the two ammeters and the wattmeter.

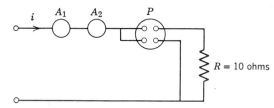

FIG. P-5.51

  a $i = 10$ amp.
  b $i = 10 \cos \omega t$ amp.
  c

For positive values of $\cos \omega t$, this wave is the same as part b, for negative values of $\cos \omega t$, $i = 0$. This wave would result if a perfect diode were placed in series with a sinusoidal voltage source and resistance, and is called a half-wave rectified wave.

d

This is a full-wave rectified wave and can be obtained by the use of diodes. This wave is the same as part b for positive values of cos ωt and the negative of part b for negative values of cos ωt.

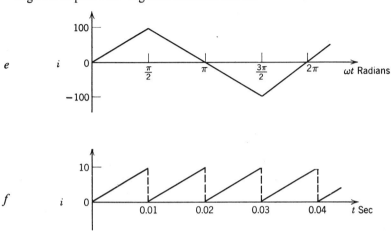

5.52 In the circuits shown in Fig. P-5.52 the meters read as shown. (All read rms. values.) For each circuit, find the reading of ammeter A_0 and/or voltmeter V_0 as appropriate.

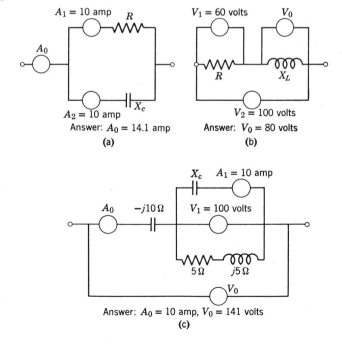

FIG. P-5.52

Steady-State Analyses with Sinusoidal Sources 281

5.53 For the circuit shown in Fig. P-5.53 the voltmeters and ammeters read rms values, the wattmeter reads average power. Voltmeter V_1 reads 223 volts, voltmeter V_2 reads 100 volts, ammeter A reads 14.14 amperes, and the wattmeter reads 1000 watts.

a Find the value of the resistance R_0.
b What is the power factor of the circuit?

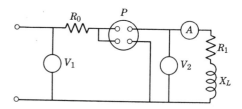

FIG. P-5.53

Partial answer: Circuit power factor is 0.95. $R_0 = 10\ \Omega$.

5.54 An American in Europe has a toaster which is rated to operate with amplitudes of 160 volts and 14 amperes. The toaster may be considered purely resistive. The available sinusoidal voltage has an amplitude of 320 volts. The American wishes to place an inductive coil in series with the toaster. We are asked to determine the voltage rating of this coil.

a Assume the coil is purely reactive. Sketch a phasor diagram using the voltage across the toaster as the reference phasor and from this phasor diagram calculate the voltage of the coil.
b Assume that the coil has a reactance ten times its resistance. Sketch a phasor diagram as before but now sketch the coil voltage at its proper angle. Using the diagram as an aid calculate the coil voltage.

5.55 The load shown in Fig. P-5.55 takes an average power of 10 watts at a power factor of 0.8 lagging. $V_{ab} = 100$ volts and the frequency is 1000 cps. It is desired to place a capacitor in parallel with the load so that the net power factor is one.

a Determine $\hat{I}_1$ using $\hat{V}_{ab}$ as reference phasor.
b Sketch $\hat{I}_1$ and then sketch $\hat{I}_2$ and $\hat{I}_0$ at their appropriate angles. From this sketch, evaluate $\hat{I}_2$.
c Calculate C.

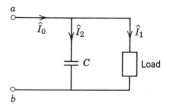

FIG. P-5.55

5.56 For the circuit shown in Fig. P-5.56, prove that the load resistance R_L must be equal to the magnitude of the generator impedance $\hat{Z}_g$ to have maximum average power supplied to R_L.

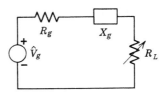

FIG. P-5.56

5.57 For the circuit shown in Fig. P-5.57 find the value of R_L for which the average power in R_L will be maximum. What is the amount of this power?

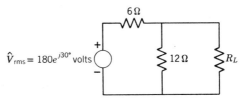

FIG. P-5.57

Partial answer: $P_{max} = 900$ watts.

5.58 For the circuit shown in Fig. P-5.58:
a Determine the value of $\hat{Z}_L$ for which the average power in $\hat{Z}_L$ will be maximum.
b Determine the value of this average power.
c Repeat parts a and b if $\hat{Z}_L$ must be purely resistive.

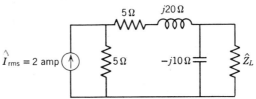

FIG. P-5.58

Partial answer: a, $\hat{Z}_L = 5 + j15\ \Omega$.
b, $P = 2.5$ watts.

5.59 In the circuit shown in Fig. P-5.59 the switch is closed at time $t = 0$. $v_{ab}(0-) = 10$ volts.
a Use the Laplace transform of a complex time function to find the complete solution for the current i.

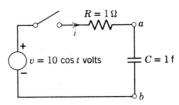

FIG. P-5.59

Answer: a, $i = -5e^{-t} + 5\sqrt{2} \cos(t + 45°)$ amp.
b Repeat part a with $v = 100 \cos 400t$, $R = 10\ \Omega$, $C = 250\ \mu f$, and $v_{ab}(0-) = 100$ volts.

Steady-State Analyses with Sinusoidal Sources 283

5.60 For the circuit of Fig. P-5.60 use the Laplace transform of a complex time function to determine the complete response v_{ab}.

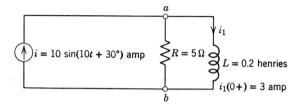

FIG. P-5.60

5.61 In the circuit of Fig. P-5.61 steady state exists. $v_{ab}(0-) = 0$ volts. The switch is closed at $t = 0$. Determine the complete solution for i by both the classical and Laplace transform methods. In each method you may use the phasor method to determine the steady state alternating-current response.

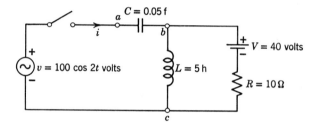

FIG. P-5.61

Answer: $i = \dfrac{20}{\sqrt{3}} e^{-t} \sin\sqrt{3}\,t + 10\sqrt{2}\cos(2t + 45°)$ amp.

5.62 In Fig. P-5.62 steady state exists with the switch K open.

a Using the nodal method, determine $\hat{V}_{bc}$. Then write the expressions for v_{bc} and i_2.

b K is closed at $t = 0$. Determine the complete solution for i_2 by both the classical and the Laplace transform methods. In each method you may use phasors to determine the steady-state alternating-current response.

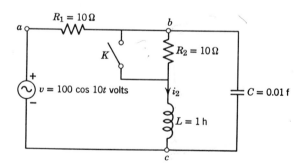

FIG. P-5.62

Partial answer: a, $i_2 = 4.47\cos(10t - 63.4°)$ amp.

284

5.63 In Fig. P-5.63 the switch is closed at $t = 0$. $i(0-) = 0$.
Determine the complete solution for i by each of the following methods.
a The method involving the Laplace transform of a complex time function.
b The classical method. You may use phasors to determine the steady state response.

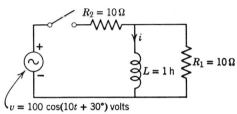

FIG. P-5.63 $v = 100 \cos(10t + 30°)$ volts

Answer: $i = -3.73e^{-5t} + 4.47 \cos(10t - 33.4°)$ amp.

6 Graphical Methods Applied to Phasor Network Functions for a Variable Element or for Variable Frequency

It is the purpose of this chapter to study the effect of the variation of one element of a network or of the variation of frequency on the steady-state sinusoidal responses. We shall do this by observing the effect of these variations on the phasor network functions.

The effect of the variation of a single element in a network is of particular importance in power transmission or in control systems. The effect of the variation of frequency is of particular importance in communications, since each communication channel requires a band of frequencies and also since there should be a minimum interference between these channels.

The variations in phasor network functions can be most clearly presented by graphical methods.

6.1 LOCI OF PHASOR NETWORK FUNCTIONS FOR A VARIABLE ELEMENT

We shall consider the effect of variations of a single element, R, L, or C, on the phasor network functions. We shall demonstrate that the loci of such network functions will lie on circles in the complex plane, a straight line being considered the special case of a circle of infinite radius. First a simple numerical example will be used to demonstrate the loci of driving-point immittances as one element is varied.

6.1.1 Loci of Driving Point Immittances for a Variable Element in a Simple Circuit

Consider the circuit of Fig. 6.1a in which a sinusoidal voltage source is applied and for which we are asked to determine the steady-state effects obtained by varying L from 0 to ∞. The arrow through L indicates that this element is variable.

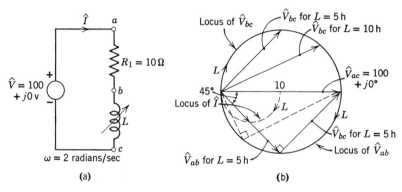

FIG. 6.1 Circuit with variable L and loci of voltages and currents.

It is fairly easy to demonstrate for this circuit that the locus of $\hat{I}$ as L varies from 0 to $+\infty$ must be a semicircle. As L is increased the magnitude of the current $\hat{I}$ decreases from 10 to 0 and its phase angle changes from 0 to $-90°$. The voltage $\hat{V}_{ab}$ must be in phase with the current, the voltage $\hat{V}_{bc}$ must lead the current by $90°$, and the sum of these phasors must add to a constant $\hat{V}_{ac}$. It follows directly from geometry that the locus of $\hat{V}_{ab}$ (the set of points representing $\hat{V}_{ab}$ for all values of L) will be a semicircle with diameter $\hat{V}_{ac}$. Since $\hat{I} = \hat{V}_{ab}/R$, the locus of $\hat{I}$ will also be a semicircle. Also $\hat{V}_{bc}$ will have the locus of a semicircle from geometry.

The network functions that are involved will have loci of straight lines or arcs of circles. $\hat{Z}_{ac}$, the driving-point impedance shown in Fig. 6.2a, may be sketched by inspection since it is a semiinfinite line parallel to the axis of imaginaries starting at $10 + j0$ for $\omega L = 0$. The locus of $\hat{Y}_{ac} = \hat{I}/\hat{V}_{ac}$ may be deduced from Fig. 6.1b since $\hat{V}_{ac}$ is a real constant and thus the locus of $\hat{I}/\hat{V}_{ac}$ differs from the locus of $\hat{I}$ by only a constant. The locus of $\hat{Y}_{ac}$ is a semicircle in the fourth quadrant as shown in Fig. 6.2b. The dotted lines shown in the first quadrant indicate values for $\hat{Y}_{ac}$ which would be obtained for negative values of L. It is often helpful to sketch the complete circular locus even though only a portion is physically realizable. The transfer functions $\hat{V}_{ab}/\hat{V}_{ac}$ and $\hat{V}_{bc}/\hat{V}_{ac}$ are shown in Figs. 6.2c and 6.2d, and are readily sketched from the loci of $\hat{V}_{ab}$ and $\hat{V}_{bc}$ since $\hat{V}_{ac}$ is a constant.

Phasor Network Functions for a Variable Element or for Variable Frequency 287

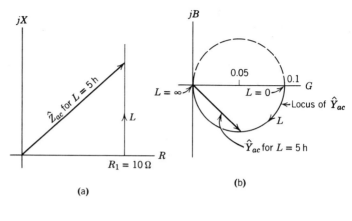

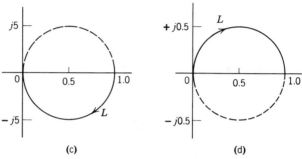

FIG. 6.2 Loci of network functions for circuit of Fig. 6.1. (a) Locus of $\hat{Z}_{ac} = \hat{V}_{ac}/\hat{I} = R_1 + j\omega L$. (b) Locus of $\hat{Y}_{ac} = \hat{I}/\hat{V}_{ac} = G + jB$. (c) Locus of $\hat{V}_{ab}/\hat{V}_{ac}$. (d) Locus of $\hat{V}_{bc}/\hat{V}_{ac}$.

It may seem strange that the reciprocal of a function which has the locus of a straight line has the locus of a circle. Let us prove this to ourselves for this particular circuit.

$$\hat{Y}_{ac} = \frac{1}{\hat{Z}_{ac}} = \frac{1}{R_1 + j\omega L} = \frac{R_1 - j\omega L}{R_1^2 + \omega^2 L^2}$$

$$= \frac{10 - j\omega L}{10^2 + \omega^2 L^2} = G + jB. \qquad (6.1)$$

We may equate reals and imaginaries in Equation 6.1 to obtain

$$G = \frac{10}{10^2 + \omega^2 L^2} \qquad (6.2)$$

and

$$B = \frac{-\omega L}{10^2 + \omega^2 L^2} \qquad (6.3)$$

We may eliminate the term, ωL, from the above equations and thus obtain a single expression relating G and B. This may be done by noting that $\dfrac{G}{B} = -\dfrac{10}{\omega L}$ and then using this relation in Equation 6.2 to obtain

$$G^2 - 0.1G + B^2 = 0. \tag{6.4}$$

If we complete the square on the left side of this equation, we obtain,

$$(G - 0.05)^2 + B^2 = (0.05)^2 \tag{6.5}$$

This is the equation of a circle with center located at $G = 0.05$, $B = 0$, and with a radius of 0.05 mhos, exactly as shown in Fig. 6.2b. This locus shows all possible values of the admittance function but does not show the magnitude or angle for a particular value of L unless this is specifically indicated. We may calculate and plot these if this is desirable. For the circuit of Fig. 6.1,

$$\hat{Y}_{ac} = \dfrac{1}{\hat{Z}_{ac}} = \dfrac{1}{R_1 + j\omega L} = \dfrac{1}{\sqrt{R_1^2 + \omega^2 L^2}} e^{-j\tan^{-1}(\omega L/R_1)}$$

Since $R_1 = 10\ \Omega$ and $\omega = 2$ radians/sec, we may calculate $\hat{Y}_{ac}$ for various values of L. For $L = 5$ henries,

$$\hat{Y}_{ac} = \dfrac{1}{\sqrt{10^2 + 10^2}} e^{-j\tan^{-1} 1} = 0.0707 e^{-j45°},$$

For $L = 10$ henries,

$$\hat{Y}_{ac} = \dfrac{1}{\sqrt{10^2 + 20^2}} e^{-j\tan^{-1} 2} = 0.045 e^{-j63.5°}.$$

The values of magnitude and phase may be plotted as shown in Fig. 6.3. We shall sometimes prefer this method for showing the variation of a network function with the variation of a network element.

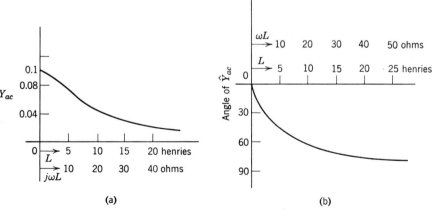

FIG. 6.3 Sketch of magnitude and angle of $\hat{Y}_{ac}$ vs. L for Fig. 6.1. (a) Y_{ac} vs. L or ωL. (b) Angle of $\hat{Y}_{ac}$ vs. L or ωL.

Thus far it appears that the loci of all network functions may be arcs of circles as one element of a network is varied. The straight line must be considered as a degenerate circle of infinite radius. Let us proceed to place another impedance in parallel with the impedance of Fig. 6.1 and determine the locus of the resulting impedance. The circuit is shown in Fig. 6.4a. Since admittances in parallel add algebraically, we calculate $\hat{Y}_2 = 1/\hat{Z}_2 = 1/(4 - j8) = 0.05 + j0.10$ mhos. This is shown in Fig. 6.4b as well as the sketch of $\hat{Y}_1 = \hat{Y}_{ac}$ which is the same as that obtained before. The locus of $\hat{Y}_{a'c'} = \hat{Y}_1 + \hat{Y}_2$ is shown directly in Fig. 6.4c. We now need to obtain the locus of the impedance $\hat{Z}_{a'c'}$ or $1/\hat{Y}_{a'c'}$, which is the reciprocal of $\hat{Y}_{a'c'}$. This means that a particular point on the $\hat{Y}_{a'c'}$ locus will become a point on the $\hat{Z}_{a'c'}$ locus with a magnitude which is the reciprocal of the magnitude of its admittance, and an angle which is the negative of the admittance angle. For convenience we label the points d, e, f, g, and h on the admittance locus. The point d with

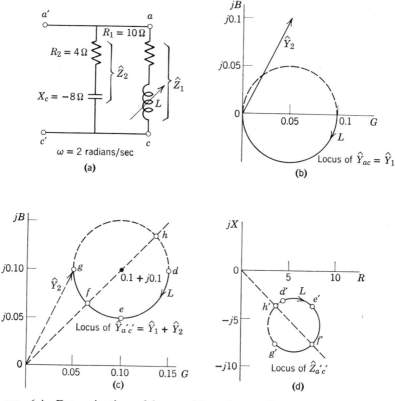

FIG. 6.4 Determination of locus of impedance of two parallel branches with one variable element. (a) Circuit. (b) Locus of $\hat{Y}_2$ and $\hat{Y}_1$. (c) Locus of $\hat{Y}_1 + \hat{Y}_2 = \hat{Y}_{a'c'}$. (d) Locus of $\hat{Z}_{a'c'}$.

coordinates $0.15 + j0.10$ will become d' with the coordinates

$$\frac{1}{0.15 + j0.10} = 4.62 - j3.08.$$

The point e having the coordinates $0.10 + j0.05$ becomes c' at $8 - j4$. The points f and h lie on a line passing through the center of the circle of admittance. This center is located at $0.141e^{j45°}$ and, since the radius of the circle is 0.05, the point f' will be at $1/(0.091e^{j45°}) = 11e^{-j45°}$ and the point h' will be at $1/(0.191e^{j45°}) = 5.25e^{-j45°}$. The point g corresponds to the value of $\hat{Y}_2$ and thus g' lies at $\hat{Z}_2 = 4 - j8$. The resulting impedance locus, sketched in Fig. 6.4d, appears to lie on the arc of a circle. In fact it does and we shall show this mathematically in the following article.

6.1.2 Proof that the Reciprocal of a Function Having a Circular Locus also has a Circular Locus

"If a function has a circular locus in the complex plane, the reciprocal of this function also has a circular locus."

Proof. Let the two complex planes be the $\hat{z}$ plane with coordinates x and jy and the $\hat{w}$ plane with coordinates u and jv as shown in Fig. 6.5. In the $\hat{z}$ plane the equation of the circle is

$$(x - a)^2 + (y - b)^2 = r^2. \tag{6.6}$$

Since we define $\hat{w} = 1/\hat{z}$, or $\hat{z} = 1/\hat{w}$,

$$x + jy = \frac{1}{u + jv} = \frac{u - jv}{u^2 + v^2} = \frac{u}{u^2 + v^2} - j\frac{v}{u^2 + v^2}. \tag{6.7}$$

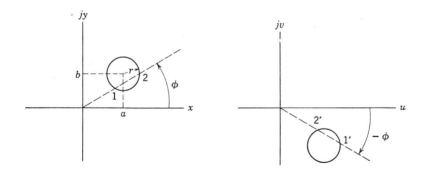

$\hat{z}$ Plane $\hat{w}$ Plane

FIG. 6.5. Circles in complex planes ($\hat{z} = 1/\hat{w}$).

From Equation 6.7 we may equate reals and imaginaries to give

$$x = \frac{u}{u^2 + v^2},$$
and
$$y = \frac{-v}{u^2 + v^2}.$$
(6.8)

Substitution of Equation 6.8 into Equation 6.6 will give the locus in the $\hat{w}$ plane of the circle in the $\hat{z}$ plane.

$$\left(\frac{u}{u^2 + v^2} - a\right)^2 + \left(\frac{-v}{u^2 + v^2} - b\right)^2 = r^2.$$
(6.9)

Simplification of Equation 6.9 gives

$$u^2 - 2ua(u^2 + v^2) + v^2 + 2vb(u^2 + v^2)$$
$$+ (a^2 + b^2 - r^2)(u^2 + v^2)^2 = 0,$$

or

$$(u^2 + v^2)[1 - 2ua + 2vb + (a^2 + b^2 - r^2)(u^2 + v^2)] = 0.$$

Ignoring the trivial solution, $u^2 + v^2 = 0$, we obtain

$$u^2 + v^2 - \frac{2ua}{a^2 + b^2 - r^2} + \frac{2vb}{a^2 + b^2 - r^2} + \frac{1}{a^2 + b^2 - r^2} = 0,$$

or

$$\left(u - \frac{a}{a^2 + b^2 - r^2}\right)^2 + \left(v + \frac{b}{a^2 + b^2 - r^2}\right)^2 = \frac{r^2}{(a^2 + b^2 - r^2)^2}.$$
(6.10)

This is the equation of a circle in the $\hat{w}$ plane and thus proves the theorem.[1]

It should be noted that a point in the $\hat{z}$ plane with an angle ϕ becomes a point in the $\hat{w}$ plane with an angle of $-\phi$. Thus the points 1 and 2 in the $\hat{z}$ plane become 1' and 2' in the $\hat{w}$ plane, respectively.

The straight line may be considered the special case of a circle having infinite radius. For the special case of a line parallel to the y axis, $x - c = 0$, we use $x = u/(u^2 + v^2)$ to obtain,

$$\frac{u}{u^2 + v^2} - c = 0, \quad \text{or} \quad u^2 - \frac{u}{c} + v^2 = 0.$$

By completing the square we get

$$\left(u - \frac{1}{2c}\right)^2 + v^2 = \left(\frac{1}{2c}\right)^2.$$
(6.11)

---

[1] One of our students, N. Stillwell, has pointed out that if a scale change of $1/(a^2 + b^2 - r^2)$ exists between the $\hat{z}$ and $\hat{w}$ planes, the circles will have the same **radius** and will be located symmetrically with respect to the axis of reals.

This is the equation of a circle with center located at $u = 1/2c$, $v = 0$, and with a radius of $1/2c$. Thus this circle will pass through the origin, which we should expect of the reciprocal of all straight lines since the reciprocal of an infinitely large number is zero.

For the straight line parallel to the x axis, $y - d = 0$, a similar development will show

$$u^2 + \left(v + \frac{1}{2d}\right)^2 = \left(\frac{1}{2d}\right)^2. \tag{6.12}$$

Example 6.1

GIVEN: The circuit in Fig. E-1.1.

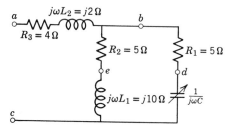

FIG. E-1.1

TO FIND: The locus of the impedance $\hat{Z}_{ac}$ with C varying from 0 to ∞.

SOLUTION: It is best to start with the impedance locus of the branch containing the variable element. The locus of $\hat{Z}_{bdc}$ is shown in Fig. E-1.2. The reciprocal of $\hat{Z}_{bdc}$, $\hat{Y}_{bdc}$ is shown in Fig. E-1.3.

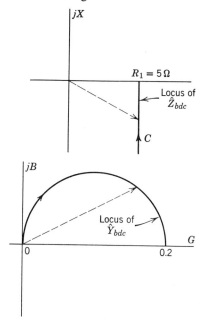

FIG. E-1.2

FIG. E-1.3

Since $\hat{Y}_{bc} = \hat{Y}_{bec} + \hat{Y}_{bdc}$, and since

$\hat{Y}_{bec} = 1/(5 + j10) = (5 - j10)/125 = 0.04 - j.08$,

the locus of $\hat{Y}_{bc}$ may be shown graphically in Fig. E-1.4. Note that for these two parallel branches there are two points, h and j, at which the admittance is real. Also note that the condition for minimum admittance is shown by the point, g, at which a line drawn from the origin to the center of the semicircle crosses the locus of $\hat{Y}_{bc}$.

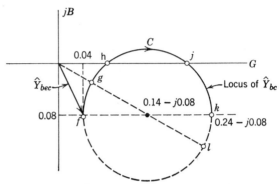

FIG. E-1.4

The locus of the impedance, $\hat{Z}_{bc}$, may be found by taking the reciprocal of points in the locus of $\hat{Y}_{bc}$. f' and k' of Fig. E-1.5 may readily be calculated. $f' = 1/(0.04 - j0.08) = 5 + j10$ ohms, and $k' = 1/(0.24 - j0.08) = 3.75 + j1.25$ ohms. The center of the circle in the $\hat{Z}$ plane **cannot** be found by taking the reciprocal of the center in the $\hat{Y}$ plane since these points are not on the loci. The center must lie halfway between the points g' and l' since these represent the maximum and minimum magnitudes of impedance. g' and l' lie on a line through the origin, making an $\tan^{-1} 0.08/0.14$ with the axis of reals. The magnitude of g is $\sqrt{0.14^2 + 0.08^2} - 0.1 = 0.061$ and thus the magnitude of g' is $1/0.061 = 16.4$ ohms. The magnitude of l is 0.261 and of $l' = 1/0.261 = 3.84$. The radius of the circle must be equal to $(16.4 - 3.84)/2 = 6.28$ ohms and its center must be located at $6.28 + 3.84 = 10.12$ ohms from the origin. This establishes the locus of $\hat{Z}_{bc}$. The dashed portions of the circle represent values on the locus if C were negative.

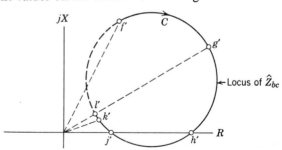

FIG. E-1.5

The center and radius of the circle could have been found from Equation 6.5. Also, if one wished to keep the same dimensions, one could locate the center of the circle in the $\hat{Z}$ plane at the same distance from the origin as in the $\hat{Y}$ plane and draw a circle of the same radius. The scale for the $\hat{Z}$ plane must be $1/(a^2 + b^2 - r^2)$ times that of the $\hat{Y}$ plane. For this example, if one inch represents 0.08 mhos in the $\hat{Y}$ plane, one inch in the $\hat{Z}$ plane must represent

$$\frac{0.08}{0.14^2 + 0.08^2 - 0.1^2} = \frac{0.08}{0.016} = 5 \text{ ohms.}$$

Finally the impedance $\hat{Z}_{ac} = \hat{Z}_{ab} + \hat{Z}_{bc} = 4 + j2 + \hat{Z}_{bc}$. The locus of this impedance is shown in Fig. E-1.6.

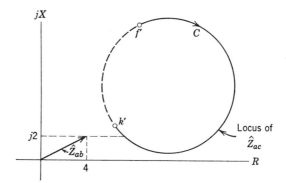

FIG. E-1.6

Example 6.2

GIVEN: The circuit of Fig. E-2.1.

TO FIND: (a) The maximum magnitude of the impedance $\hat{Z}_{ac}$ as R_1 is varied from 0 to ∞.

(b) The value(s) of R_1 for which the phase angle of $\hat{Z}_{ac}$ is zero.

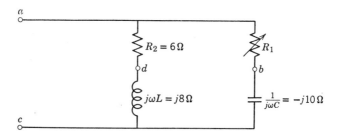

FIG. E-2.1

SOLUTION: (a) The locus of $\hat{Z}_{abc}$ is shown in Fig. E-2.2 and the reciprocal of $\hat{Z}_{abc}$, $\hat{Y}_{abc}$, is shown in Fig. E-2.3. $\hat{Y}_{adc}$ is also sketched in Fig. E-2.3, $\hat{Y}_{adc}$ being $1/(6 + j8) = 0.06 - j0.08$ mhos. The resultant $\hat{Y}_{ac} = \hat{Y}_{abc} + \hat{Y}_{adc}$ is shown in Fig. E-2.4.

FIG. E-2.2

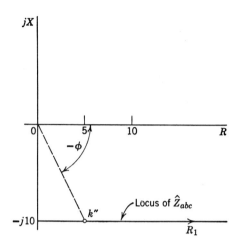

FIG. E-2.3

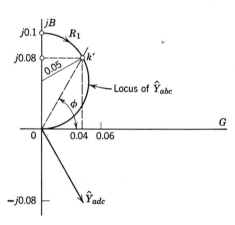

FIG. E-2.4

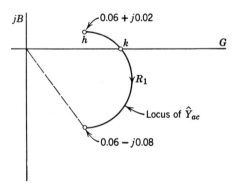

In this particular case the minimum magnitude of admittance is observed to occur at the point h (where $R_1 = 0$), and is equal to $\sqrt{0.06^2 + 0.02^2} = 0.0632$. Then $Z_{ac_{\max}} = 1/0.0632 = 15.9$ ohms.

(b) We now proceed to find the value(s) of R_1 for which the phase angle of $\hat{Z}_{ac}$ is zero. This could be done analytically; however, we shall show how this can be obtained graphically from the sketches we have already made. From Fig. E-2.4 there is only one value of R_1 for which the power factor of $\hat{Z}_{ac}$ is unity. This corresponds to the point k on the locus of $\hat{Y}_{ac}$, and also to the point k' on the locus of $\hat{Y}_{abc}$ in Fig. E-2.3. The vertical projection of k' is 0.08; the horizontal projection can readily be calculated as $\sqrt{(0.05)^2 - (0.03)^2} = 0.04$. This means that $\hat{Y}_{abc} = 0.04 + j0.08 = \sqrt{0.04^2 + 0.08^2}\, e^{j\,\tan^{-1} 2}$. To find R_1 we could now take $\mathrm{Re}(\hat{Z}_{abc})$. However, let us note that we could use Fig. E-2.2. The value of R_1 must lie at k'' which lies at the intersection of the locus of $\hat{Z}_{abc}$ and the line drawn through the origin with a negative angle whose tangent is 2. By inspection $R_1 = 5$ ohms in order that the angle of $\hat{Z}_{ac}$ be zero.

6.1.3 General Proof that Loci of Phasor Network Functions are Circular for a Single Variable Element

We have demonstrated that the loci of certain phasor network functions are arcs of circles as one element in the network is varied. However, we have not proved this for the general case. One approach is to use the general phasor voltage equations for the loop method, similar to Equation 4.68, page 179. We shall be careful to choose the current through the variable element as one of the independent unknowns and for convenience we shall also choose the current through the **single** voltage source as the independent unknown current, I_1. Thus the equations may be written:

$$\begin{aligned}
\hat{Z}_{11}\hat{I}_1 + \hat{Z}_{12}\hat{I}_2 + \hat{Z}_{13}\hat{I}_3 + \ldots \hat{Z}_{1n}\hat{I}_n &= \hat{V}_1 \\
\hat{Z}_{21}\hat{I}_1 + \hat{Z}_{22}\hat{I}_2 + \hat{Z}_{23}\hat{I}_3 + \ldots \hat{Z}_{2n}\hat{I}_n &= 0 \\
\hat{Z}_{31}\hat{I}_1 + \hat{Z}_{32}\hat{I}_2 + \hat{Z}_{33}\hat{I}_3 + \ldots \hat{Z}_{3n}\hat{I}_n &= 0 \\
\vdots \\
\hat{Z}_{n1}\hat{I}_1 + \hat{Z}_{n2}\hat{I}_2 + \hat{Z}_{n3}\hat{I}_3 \ldots \ldots \hat{Z}_{nn}\hat{I}_n &= 0
\end{aligned} \qquad (6.13)$$

To prove that the loci of driving-point immittances will be circular we solve for $\hat{I}_1$ using Cramer's rule.

$$\hat{I}_1 = \frac{\begin{vmatrix} \hat{V}_1 & \hat{Z}_{12} & \hat{Z}_{13} & \hat{Z}_{14} & \cdots & \hat{Z}_{1n} \\ 0 & \hat{Z}_{22} & \hat{Z}_{23} & & \cdots & \hat{Z}_{2n} \\ 0 & \hat{Z}_{32} & \hat{Z}_{33} & & \cdots & \hat{Z}_{3n} \\ \vdots & & & & & \\ 0 & \hat{Z}_{n2} & \hat{Z}_{n3} & & \cdots & \hat{Z}_{nn} \end{vmatrix}}{\Delta} = \frac{\hat{V}_1 \Delta_{11}}{\Delta}, \qquad (6.14)$$

in which Δ is the determinant of the impedance matrix and Δ_{11} is the cofactor with the first row and first column omitted.

The driving point admittance, $\hat{Y}_1$, is

$$\hat{Y}_1 = \frac{\hat{I}_1}{\hat{V}_1} = \frac{\Delta_{11}}{\Delta}. \tag{6.15}$$

Let us consider the variable element to be part of $\hat{Z}_{33}$ so that

$$\hat{Z}_{33} = \hat{Z}_0 + \hat{Z}'_{33}, \tag{6.16}$$

in which $\hat{Z}_0$ may be **one** of the following: R, $j\omega L$, or $1/j\omega C$.

Δ may be evaluated by expanding along its third row,

$$\Delta = \hat{Z}_{31}\Delta_{31} + \hat{Z}_{32}\Delta_{32} + (\hat{Z}_0 + \hat{Z}'_{33})\Delta_{33} + \ldots \hat{Z}_{3n}\Delta_{3n}. \tag{6.17}$$

Similarly Δ_{11} may be evaluated by expanding along the row containing $\hat{Z}_{33}$. Thus

$$\Delta_{11} = \hat{Z}_{32}\Delta_{1132} + (\hat{Z}_0 + \hat{Z}'_{33})\Delta_{1133} + \ldots \hat{Z}_{3n}\Delta_{113n}, \tag{6.18}$$

in which Δ_{1132} is the cofactor of the cofactor Δ_{11} with the original third row and second column omitted. We have been careful to choose our independent currents so that $\hat{Z}_0$ is not part of the cofactors in either Equation 6.17 or Equation 6.18. Thus Δ_{21}, Δ_{32}, Δ_{1133}, etc. are simply complex quantities and we may write Equation 6.15 as

$$\hat{Y}_1 = \frac{\Delta_{11}}{\Delta} = \frac{\hat{A}\hat{Z}_0 + \hat{B}}{\hat{C}\hat{Z}_0 + \hat{D}} \tag{6.19}$$

in which $\hat{A}$, $\hat{B}$, $\hat{C}$, and $\hat{D}$ are complex quantities which do not involve $\hat{Z}_0$. Equation 6.19 is known as a "bilinear transformation," which is the most general form for which one and only one value of $\hat{Y}_1$ corresponds to each value of $\hat{Z}_0$ and conversely. Equation 6.19 may be written

$$\hat{Y}_1 = \frac{\hat{A}\hat{Z}_0 + \hat{B}}{\hat{C}\hat{Z}_0 + \hat{D}} = \frac{\hat{A}}{\hat{C}} + \frac{\hat{B} - \hat{A}\hat{D}/\hat{C}}{\hat{C}\hat{Z}_0 + \hat{D}}$$

$$= \frac{\hat{A}}{\hat{C}} + \frac{(\hat{B}\hat{C} - \hat{A}\hat{D})/\hat{C}^2}{\hat{Z}_0 + \hat{D}/\hat{C}}. \tag{6.20}$$

The locus of the quantity $(\hat{Z}_0 + \hat{D}/\hat{C})$ will be a straight line since $\hat{Z}_0$ is limited to one of the quantities: R, $j\omega L$, or $1/j\omega C$, in which R, L, or C may vary from 0 to ∞. The locus of $1/[\hat{Z}_0 + (\hat{D}/\hat{C})]$ will therefore be circular. The quantity $(\hat{B}\hat{C} - \hat{A}\hat{D})/\hat{C}^2$ is a complex number having a magnitude and angle; the effect of multiplying a circular locus by some magnitude is simply to change the scale; the effect of multiplying a circular locus by $e^{j\phi}$ is simply to rotate the locus through the angle ϕ. The effect of adding a complex number, $\hat{A}/\hat{C}$, to a circular locus is simply to shift the location of the locus. Thus the driving point admittance, $\hat{Y}_1$ will always have the locus of a circle, a straight line being the special case of a circle of infinite radius. Since the driving point impedance is the reciprocal of the driving point admittance, its locus will also be circular.

With regard to the locus of transfer functions, we may use Equation 6.13 to determine the current $I_k(k \neq 1)$; and thus obtain a transfer admittance

$$\hat{Y}_{k1} = \frac{\hat{I}_k}{\hat{V}_1} = \frac{\Delta_{1k}}{\Delta}, \tag{6.21}$$

in which the form of Δ_{1k} is similar to that of Δ_{11} of our illustration except that Δ_{1k} is not a function of $\hat{Z}_0$ when $k = 3$. This means that $\hat{Y}_{k1}$ can also be expressed as a bilinear function of $\hat{Z}_0$ and that its locus will be circular. A transfer function may be a gain function corresponding to the ratio of a "response" voltage to the driving function $\hat{V}_1$. This corresponds simply to some impedance times the transfer admittance, and thus the locus of such a transfer function will also be circular.

Some transfer functions are based on having an independent current source instead of an independent voltage source, and consist of either the ratio of some voltage response to this current source or the ratio of some current response to this current source. The loci of these transfer functions will also be circular. We may deduce this by changing our voltage source to an equivalent current source; or we may prove this independently by starting with the general current equations for the node voltage method (see Equation 4.58).

Hence the locus of any network function is circular if one element in the network is variable.

Example 6.3

GIVEN: The circuit of Fig. E-3.1.

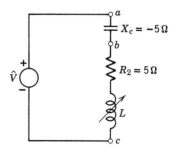

FIG. E-3.1

TO FIND: The locus of the transfer voltage $\hat{V}_{bc}/\hat{V}_{ac}$.

SOLUTION: By voltage ratio,

$$\frac{\hat{V}_{bc}}{\hat{V}_{ac}} = \frac{j\omega L + R_2}{j\omega L + R_2 + \dfrac{1}{j\omega C}}$$

$$= \frac{j\omega L + 5}{j\omega L + 5 - j5}$$

$$= 1 + \frac{j5}{j\omega L + 5 - j5} \tag{1}$$

The locus of $j\omega L + 5 - j5$ is shown in Fig. E-3.2, and the locus of its reciprocal is shown in Fig. E-3.3.

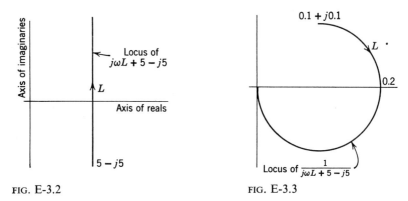

FIG. E-3.2

FIG. E-3.3

Multiplication by j is equivalent to rotating all points on the locus 90° in a counterclockwise sense. Multiplication by 5 changes the magnitude

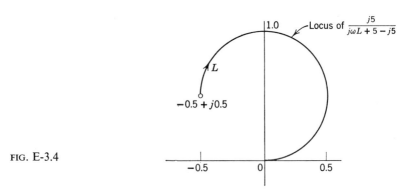

FIG. E-3.4

of each point by the factor of 5. The locus of $j5/(j\omega L + 5 - j5)$ is then shown in Fig. E-3.4, and the final locus of $\hat{V}_{bc}/\hat{V}_{ac}$ is shown in Fig. E-3.5.

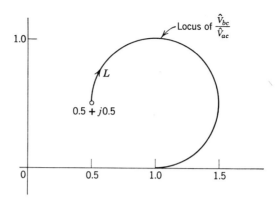

FIG. E-3.5

6.2 FREQUENCY CHARACTERISTICS OF NETWORK FUNCTIONS

We have found that the locus of a phasor network function will be circular if one element of the network is varied. What is the effect on the network function if the frequency is varied? We would expect that this variation would be more complex since the immittance of each inductance and capacitance is affected by frequency, and that the locus would not be circular except in the simplest cases. We would also expect the variation to be important physically since we usually require networks to operate over a range of frequencies.

It is possible to use the approach of the previous section in which the network function was written as a function of angular frequency ω; then the magnitude and phase of the network function could be evaluated for various values of ω. This direct approach is really not as simple as a more sophisticated approach in which we first write the transform network function as the ratio of two factored polynomials and then obtain the phasor network function by letting $s = j\omega$. This latter method permits a graphical display of the roots of the polynomials such that one can obtain a clear picture of the variation of the network function with frequency. This method also shows some of the relationships between the transform domain, the phasor domain and the time domain.

6.2.1 Poles and Zeros of Network Functions

In the transform domain our passive elements are expressed in the form of Ls, R, $1/Cs$, Ms, or their reciprocals. If we combine these terms in any arrangement which results in a network function, we find that this combination can always be written as the ratio of two polynomials. For an illustration the driving point impedance of a series *RLC* circuit is

$$Z(s) = Ls + R + \frac{1}{Cs} = \frac{LCs^2 + RCs + 1}{Cs}$$

$$= L\left(\frac{s^2 + \frac{R}{L}s + \frac{1}{LC}}{s}\right) \qquad (6.22)$$

A particular transfer function, which is the ratio of the voltage across the RL portion to the total voltage, is

$$T(s) = \frac{Ls + R}{Ls + R + 1/Cs} = \frac{Cs(Ls + R)}{LCs^2 + RCs + 1}$$

$$= \frac{s(s + R/L)}{s^2 + (R/L)s + 1/LC}. \qquad (6.23)$$

Both of these functions are ratios of two polynomials in s. As a matter of fact any network function for a linear network can be so expressed. One way of proving this would be to start with the general node or loop method and show that any network function is an algebraic term times the ratio of two determinants the terms of which are algebraic. We shall not go through the proof but shall start with the proposition that any network function may be written as

$$H(s) = A\frac{s^n + a_{n-1}s^{n-1} + \ldots + a_1s + a_0}{s^m + b_{m-1}s^{m-1} + \ldots + b_1s + b_0}. \qquad (6.24)$$

If the numerator polynomial is factored into its n roots and the denominator polynomial is factored into its m roots, Equation 6.24 may be written

$$H(s) = A\frac{(s - s_1)(s - s_2) \ldots (s - s_n)}{(s - s_a)(s - s_b) \ldots (s - s_m)}. \qquad (6.25)$$

Equation 6.24 is known as the "standard form" for writing the ratio of two polynomials, the coefficient of the highest degree term being made unity in both numerator and denominator. Equation 6.25 is known as the "standard factored form" for writing the ratio of two polynomials. A is a constant known as the scale factor. The roots s_1, $s_2, s_3 \ldots s_n$ are known as the "zeros" of the network function since the function is zero for these values of s. The roots $s_a, s_b, s_c \ldots s_m$ are known as the "poles" of the network function since the function is infinite for these values of s. It is interesting to observe that a network function is completely specified by its poles, zeros, and scale factor.

In addition to the finite poles and zeros the network function may have poles or zeros at $s = \infty$. If $n > m$ in Equation 6.24, the network function has a pole of order $(n - m)$ at $s = \infty$. If $n < m$ in Equation

6.24, the network function has a zero of order $(m - n)$ at $s = \infty$. Thus if we include the poles or zeros at $s = \infty$, the total number of poles will always equal the total number of zeros for a network function. However when reference is made to the poles and zeros of a function, it is understood that these mean only the finite poles and zeros.

Our objectives are to obtain two characteristic curves of network functions; one is a curve of magnitude versus angular frequency, ω; the other is a curve of phase angle versus ω. This is done as follows: locate the poles and zeros in a diagram of the s plane, then let $s = j\omega$ for the network function and evaluate magnitude and angle as ω varies from 0 to ∞ by use of the pole-zero diagram and the scale factor. Let us consider first a very simple network with a single pole;

$$H(s) = \frac{A}{s - s_a}. \tag{6.26}$$

A pole-zero diagram is shown in Fig. 6.6a, the single pole, s_a, being represented by the cross, $\times$. The phasor network function, $H(j\omega) = \hat{H}$

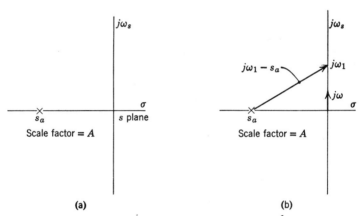

(a)   (b)

FIG. 6.6 Pole-zero diagram for $H(s) = A/(s - s_a)$ and for $\hat{H}$. (a) Pole-zero diagram for $H(s) = A/(s - s_a)$. (b) Diagram for evaluating $\hat{H}$ as function of ω.

may be obtained directly from $H(s)$, as shown in Section 5.9.

$$H(j\omega) = \hat{H} = H(s)\bigg|_{s=j\omega} = \frac{A}{s - s_a}\bigg|_{s=j\omega} = \frac{A}{j\omega - s_a} \tag{6.27}$$

When we set $s = j\omega$, we are setting $\sigma = 0$, and $\omega_s = \omega$. Since ω is physically a positive real quantity we are restricting s to the upper portion of the imaginary axis. As ω is varied from zero to infinity we can picture s of the transform function as varying from the origin upward along the ω_s axis. For a particular value of ω, such as ω_1, the factor, $j\omega_1 - s_a$, may be sketched as a directed line segment as shown in

Fig. 6.6b. The fact that the difference of two complex quantities may be shown graphically was demonstrated in Chapter 5 but is shown again in Fig. 6.7. The term $\hat{A} - \hat{B}$ is a directed line segment drawn from $\hat{B}$ to $\hat{A}$. In the same manner, the factor, $(j\omega_1 - s_a)$, is a directed line segment drawn from s_a to $j\omega_1$.

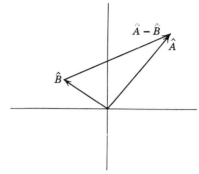

FIG. 6.7 Graphical difference of two complex numbers.

For our particular $\hat{H}$ as represented in Fig. 6.6b, the factor $(j\omega - s_a)$ has a magnitude of $|s_a|$ and an angle of $0°$ at $\omega = 0$. As ω increases, the magnitude and the angle both increase such that at $\omega = |s_a|$, the magnitude has increased by a factor of $\sqrt{2}$ and the angle has increased

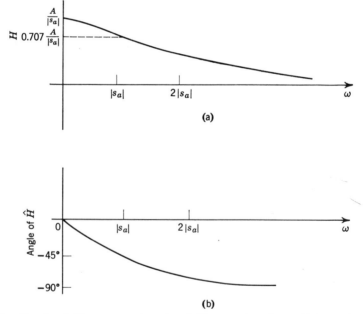

FIG. 6.8 Sketch of H vs. ω and angle of $\hat{H}$ vs. ω for $\hat{H} = A/(j\omega - s_a)$. (a) H vs. ω. (b) Angle of $\hat{H}$ vs. ω.

to 45°. As ω increases toward infinity the magnitude of $(j\omega - s_a)$ also increases indefinitely and the angle approaches 90°. Remembering that the factor $(j\omega - s_a)$ is in the denominator and that a scale factor of A is present, we can sketch H, the magnitude of $\hat{H}$, versus ω and the angle of $\hat{H}$ versus ω. This is shown in Fig. 6.8.

Let us demonstrate with a more complex function. Let

$$H(s) = A \frac{(s - s_1)}{(s - s_a)(s - s_b)}, \tag{6.28}$$

in which $A = 10$, $s_1 = 0$, $s_a = -2 + j2$ and $s_b = -2 - j2$. Recall from algebra that, if the roots of a polynomial are complex, they occur in pairs as complex conjugates.

The pole-zero diagram of $H(s)$ is shown in Fig. 6.9a, the factors for determining $\hat{H}$ at $\omega = \omega_1$ are shown in Fig. 6.9b, the magnitude versus frequency characteristic for $\hat{H}$ is shown in Fig. 6.9c, and the phase

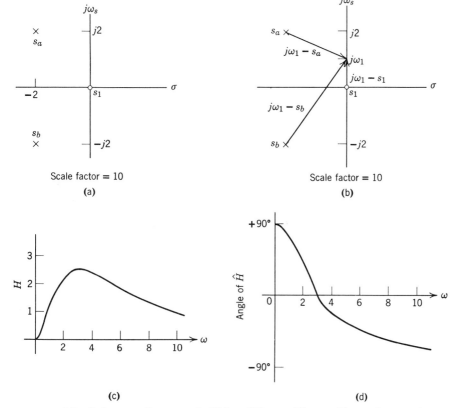

FIG. 6.9 Pole-zero diagram of $H(s) = 10(s - s_1)/(s - s_a)(s - s_b)$, magnitude and phase characteristics. (a) Pole-zero diagram. (b) Factors for $\omega = \omega_1$. (c) Magnitude characteristic. (d) Phase characteristic.

versus frequency characteristic is shown in Fig. 6.9d. The zero at $s_1 = 0$ is indicated by a circle; since this zero lies on the imaginary axis it means that the magnitude of the function is zero at $\omega = 0$. In evaluating magnitude and phase for various values of ω, we need to keep in mind that

$$H = H(j\omega) = H(s)\Big|_{s=j\omega} = A\frac{j\omega - s_1}{(j\omega - s_a)(j\omega - s_b)} \tag{6.29}$$

Each factor, $(j\omega - s_n)$, has a magnitude and phase angle and may be considered a phasor. Thus for any particular value of ω the magnitude of $\hat{H}$ is the scale factor times the products of the magnitudes of all the phasors in the numerator divided by the products of the magnitudes of all the phasors in the denominator. The phase angle of the function $\hat{H}$ is the sum of the angles of all the phasors in the numerator minus the sum of all the angles of the phasors in the denominator. In using the pole-zero diagram, the phasors from the zeros are associated with the numerator, the phasors from the poles are associated with the denominator.

With regard to the variation of the magnitude of $\hat{H}$ as ω varies from 0 to ∞, the magnitude must start at zero, have a maximum near $\omega = 2$ or 3 (since this is close to a pole), and become zero at $\omega = \infty$. As ω becomes very large the distance to a finite pole is essentially the same as the distance to a finite zero; in this case the number of poles exceeds the number of zeros, so the magnitude becomes zero as ω approaches ∞. With regard to the variation of the angle of $\hat{H}$ as ω varies from 0 to ∞, at $\omega = 0+$ the angle in the numerator is $+90°$, the net angle in the denominator is zero, and thus the angle of $\hat{H}$ is $+90°$. The angle of H at $\omega = \infty$ is $-90°$ since the number of poles exceed the number of zeros by one. The angle of $\hat{H}$ must be zero somewhere near $\omega = 3$ since it is here that the net angle in the denominator will equal the $90°$ of the numerator. We calculate $\hat{H}$ at $\omega = 2$ using Fig. 6.9b to guide us.

$$\hat{H} = A\frac{(j\omega - s_1)}{(j\omega - s_a)(j\omega - s_b)}\Big|_{\omega=2} = 10\frac{2e^{j90°}}{(2e^{j0°})(\sqrt{2^2 + 4^2}\,e^{j\tan^{-1}2})}$$

$$= 2.23e^{j26.5°}.$$

We also calculate $\hat{H}$ at $\omega = 3$.

$$\hat{H} = 10\frac{(3e^{j90°})}{(\sqrt{2^2 + 1^2}\,e^{j\tan^{-1}\frac{1}{2}})(\sqrt{2^2 + 5^2}\,e^{j\tan^{-1}\frac{5}{2}})}$$

$$= 2.5\frac{e^{j90°}}{e^{j26.5°}e^{j68.2°}} = 2.5e^{-j4.7°}$$

The sketches of Figs. 6.9c and 6.9d are obtained with a minimum of calculation. This is usually adequate. If one wants detailed data at various frequencies one can perform additional calculations.

Example 6.4

GIVEN: The circuit of Fig. E-4.1 with $R = 3$ ohms, $C = \frac{1}{12} f$.

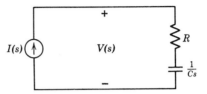

FIG. E-4.1

TO FIND: The pole-zero diagram for $Z(s) = [V(s)]/[I(s)]$ and the magnitude and phase characteristics as a function of frequency for $\hat{Z}$.

SOLUTION: The transform driving point impedance is first put in factored form.

$$Z(s) = R + \frac{1}{Cs} = \frac{RCs + 1}{Cs} = R\left(\frac{s + 1/RC}{s}\right). \tag{1}$$

Substitution of numbers results in

$$Z(s) = 3\left(\frac{s + 4}{s}\right). \tag{2}$$

$Z(s)$ has a scale factor of 3, a zero at $s = -4$ and a pole at $s = 0$. A pole-zero diagram is shown in Fig. E-4.2 and from this diagram

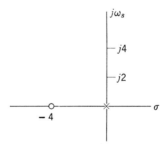

FIG. E-4.2

sketches of magnitude and phase of $\hat{Z}$ versus frequency are made and shown in Figs. E-4.3 and E-4.4.

$$\hat{Z} = Z(s)\bigg|_{s=j\omega} = 3\left(\frac{j\omega + 4}{j\omega}\right)$$

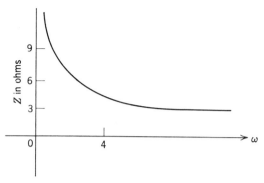

FIG. E-4.3

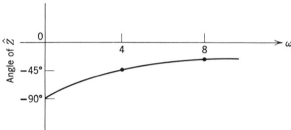

FIG. E-4.4

The magnitude of $\hat{Z}$ is infinite at $\omega = 0$ because of the pole at this location, and approaches the scale factor, 3, as ω approaches infinity. At $\omega = 4$ the numerator has a magnitude equal to $\sqrt{2}$ times the denominator or $Z = 3\sqrt{2} = 4.24$ ohms. The phase of $\hat{Z}$ is $-90°$ at $\omega = 0+$ and then decreases to zero as ω approaches infinity. At $\omega = 4$, the angle in the numerator is $45°$; since the angle in the denominator is a constant of $90°$, the net angle is $-45°$.

Example 6.5

GIVEN: The circuit of Fig. E-5.1 with $L = 0.5$ henries, $C = \frac{1}{8}$ farad.
TO FIND: The pole-zero diagram for $Y(s) = I(s)/V(s)$, and the magnitude and phase characteristics for $\hat{Y}$ as a function of frequency.

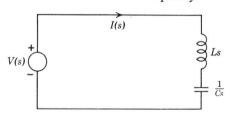

FIG. E-5.1

SOLUTION: The transform driving point admittance is

$$Y(s) = \frac{1}{Ls + 1/Cs} = \frac{Cs}{LCs^2 + 1} = \frac{1}{L}\left(\frac{s}{s^2 + 1/LC}\right). \tag{1}$$

The substitution of numbers into Equation 1 gives

$$Y(s) = 2\left(\frac{s}{s^2 + 16}\right) = 2\frac{s}{(s + j4)(s - j4)}. \tag{2}$$

The pole-zero diagram is shown in Fig. E-5.2. The poles in this case are a conjugate pair on the imaginary axis. The magnitude and phase

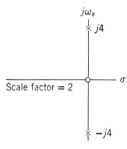

FIG. E-5.2

characteristics of $\hat{Y}$ are shown in Figs. E-5.3 and E-5.4 respectively. The discontinuity of 180° in the phase angle is caused by the pole on the imaginary axis.

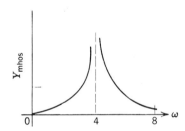

FIG. E-5.3

This is a purely reactive circuit and thus it is possible to combine the magnitude and phase characteristics in one curve. Since $\hat{Y} = G + jB$

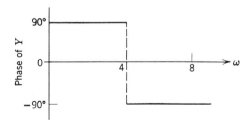

FIG. E-5.4

and $G = 0$, a sketch of B versus ω (see Fig. E-5.5) gives all the information contained in the previous two figures.

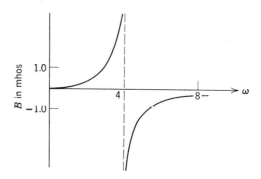

FIG. E-5.5

Example 6.6

GIVEN: The circuit of Fig. E-6.1 with $R_1 = R_2 = 1$ ohm, $C = 0.5 f$, and $L = \frac{1}{8}$ henries.

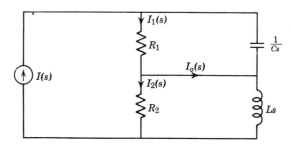

FIG. E-6.1

TO FIND: Magnitude and phase characteristics of the transfer function, $\hat{I}_0/\hat{I}$.

SOLUTION: $I_0(s) = I_1(s) - I_2(s)$. (1)

Since by current ratio,

$$I_1(s) = \left(\frac{1/Cs}{R_1 + 1/Cs}\right) I(s), \tag{2}$$

and

$$I_2(s) = \left(\frac{Ls}{Ls + R_2}\right) I(s), \tag{3}$$

$$I_0(s) = \left(\frac{1}{R_1 Cs + 1} - \frac{Ls}{Ls + R_2}\right) I(s),$$

or

$$T(s) = \frac{I_0(s)}{I(s)} = \frac{R_2 - LR_1 Cs^2}{(R_1 Cs + 1)(Ls + R_2)}$$

$$= -\frac{s^2 - R_2/LR_1 C}{(s + 1/R_1 C)(s + R_2/L)}. \tag{4}$$

Substitution of numbers results in

$$T(s) = \frac{I_0(s)}{I(s)} = -\frac{s^2 - 16}{(s+2)(s+8)}$$
$$= -\frac{(s-4)(s+4)}{(s+2)(s+8)}. \tag{5}$$

The scale factor is -1. This is treated as $e^{j180°}$ as far as evaluating the angle of $\hat{T}$ is concerned. A pole-zero diagram is shown in Fig. E-6.2. In this particular case the two zeros are symmetrically located with respect to the imaginary axis, so there is no net phase angle in the numerator. The magnitude and phase characteristics are shown in Figs. E-6.3 and E-6.4.

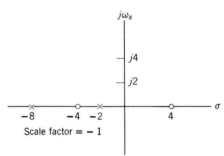

FIG. E-6.2

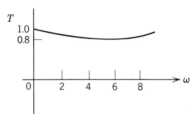

FIG. E-6.3

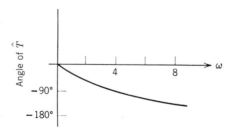

FIG. E-6.4

6.2.2 Relation of Poles of Network Functions to Free Response in Time Domain (Restrictions on Pole and Zero Locations)

In the previous examples all poles of the network functions appeared either in the left half of the s plane or on the imaginary axis; this was also true of the zeros of driving-point functions. There are good reasons for this. Let us first consider the transfer function which is always the ratio of a response function to a driving function.

$$T(s) = \frac{R(s)}{D(s)} \tag{6.30}$$

Then

$$R(s) = T(s)D(s), \tag{6.31}$$

and

$$r(t) = \mathscr{L}^{-1}R(s) \tag{6.32}$$

In considering the expansion of Equation 6.31 by partial fractions prior to obtaining the inverse Laplace transform, it is clear that the poles of $T(s)$ correspond to the poles of the free response terms and that the poles of $D(s)$ correspond to the poles of the forced or steady state response terms. Thus a factor of $(s - s_a)$ in the denominator of $T(s)$ will become the term, $A/(s - s_a)$, by partial fraction expansion and this will transform to $Ae^{s_a t}$ in the time domain. If our networks have a resistance element it is necessary that the free response go to zero as time increases and thus s_a must have a negative real term, thus forcing the poles to be in the left half plane. It is possible to have multiple roots since $(s - s_a)^2$ in the denominator of the transfer function means a free response of $A_1 e^{s_a t} + A_2 t e^{s_a t}$, which goes to zero as time increases without limit if the real part of s_a is negative. If our network has only reactive elements, as illustrated in Example 6.5, the poles may lie on the imaginary axis. In this case only simple roots are permitted since in the time domain double roots would mean terms of the form, $A_1 \cos \omega t + A_2 t \cos \omega t$, which are not possible for the free response of a passive network.

As for driving point immittances, we may treat these as either the ratio of $R(s)/D(s)$ or the reciprocal of this ratio. Because of the reciprocal feature of the definition, the zeros of driving-point immittances have the same restrictions of being either in the left half of the s plane or on the imaginary axis.

6.2.3 Resonance

If the magnitude and phase angle of a network function vary markedly over a limited range of frequencies as ω is varied from 0 to ∞, the phenomenon is called resonance and the frequency about which this effect is centered is said to be the resonant frequency. This will occur if there is an isolated pair of complex conjugate roots for which the real part is small in comparison with the imaginary part. "Isolated" in this case means that the distance between these roots and other roots of the function is large in comparison to the distance between these roots and the imaginary axis.

Figure 6.10 demonstrates this situation for an admittance function which shows resonance near the pole s_a. Such poles are the roots of a quadratic which may be written

$$s^2 + Bs + \omega_0^2 = 0, \tag{6.33}$$

in which B and ω_0 have the dimensions of frequency, and $\omega_0 > B/2$. The roots are

$$s_a = -\frac{B}{2} + j\sqrt{\omega_0^2 - (B/2)^2},$$

and

$$s_b = -\frac{B}{2} - j\sqrt{\omega_0^2 - (B/2)^2}.$$

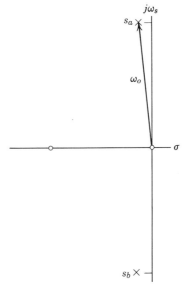

FIG. 6.10 Pole-zero diagram for admittance function which shows resonance.

An expanded view of the region near s_a is shown in Fig. 6.11. If the variable frequency ω is equal to $\sqrt{\omega_0^2 - (B/2)^2}$, the factor $(j\omega - s_a)$ will have minimum magnitude and zero angle. For values of ω which differ from $\sqrt{\omega_0^2 - (B/2)^2}$ by the frequency of $B/2$, the factor, $(j\omega - s_a)$, will be larger in magnitude by $\sqrt{2}$ and will change in angle by 45°. If the distance to other poles or zeros is large in comparison to $B/2$, we are justified in ignoring the change in other factors of the network function. As an approximation, we assume that the admittance function of Fig. 6.10 has its maximum value at ω_0, the resonant frequency. With a voltage source of fixed magnitude, the current and the average power also have their maximum values at this frequency. For a change in frequency of

$B/2$, the current has decreased by a factor of $1/\sqrt{2}$, and the average power by a factor $1/2$. The frequency difference between the two half-power points is known as the "bandwidth." The bandwidth in this case is B. The ratio of the resonant frequency to the bandwidth, ω_0/B, is a measure of the quality of the resonant circuit and is given the symbol, Q. The student will learn more about Q in his electronics courses; we shall not develop this subject.

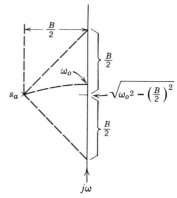

FIG. 6.11 Expanded view near pole, s_a, of Fig. 6.10.

It is interesting to note that the magnitude of s_a or s_b is not a function of B. Since

$$s_a = -\frac{B}{2} + j\sqrt{\omega_0^2 - (B/2)^2}, \quad |s_a| = \omega_0,$$

and therefore if B is varied, with ω_0 held constant, the locus of s_a is a circle with radius ω_0.

Example 6.7

GIVEN: The series RLC circuit of Fig. E-7.1 with $R = 2$ ohms, $L = 1$ henry and $C = 1/101$ farads.

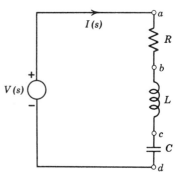

FIG. E-7.1

314

TO FIND: (a) The magnitude and phase characteristics of the admittance, $\hat{Y} = \hat{I}/\hat{V}$; and (b) a sketch of the locus of the poles of $Y(s)$ as R is varied.

SOLUTION: (a) The expression for the admittance, $Y(s)$, is

$$Y(s) = \frac{I(s)}{V(s)} = \frac{1}{Ls + R + 1/Cs} = \frac{1}{L}\left(\frac{s}{s^2 + (R/L)s + 1/LC}\right). \quad (1)$$

Substitution of numbers gives

$$Y(s) = \frac{s}{s^2 + 2s + 101} = \frac{s}{(s + 1 - j10)(s + 1 + j10)}. \quad (2)$$

A pole-zero diagram is shown in Fig. E-7.2. A sketch of the magnitude characteristic is shown in Fig. E-7.3, and a sketch of the phase characteristic is shown in Fig. E-7.4.

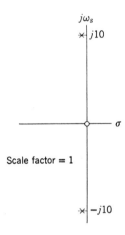

FIG. E-7.2

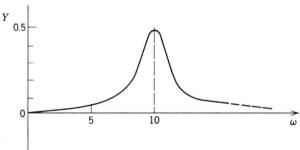

FIG. E-7.3

For this circuit the bandwidth, B, is 2 radians per second while the resonant frequency, $\omega_0 = \sqrt{101} = 10.05$ radians per second. This means the Q of the circuit is $\omega_0/B \cong 5$. The free response frequency, $\omega = 10$, differs slightly from the resonant frequency.

Phasor Network Functions for a Variable Element or for Variable Frequency 315

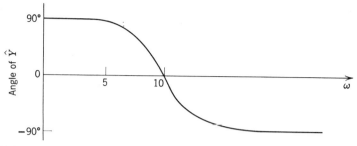

FIG. E-7.4

By inspection of the circuit the maximum admittance occurs when $\omega L = 1/\omega C$ or $\omega^2 = 1/LC = 101$. In this case the resonant frequency is the same as that which gives unity power factor or maximum magnitude of admittance. If we had been asked for the transfer function, $V_{cd}(s)/V(s)$, this would be

$$T(s) = \frac{V_{cd}(s)}{V(s)} = \frac{1/Cs}{Ls + R + 1/Cs} = \frac{1}{LC}\left(\frac{1}{s^2 + (R/L)s + 1/LC}\right). \quad (3)$$

This transfer function does not have its maximum value at the resonant frequency, but at a slightly lower frequency.

(b) Since

$$Y(s) = \frac{1}{L}\left[\frac{s}{s^2 + (R/L)s + 1/LC}\right]$$

the roots of the denominator are

$$s_a = -\frac{R}{2L} + j\sqrt{\frac{1}{LC} - \left(\frac{R}{2L}\right)^2}$$

and

$$s_b = -\frac{R}{2L} - j\sqrt{\frac{1}{LC} - \left(\frac{R}{2L}\right)^2}.$$

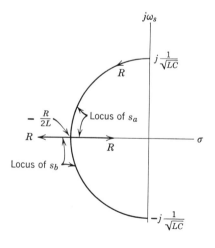

FIG. E-7.5

For $R = 0$, $s_a = j(1/\sqrt{LC})$ and $s_b = -j(1/\sqrt{LC})$. As R increases from zero, the magnitudes of s_a and s_b stay equal to $1/\sqrt{LC}$ until $R/2L = 1/\sqrt{LC}$. The locus of the roots is circular as shown in Fig. E-7.5, until $(R/2L)^2 = 1/LC$. At this value of R, the poles s_a and s_b become real and equal. For larger values of R, one root, s_a, approaches 0 as R increases indefinitely; the other root, s_b, approaches $-(R/L)$ as R increases indefinitely.

6.2.4 Driving-Point Immittances of Reactive Networks

The study of reactive circuits is helpful in giving us an approximate view of the operation of actual circuits with high Q. It also permits us to deal with fairly complex circuits with a minimum of algebra.

If a network is purely reactive, having no resistance elements, the free response will not attenuate nor will it increase indefinitely with time. The free response will consist only of constant terms plus sinusoidal terms. Thus, the poles and zeros of the driving point immittances must all be simple roots on the imaginary axis. The poles and zeros must alternate; that is, two zeros are always separated by a pole and two poles are always separated by a zero. This latter statement can be shown to be true by considering the poles and zeros to be slightly to the left of the imaginary axis as they would be in a circuit with some dissipation. The presence of one pole or one zero causes a change of nearly 180° in the phase of the driving point immittance as ω is varied past the frequency of the root. In order to limit the phase of the driving point immittance to the range from $-90°$ to $+90°$, it is essential that a pole be followed by a zero and conversely. This means also that the degrees of the polynomials in the numerator and denominator can differ only by one.

In Fig. 6.12 we give the pole zero diagrams and frequency characteristics for the L, the C, and the LC series circuit. The latter is said to be in series resonance at $\omega = 1/\sqrt{LC}$. It is interesting to see that the reactance characteristic of L is identical in form to the susceptance characteristic of C.

In Fig. 6.13 are shown the characteristics of slightly more complex circuits. In Fig. 6.13a is a parallel LC circuit which has an infinite impedance or zero admittance at $\omega = 1/\sqrt{LC}$. This is a resonant frequency although this type is sometimes called antiresonant frequency to distinguish it from zero impedance frequencies. Parts b and c of Fig. 6.13 demonstrate that the same frequency characteristics may be obtained with two different circuits. Also note that the circuit of Fig. 6.13b is an inductance in parallel with a series LC circuit and that

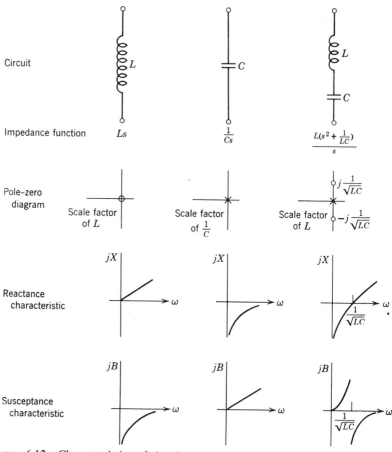

FIG. 6.12 Characteristics of simple reactive circuits.

therefore its susceptance characteristic is the sum of the susceptance of the inductance plus the susceptance of the series LC circuit.

A general expression for the transform impedance of a reactive circuit may be written

$$Z(s) = A \frac{s(s^2 + \omega_1^2)(s^2 + \omega_3^2)(\quad) \cdots}{(s^2 + \omega_2^2)(s^2 + \omega_4^2)(\quad) \cdots} \tag{6.34}$$

If we limit ourselves to three resonant frequencies the impedance will be of the form

$$Z(s) = A \frac{s(s^2 + \omega_1^2)}{(s^2 + \omega_2^2)(s^2 + \omega_4^2)}, \tag{6.35}$$

or

$$Z(s) = A \frac{(s^2 + \omega_1^2)(s^2 + \omega_3^2)}{s(s^2 + \omega_2^2)}. \tag{6.36}$$

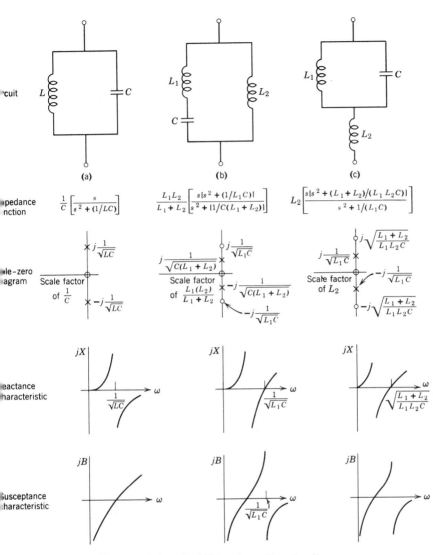

FIG. 6.13 Characteristics of additional reactive circuits.

The form of Equation 6.36 is obtained from Equation 6.34 by considering $\omega_4 = 0$. This is necessary since the degrees of the polynomials in numerator and denominator must differ only by one.

6.3 SUMMARY

The locus of any phasor network function for a passive network is circular if *one* element is varied. If frequency is varied the magnitude and phase characteristics may be conveniently obtained from a pole-zero diagram. The poles of all network functions and the zeros of driving point immittances must lie on left half of s plane or on the imaginary axis. Resonance occurs in circuits in which the effect of resistance is relatively small and is characterized by relatively large variations in response with small variations in frequency.

FURTHER READING

For further reading on the subjects of this chapter, see Brenner and Javid's *Analysis of Electric Circuits*, Chapter 15, McGraw-Hill, New York, 1959. For additional material on frequency characteristics see Clement and Johnson's *Electrical Engineering Science*, Chapters 14 and 15, McGraw-Hill, New York, 1960, or M. E. Van Valkenburg, *Network Analysis*, Chapters 11 and 12, Prentice-Hall, Englewood Cliffs, N.J., 1955.

PROBLEMS

6.1 For each of the circuits shown in Fig. P-6.1:
 a Sketch on cross-section paper the driving-point impedance. (Let $1'' = 5\,\Omega$.)
 b Sketch the driving-point admittance. (Let $1'' = 0.1$ mhos.)
On each locus indicate with an arrow the direction of increasing R, L, or C as appropriate.

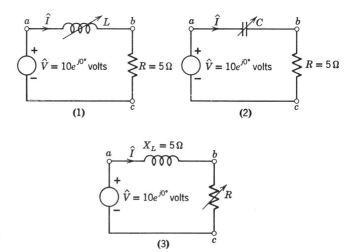

FIG. P-6.1

6.2 For each of the circuits of Fig. P-6.1:
 a Sketch the locus of the current $\hat{I}$.
 b Sketch the locus of $\hat{V}_{ab}/\hat{V}_{ac}$.
 c Sketch the locus of $\hat{V}_{bc}/\hat{V}_{ac}$.

6.3 For each of the circuits of Fig. P-6.1:
  a Sketch the magnitude of $\hat{Y}_{ac}$ versus the variable element (R, L, or C as appropriate).
  b Sketch the angle of $\hat{Y}_{ac}$ versus the variable element.

6.4 Clearly showing the steps used, determine the locus of $\hat{Z}_{ab}$ and $\hat{Y}_{ab}$ for the circuit shown in Fig. P-6.4. Show numerical values at key points and indicate the direction of the increasing value of the variable element.

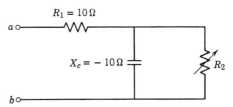

FIG. P-6.4

6.5 Repeat problem 6.4 for the circuit shown in Fig. P-6.5.

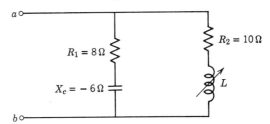

FIG. P-6.5

6.6 Repeat problem 6.4 for the circuits shown in Fig. P-6.6.

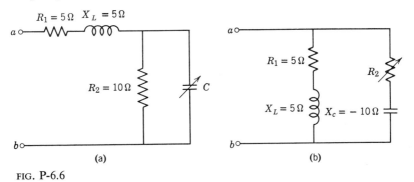

FIG. P-6.6

6.7 Prove that the locus of the line $y = d$ in the $\hat{z}$ plane becomes a circle in the $\hat{w}$ plane.

6.8 For the straight line, $ax + by = 1$, in the $\hat{z}$ plane, show that the locus in the $\hat{w}$ plane is the circle,

$$\left(u - \frac{a}{2}\right)^2 + \left(v + \frac{b}{2}\right)^2 = \left(\frac{a}{2}\right)^2 + \left(\frac{b}{2}\right)^2.$$

Phasor Network Functions for a Variable Element or for Variable Frequency

6.9 Clearly showing all steps, determine the locus of the current $\hat{I}$ for the circuit shown in Fig. P-6.9.

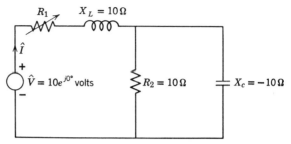

FIG. P-6.9

6.10 Sketch the locus of the current $\hat{I}$ if $\hat{V} = 10e^{j45°}$ volts for the circuit of Fig. P-6.9.

6.11 For the circuit shown in Fig. P-6.11, show all steps to:
 a Determine the locus of $\hat{Z}_{ab}$. (Indicate values at key points and direction of increasing magnitude of the variable element C.)
 b Determine, from the locus diagrams, the value(s) of C in farads that will make $\hat{Z}_{ab}$ real (purely resistive); $\omega = 2$ radians/sec.
 c Determine the value(s) of $\hat{Z}_{ab}$ for part b.

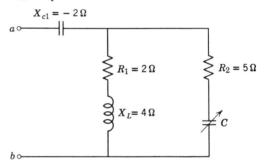

FIG. P-6.11

6.12 For the circuit shown in Fig. P-6.12:
 a Sketch the loci of the impedance, $\hat{Z}_{ab}$, and the admittance, $\hat{Y}_{ab}$, showing values at key points.
 b Sketch the locus of the current $\hat{I}$, showing values at key points.
 c For what value(s) of C will the power factor be unity?
 d For what value(s) of C will the phase angle of $\hat{Y}_{ab}$ be $+60°$?

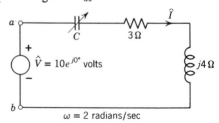

FIG. P-6.12
Partial Answer: (c) $C = 0.125$ f.

6.13 For the circuit of Fig. P-6.13:
  a Sketch the loci of admittance $\hat{Y}_{ab}$ and impedance $\hat{Z}_{ab}$, showing values at key points.
  b Sketch the locus of the current $\hat{I}$.
  c Determine the value(s) of L for which $\hat{I}$ and $\hat{V}$ are in phase.
  d Determine the magnitude and phase angle for $\hat{I}$ for minimum value of I. Also determine the value of L at this point.
  e Determine the phase angle of $\hat{I}$ for maximum value of I.

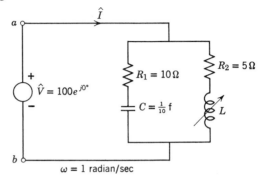

FIG. P-6.13

$\omega = 1$ radian/sec

Partial Answer: (c) $L = 1.34, 18.7$ henries.

6.14 For the circuit of Fig. P-6.14:
  a Sketch the impedance, $\hat{Z}_{ab}$, and admittance, $\hat{Y}_{ab}$, loci.
  b Sketch the locus of the current, $\hat{I}$.
  c For what value(s) of R will the power factor be unity?
  d Calculate the minimum and maximum values of I and the value of R to produce each.

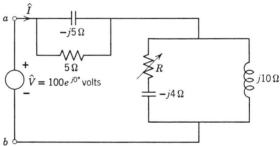

FIG. P-6.14

6.15 For the circuit shown in Fig. P-6.15:
  a Showing all steps, sketch the locus of $\hat{V}_{ab}$.
  b Determine the minimum value of V_{ab}.

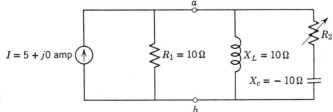

FIG. P-6.15

Phasor Network Functions for a Variable Element or for Variable Frequency   323

6.16 For the circuit shown in Fig. P-6.16:
   a Showing all steps, sketch the locus of $\hat{I}$.
   b Find the maximum and minimum values of I.
   c Find the maximum and minimum power factor ($\cos \theta$).
   d Find the value of R to give minimum current.

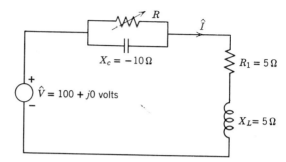

FIG. P-6.16

6.17 In Fig. P-6.16 let X_c be variable from 0 to $-\infty$, and let $R = 10\ \Omega$, $R_1 = 3\ \Omega$, and $X_L = 4\ \Omega$.
   a Showing all steps, sketch the locus of $\hat{I}$.
   b Find the maximum and minimum values of I.
   c Find the maximum and minimum power factor.
   d Find the magnitude of $\hat{I}$ at the power factors of part *c*.

6.18 For the circuit shown in Fig. P-6.18:
   a Showing all steps, sketch the locus of $\hat{V}_{ab}$.
   b Find the maximum and minimum values of V_{ab}.
   c For what values of X_L will the power factor by unity? 0.707?
   d For what values of X_L will the power factor be leading?

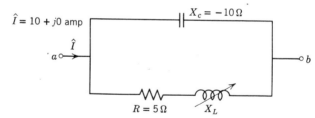

FIG. P-6.18

6.19 For the circuit of Fig. P-6.19:

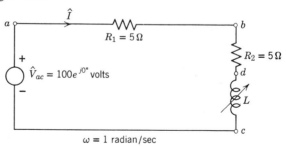

FIG. P-6.19

a Sketch the loci of $\hat{I}$, $\hat{V}_{ad}$, $\hat{V}_{dc}$ and from these determine the locus of the voltage $\hat{V}_{bc}$.
b Determine the value of L which makes the angle between $\hat{V}_{ac}$ and $\hat{V}_{bc}$ a maximum.

6.20 For Fig. P-6.20, with ω constant, what conditions must be satisfied for a variation in R to permit a net admittance, $\hat{Y}_{ab}$, which has no susceptance term? (Unity power factor.) Give both an analytical and graphical proof.

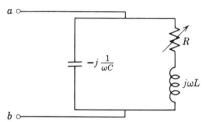

FIG. P-6.20

6.21 For the circuit shown in Fig. P-6.21 use the techniques of Example 6.3, page 299, to determine the locus of the **transfer function** $\hat{I}_1/\hat{I}_0$.

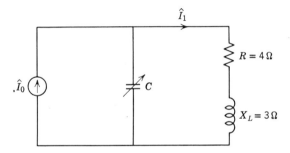

FIG. P-6.21

6.22 For the circuit shown in Fig. P-6.22:
a Determine the locus of the transfer voltage $\hat{V}_{cd}/\hat{V}_{ad}$.
b If $\hat{V}_{ad} = 1e^{j0°}$ volts, find the maximum value of V_{cd}.

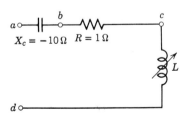

FIG. P-6.22

6.23 For the circuit of Fig. P-6.23, prove that if $R^2 = L/C$, V_{cd} will be zero for all values of ω. A graphical proof is desired as well as an analytic proof.

Phasor Network Functions for a Variable Element or for Variable Frequency 325

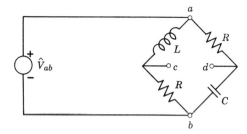

FIG. P-6.23

6.24 In problem 6.13 it is possible that if R_2 is made too large no value of L will produce unity power factor. Use a phasor locus diagram to determine the critical value of R_2 for unity power factor (the largest possible value that R_2 may have and still have unity power factor by adjusting L). With R_2 equal to its critical value, determine the value of L to produce unity power factor. Determine the value of L (with R_2 equal to its critical value) that will cause I to have its minimum value. What is the minimum value of I? What is the phase angle of $\hat{I}$ at minimum value?

6.25 In the circuit of problem P-6.23, $R = 2\ \Omega$, L is variable; a capacitance of unknown value is connected between terminals d and b; and an ideal voltmeter is connected between terminals d and c. L is adjusted until the voltmeter reads zero, at which time L is known to be 0.004 henries. What is the value of C?

6.26 For the circuit shown in Fig. P-6.26:
 a Sketch the locus of the network function $\hat{I}_1/\hat{I}$.
 b Find the value(s) of X_L such that $\hat{I}_1$ and $\hat{I}$ will be out of phase by an angle, $\tan^{-1} 2/3$.

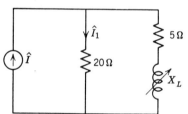

FIG. P-6.26
Answer: $X_L = 5\ \Omega, 25\ \Omega$

6.27 In 60-cps power transmission it is important to have the load voltage stay relatively constant under various load conditions. Figure P-6.27 represents a simplified circuit of a transmission system in which the generator voltage, $\hat{V}_g$, is assumed to be the reference phasor, $\hat{V}_g = 1000 + j0$ volts, the current $\hat{I}$ is assumed to vary in magnitude from 0 to 50 amperes and to vary in phase from $0°$ to $\pm 90°$. The impedance of the transmission line is approximated by the inductance ($X_L = 2$ ohms). It is desirable to determine the maximum percentage variation in the load voltage. Use a phasor diagram showing $\hat{V}_g = \hat{V}_{ab}$, $\hat{I} = 50e^{j\phi}$ for various values of ϕ, $\hat{V}_{ac} = \hat{I}jX_L$. From a phasor locus diagram calculate the maximum and minimum values for V_{cb}.

FIG. P-6.27

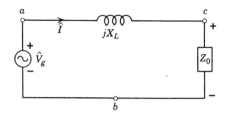

6.28 For each of the circuits shown in Fig. P-6.28, write the expressions for the driving point impedance as a ratio of polynomials in s. Give answers in standard factored form.

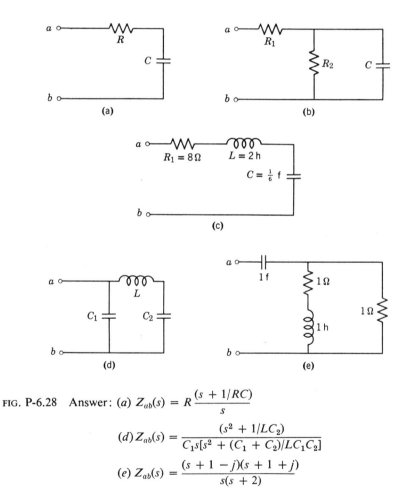

FIG. P-6.28 Answer: (a) $Z_{ab}(s) = R\dfrac{(s + 1/RC)}{s}$

(d) $Z_{ab}(s) = \dfrac{(s^2 + 1/LC_2)}{C_1 s[s^2 + (C_1 + C_2)/LC_1 C_2]}$

(e) $Z_{ab}(s) = \dfrac{(s + 1 - j)(s + 1 + j)}{s(s + 2)}$

6.29 Arrange each of the network functions given in standard factored form and plot a pole-zero diagram. From the pole-zero diagram calculate $\hat{H}$ for frequencies of $\omega = 0, 1, 2,$ and 3 radians/sec.

a $H(s) = \dfrac{10}{5s + 10}$

b $H(s) = \dfrac{10s}{5s + 5}$

c $H(s) = \dfrac{s + 1}{s^2 + 2s}$

d $H(s) = \dfrac{s^2 + 3s + 2}{s}$

e $H(s) = \dfrac{2s}{s^2 + 2s + 2}$

6.30 For each of the network functions given, plot a pole-zero diagram. From the pole-zero diagram sketch the curves of H versus ω and angle of $\hat{H}$ versus ω.

(a) $H(s) = \dfrac{1}{s + 1}$

(b) $H(s) = \dfrac{s}{s + 1}$

(c) $H(s) = \dfrac{s + 1}{s}$

(d) $H(s) = \dfrac{s}{(s + 1)(s + 2)}$

(e) $H(s) = 5 \dfrac{s}{s^2 + 2s + 26}$

6.31 For the circuit shown in Fig. P-6.31:
a Write the expression for $Z_{ab}(s)$ and plot the pole-zero diagram.
b Sketch the curves of Z_{ab} vs. ω and angle of $\hat{Z}_{ab}$ vs. ω.
c From the pole-zero diagram calculate $\hat{Z}_{ab}$ for $\omega = 0, 1,$ and 2 radians/sec.

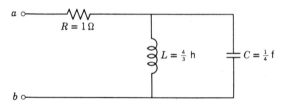

FIG. P-6.31

6.32 For the circuit shown in Fig. P-6.32:
a Write the expression for the driving-point admittance $Y_{ab}(s)$ and plot the pole-zero diagram.
b Sketch the curves of Y_{ab} vs. ω and angle of $\hat{Y}_{ab}$ vs. ω.
c From the pole-zero diagram calculate $\hat{Y}_{ab}$ for $\omega = 1$ radians/sec and $\omega = 2$ radians/sec.

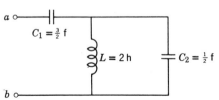

FIG. P-6.32

6.33 For the circuit shown in Fig. P-6.33:
  a Write the expression in literal form for the transfer function $H(s) = V_{cb}(s)/V_{ab}(s)$.
  b Using $R_1 = R_2 = 1\,\Omega$, $L = 1$ h and $C = 1$ f, plot the pole-zero diagram.
  c Sketch the curves of H vs. ω and angle of $\hat{H}$ vs. ω.

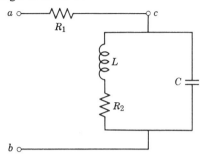

FIG. P-6.33

6.34 For the circuit shown in Fig. P-6.34:
  a Show that if $R^2 = L/C$ the voltage $\hat{V}_{cd}$ is a constant in magnitude but changes in phase angle as ω varies.
  b Sketch the variation of phase angle versus ω.

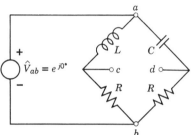

FIG. P-6.34

6.35 Figure P-6.35 shows the pole-zero diagram for the driving point impedance of a certain network.
  a If this circuit is driven by a voltage source, give the general form of the free response.
  b If the circuit is driven by a current source, give the form of the free response.

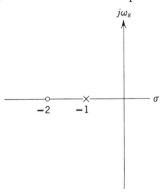

FIG. P-6.35

6.36 A certain network's driving-point impedance function has a pole at -1 and a pole at infinity and a pair of complex conjugate zeros at $-1 \pm j1$. Give the general form of the free response if:
  a The driving function is a voltage source.
  b The driving function is a current source.

6.37 Figure P-6.37a shows the pole-zero diagram of a certain transfer function:
  a Give the nature of the free response and show a rough sketch of it.
  b The network elements are now adjusted such that the pole-zero diagram is that shown in Fig. P-6.37b. Again, give the nature of the free response and show a rough sketch.
  c If the elements are again adjusted such that the poles of Fig. P-6.37b are relocated 10 units to the left, give the nature of the free response and show a rough sketch.

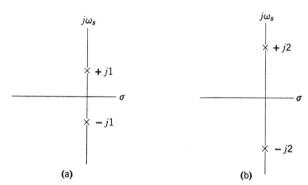

FIG. P-6.37 (a) (b)

6.38 For the circuit shown in Fig. P-6.38:
  a Write, in literal form, the expression for the driving-point impedance $Z_{ab}(s)$.
  b For $R = 1\ \Omega$, $L = \frac{1}{26}$ henry and $C = \frac{1}{4}$ farad, plot the pole-zero diagram and then sketch curves of Z_{ab} vs. ω and angle of $\hat{Z}_{ab}$ vs. ω.

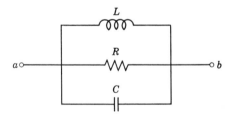

FIG. P-6.38

6.39 From the pole-zero diagram for the circuit of Fig. P-6.38:
  a Determine the approximate frequency of resonance.
  b Determine the impedance at the resonant frequency.
  c Determine the bandwidth.
  d Calculate the percent error of the approximation of part a.

6.40 If a current source of $\hat{I} = 10e^{j0°}$ amperes is connected to the terminals a, b of Fig. P-6.38, with the data of Prob 6.38b, use a pole zero-diagram to:
  a Determine the maximum value of average power input to this circuit.
  b Determine at what frequency this will occur.
  c Determine at what frequencies the power will be one-half maximum.
  d What is the Q of this circuit?

6.41 A series RLC circuit has the pole-zero diagram shown in Fig. P-6.41 for its driving-point admittance. Find the values for R, L, and C.
Partial answer: $C = 0.008$ f.

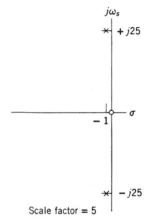

FIG. P-6.41   Scale factor = 5

6.42 For each of the reactive networks shown in Fig. P-6.42, write the transform impedance, show a pole-zero diagram, and sketch reactance versus angular frequency. Follow the form indicated in Fig. 6-13.

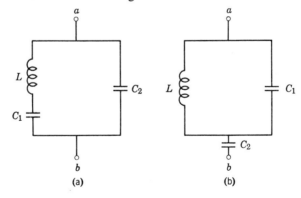

FIG. P-6.42   (a)   (b)

7 Nonsinusoidal Periodic Waves—Steady-State Response

Nature seldom smiles on her students; she guards her secrets well. Experimental engineers refer to a law known as Murphy's law which has various versions, one of them being "If something can go wrong, it will." However, there are rare occasions when study of one particular phenomenon sheds light on other phenomena and by chance or good fortune one gains something one was not seeking directly.[1] Such is the case with the study of sinusoidal sources.

In this chapter we shall find that any nonsinusoidal periodic wave of physical significance can be represented by the sum of a number of sinusoidal waves of various frequencies. Since superposition applies to linear systems, the steady-state response to each frequency can be determined independently and the net steady-state response can be determined as the sum of these individual responses. Thus our knowledge of the response to a single sinusoidal source can readily be extended to give us the response to any periodic function.

As a matter of fact our good fortune is greater than this. We shall find in Chapter 9 that any single pulse or group of pulses can be considered as having a frequency spectrum. Thus the response to any physical driving function can be inferred from the knowledge of the response to a sinusoidal source.

We shall limit ourselves in this chapter to the steady-state response to nonsinusoidal periodic functions. A periodic function of time is defined as one for which $f(t) = f(t + T)$ for all time, the period T being the **smallest** interval for which the expression holds. Figure 7.1 shows several periodic nonsinusoidal waves.

---

[1] This type of good fortune is known as serendipity from a tale about the three princes of Serendip (Ceylon).

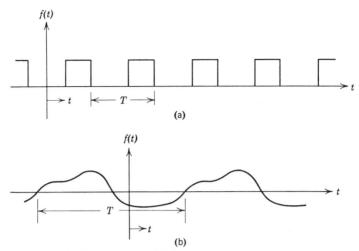

FIG. 7.1 Several periodic nonsinusoidal functions.

7.1 INTRODUCTION TO FOURIER SERIES

A remarkable theorem known as the Fourier[2] theorem states that nonsinusoidal periodic functions may be represented, with some restrictions, by the infinite series,

$$f(t) = A_0 + A_1 \cos \omega_1 t + A_2 \cos 2\omega_1 t + A_3 \cos 3\omega_1 t + \ldots$$
$$+ B_1 \sin \omega_1 t + B_2 \sin 2\omega_1 t + B_3 \sin 3\omega_1 t + \ldots$$
$$= A_0 + \sum_{n=1}^{\infty}(A_n \cos n\omega_1 t + B_n \sin n\omega_1 t). \tag{7.1}$$

An alternate way of writing the series is

$$f(t) = A_0 + \sum_{n=1}^{\infty} C_n \cos(n\omega_1 t + \alpha_n), \tag{7.2}$$

in which $C_n = \sqrt{A_n^2 + B_n^2}$, and $\alpha_n = -\tan^{-1}(B_n/A_n)$. The term $C_1 \cos(\omega_1 t + \alpha_1)$ is called the fundamental component of the series and is a sinusoid with the same period as the periodic function. The fundamental angular frequency, ω_1, is thus equal to $2\pi/T$. The terms $C_2 \cos(2\omega_1 t + \alpha_2)$, $C_3 \cos(3\omega_1 t + \alpha_3)$, etc., are known as the harmonic components of the series.

For purposes of analysis it is easier to express the series as functions of $\omega_1 t = x$ rather than as functions of t. Equation 7.1 becomes

$$f(x) = A_0 + A_1 \cos x + A_2 \cos 2x + A_3 \cos 3x + \ldots$$
$$+ B_1 \sin x + B_2 \sin 2x + B_3 \sin 3x - \ldots$$
$$= A_0 + \sum_{n=1}^{\infty}(A_n \cos nx + B_n \sin nx). \tag{7.3}$$

---
[2] Named after the French mathematician, J. B. J. Fourier.

In order for us to write $f(x)$ as a Fourier series we need to learn how to obtain the constant A_0 and the coefficients A_n and B_n of Equation 7.3. Before doing this, let us consider the conditions which are sufficient for the Fourier series to exist and the conditions which are necessary for the series to converge.

A periodic function must satisfy certain conditions in order that it may be represented by the Fourier series. The sufficient conditions, known as the Dirichlet conditions, are that the function must either be piecewise continuous throughout the period, or it may have a finite number of infinite discontinuities and be piecewise continuous between them, but then the integral $\int_{K_1}^{K_1+2\pi} f(x)\,dx$ must exist. A function is piecewise continuous throughout the period if there are a finite number of finite discontinuities between which the function is continuous. The necessary conditions that a periodic function must satisfy if it is to be represented by a Fourier series have not been determined and are still the subject of mathematical research.

All periodic driving and response functions which are physically realizable satisfy the Dirichlet conditions and thus may be expressed in a Fourier series. These conditions are also sufficient to ensure the existence of the Laplace transform of these functions.

With regard to the matter of convergence, the Fourier series converges to $f(x)$ at all points for which $f(x)$ is continuous. At those points for which $f(x)$ is discontinuous the Fourier series converges to the midpoint of the discontinuity. Also at the discontinuities there are "overshoots" known as the Gibbs phenomenon. If an infinite number of terms is used in the series this overshoot occurs at the discontinuity and thus is of no physical significance.

How many terms of the Fourier series must we consider in order to have an adequate representation of the wave for determining the network response? This is a matter of engineering judgment for each case. In general the smoother the wave the fewer terms will be needed. If we use a finite number of terms to represent a wave which has a discontinuity we will find that the Gibbs phenomenon causes overshoots in the neighborhood of the discontinuity. We shall not concern ourselves with these overshoots in this text. For further reading see E. A. Guillemin's *The Mathematics of Circuit Analysis*, Chapter 7, M.I.T. Press, Cambridge, 1949.

There is another consideration in determining the terms to be included in writing the Fourier series of a driving function. This is the frequency characteristic of the network function which is the ratio of the response function to the driving function, Specifically, if resonance occurs at a high harmonic, this harmonic should be included in the series of the driving function.

The expressions for the coefficients of the terms in the Fourier series may be readily obtained by using the "orthogonality properties" of the trigonometric functions. If m and n are integers $1, 2, 3, \ldots$, and K_1 is real, the orthogonality properties are:

$$\int_{K_1}^{K_1+2\pi} \cos mx \cos nx \, dx = \begin{cases} 0, & \text{if } m \neq n \\ \pi, & \text{if } m = n \end{cases}, \qquad (7.4)$$

$$\int_{K_1}^{K_1+2\pi} \sin mx \sin nx \, dx = \begin{cases} 0, & \text{if } m \neq n \\ \pi, & \text{if } m = n \end{cases}, \qquad (7.5)$$

and

$$\int_{K_1}^{K_1+2\pi} \sin mx \cos nx \, dx = 0 \text{ for all values of } m \text{ or } n. \qquad (7.6)$$

The proof of these properties is readily obtained by using the trigonometric identities, $\cos(\alpha \pm \beta) = \cos \alpha \cos \beta \mp \sin \alpha \sin \beta$, and $\sin(\alpha \pm \beta) = \sin \alpha \cos \beta \pm \cos \alpha \sin \beta$, to change the integrand into terms that integrate easily.

The A_0 coefficient can be obtained by integrating Equation 7.3 over one period. The application of the orthogonality properties leads to

$$A_0 = \frac{1}{2\pi} \int_{K_1}^{K_1+2\pi} f(x) \, dx. \qquad (7.7)$$

The student will be asked to prove that

$$A_n = \frac{1}{\pi} \int_{K_1}^{K_1+2\pi} f(x) \cos nx \, dx, \qquad (7.8)$$

and

$$B_n = \frac{1}{\pi} \int_{K_1}^{K_1+2\pi} f(x) \sin nx \, dx, \qquad (7.9)$$

for $n = 1, 2, 3, \ldots$.

We observe that A_0 is the average value of the function, $f(x)$, for one period; whereas A_n is twice the average value of the function, $f(x) \cos nx$, for one period, and B_n is twice the average value of the function, $f(x) \sin nx$, for one period.

7.2 SYMMETRY

To minimize the work in evaluating coefficients it is helpful to note that certain types of symmetry mean that some of these coefficients are zero "by inspection." The most common types of symmetry are called "half-wave symmetry," "sine symmetry," and "cosine symmetry." Half-wave symmetry means that even harmonics are missing or that the following coefficients are zero: A_2, B_2, A_4, B_4, A_6, B_6, etc. Sine symmetry means that all cosine terms are missing or that the following coefficients are zero: A_1, A_2, A_3, A_4, etc. Cosine symmetry means that all sine terms are missing or that the following coefficients are zero: B_1, B_2, B_3, B_4, etc.

An insight into symmetry is obtained by observing the effect of adding harmonic components to a fundamental component. In Fig. 7.2a is shown a sinusoid which would be expressed as a sine function if the origin were chosen at x_1 or x_2 and which would be expressed as a cosine function if the origin were chosen at x_3 or x_4. Mathematically, we would say it is an odd function, $f(x) = -f(-x)$, for x_1 or x_2, while it is an even function, $f(x) = f(-x)$, for x_3 or x_4. In addition the sinusoid has the property that for any value of x, $f(x) = -f(x + \pi)$.

In Fig. 7.2b a second harmonic is added to the fundamental component in such a manner that each wave would be written as a cosine function for x_3 or x_4. We observe that the resultant wave is an even function for x_3 and x_4. In Fig. 7.2d a similar result is observed for the addition of a third harmonic component to a fundamental component. If a constant term, A_0, were added, the resultant wave would still have even function symmetry or cosine symmetry about two points in any one period, these two points being one-half period apart.

In Fig. 7.2c a second harmonic component is added to a fundamental component is such a manner that each wave would be written as a sine function for x_1 or x_2. The resultant wave is an odd function for x_1 or x_2. In Fig. 7.2d a similar result is shown for the addition of a third harmonic component to a fundamental component. If now a constant term A_0 is added, the resultant wave would no longer show odd function symmetry. A periodic wave, $f(x)$, is said to have sine wave symmetry if $[f(x) - A_0]$ has odd function symmetry.

We observe that the fundamental component and any odd harmonic component has the symmetry of $f(x) = -f(x + \pi)$, while this is not true for any even harmonic component. The fact that the addition of a second harmonic component destroys this symmetry is shown in Figs. 7.2b and 7.2c whereas the fact that the addition of a third harmonic component maintains this symmetry is shown in Figs. 7.2d and 7.2e. The addition of a constant term, A_0, tends to destroy this symmetry and thus we define half-wave symmetry as existing for $f(x)$ if

$$f(x) - A_0 = -[f(x + \pi) - A_0].$$

This type of symmetry is not dependent on the choice of the origin.

In Fig. 7.3a the wave has half-wave symmetry. If either x_1 or x_2 were chosen as the origin, the wave would have sine symmetry, $A_n = 0$, and the Fourier series could be written

$$f(x) = A_0 + B_1 \sin x + B_3 \sin 3x + B_5 \sin 5x \ldots.$$

If either x_3 or x_4 were chosen as the origin, the wave would have cosine symmetry and the series could be written

$$f(x) = A_0 + A_1 \cos x + A_3 \cos 3x \ldots.$$

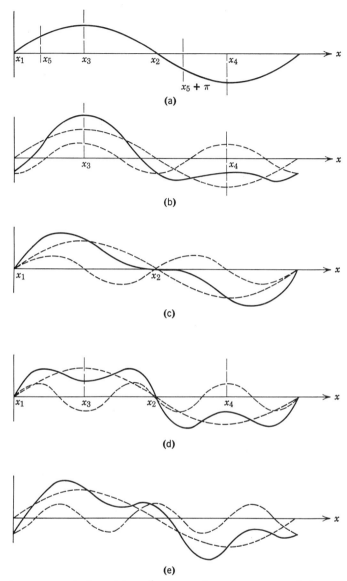

FIG. 7.2 Symmetry of sinusoid and symmetries of wave consisting of fundamental component plus one harmonic component (shown for one period only). (*a*) Fundamental component. (*b*) Fundamental component plus second harmonic component with cosine symmetry about x_3 and x_4. (*c*) Fundamental component plus second harmonic with sine symmetry about x_1 and x_2. (*d*) Fundamental plus third harmonic with sine symmetry about x_1 and x_2, cosine symmetry about x_3 and x_4. (*e*) Fundamental plus third harmonic.

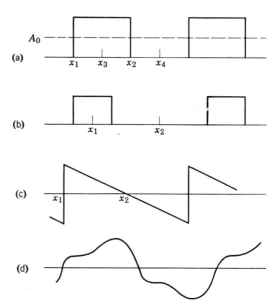

FIG. 7.3 Some periodic functions.

Figure 7.3b shows a wave which has cosine symmetry if either x_1 or x_2 is chosen as the point, $x = 0$.

Figure 7.3c shows a wave which has sine symmetry about the points x_1 and x_2.

Figure 7.3d shows a wave which has half-wave symmetry.

Example 7.1

GIVEN: The periodic function known as a square wave which is shown in Fig. E-1.1.

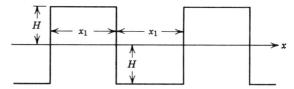

FIG. E-1.1

TO FIND: The Fourier series which represents this wave.

SOLUTION: This wave is similar to that of Fig. 7.3a except that A_0 is zero by inspection. We choose to use cosine symmetry and select the origin as shown in Fig. E-1.2. Since $A_n = \dfrac{1}{\pi} \int_{K_1}^{K_1+2\pi} f(x) \cos nx\, dx$, we may choose the lower limit of integration at any point. We choose

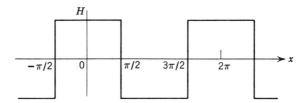

FIG. E-1.2

$K_1 = -\pi/2$ in order that the complete integral can be performed in two operations.

$$A_n = \frac{1}{\pi}\left[\int_{-\pi/2}^{\pi/2} H \cos nx\, dx + \int_{\pi/2}^{3\pi/2} -H \cos nx\, dx\right]. \tag{1}$$

$$A_n = \frac{1}{\pi}\left[\frac{H \sin nx}{n}\bigg|_{-\pi/2}^{\pi/2} - \frac{H \sin nx}{n}\bigg|_{\pi/2}^{3\pi/2}\right]$$

$$= \frac{4H}{n\pi}\sin(n\pi/2). \tag{2}$$

Since the wave had half-wave symmetry, the even terms must be zero; this is borne out by Equation 2 if we solve for even coefficients of n. For odd values of n we obtain

$$A_1 = \frac{4H}{\pi}, \quad A_3 = -\frac{4H}{3\pi}, \quad A_5 = \frac{4H}{5\pi}, \quad A_7 = -\frac{4H}{7\pi}, \text{ etc.}$$

Then

$$f(x) = \frac{4H}{\pi}\cos x - \frac{4H}{3\pi}\cos 3x + \frac{4H}{5\pi}\cos 5x - \frac{4H}{7\pi}\cos 7x \ldots \tag{3}$$

or

$$f(x) = \sum_{n=1,3,5,7} \frac{4H}{n\pi}\sin\frac{n\pi}{2}\cos nx. \tag{4}$$

Integration of a square wave results in a triangular wave. Let us integrate Equation 3 term by term to obtain

$$f(x) = \frac{4H}{\pi}\sin x - \frac{4H}{9\pi}\sin 3x + \frac{4H}{25\pi}\sin 5x - \frac{4H}{49\pi}\sin 7x \ldots \text{ etc.} \tag{5}$$

We observe that the relative magnitude of the harmonics is less for the triangular wave than for the square wave; that is, the relative amplitude of the third harmonic for the triangular wave is one third that of the square wave, etc. This is in accord with the fact that the triangular wave is smoother than the square wave.

Example 7.2

GIVEN: The periodic wave shown in Fig. E-2.1, for which the voltage is $+100$ volts for 0.001 seconds, 0 volts for 0.002 seconds, $+100$ volts for 0.001 seconds, 0 for 0.002 seconds, etc.

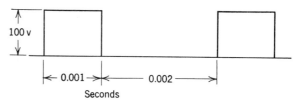

FIG. E-2.1

TO FIND: The Fourier series through the fourth harmonic.

SOLUTION: Inspection of the waveshape tells us the period is 0.003 seconds. (The wave repeats itself every 0.003 seconds.) The value of A_0 (the average or d-c value of the wave) is

$$A_0 = \frac{\text{net area over one period}}{\text{period}} = \frac{(100)(0.001) + (0)(0.002)}{0.003}$$

$$= \frac{100}{3} = 33.3 \text{ volts}$$

We now sketch A_0 on the plot of the wave and examine for symmetry. Examination shows no half wave or sine symmetry. There are two choices of $t = 0$ axis that will further help to reduce our work. If we choose either point M or N as shown in Fig. E-2.2 our wave satisfies the conditions for cosine symmetry and all B_n terms will be zero. We shall use the M position for $t = 0$. One expression for A_n is

$$A_n = \frac{1}{\pi} \int_0^{2\pi} f(x) \cos nx \, dx \qquad (1)$$

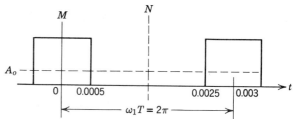

FIG. E-2.2

Since our limits are expressed in terms of angle, and it is generally easier to work with angles rather than the variable t, we shall convert our variable as follows. One period is 2π radians, then for

$t = 0$	$x = 0$ radians or $0°$
$t = 0.0005$	$x = \pi/3$ radians or $60°$
$t = 0.0025$	$x = 10\pi/6$ radians or $300°$
$t = 0.003$	$x = 2\pi$ radians or $360°$

The equations for our wave then may be written:
For

$0 < x < \dfrac{\pi}{3}$,   for $\dfrac{\pi}{3} < x < \dfrac{10\pi}{6}$,   for $\dfrac{10\pi}{6} < x < 2\pi$,

$f(x) = 100.$ $f(x) = 0,$ $f(x) = 100.$

Then

$$A_n = \frac{1}{\pi}\left[\int_0^{\pi/3}(100)\cos nx\,dx + \int_{\pi/3}^{10\pi/6}(0)\cos nx\,dx\right.$$
$$\left. + \int_{10\pi/6}^{2\pi} 100\cos nx\,dx\right]$$

$$= \frac{100}{n\pi}\left\{\left[\sin nx\Big|_0^{\pi/3} + \sin nx\Big|_{10\pi/6}^{2\pi}\right]\right\} = \frac{100}{n\pi}\sin nx\Big|_{10\pi/6}^{\pi/3}$$

$$A_1 = \frac{100}{\pi}\left[\frac{\sqrt{3}}{2} + \frac{\sqrt{3}}{2}\right] = 55,$$

$$A_2 = \frac{100}{2\pi}\left[\frac{\sqrt{3}}{2} + \frac{\sqrt{3}}{2}\right] = 27.6,$$

$$A_3 = \frac{100}{3\pi}[0] = 0, \text{ and}$$

$$A_4 = \frac{100}{4\pi}\left[-\frac{\sqrt{3}}{2} - \frac{\sqrt{3}}{2}\right] = -13.8.$$

These coefficients could also be obtained by having other limits for the integral of Equation 1, provided that the difference between these limits was 2π. Since we desire to write the series in terms of $\omega_1 t$,

$$\omega_1 = \frac{2\pi}{T} = \frac{2\pi}{0.003} = 2090 \text{ radians/sec.}$$

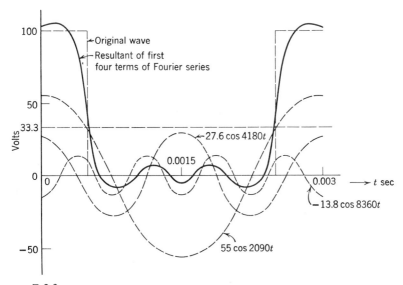

FIG. E-2.3

Our series may then be written

$v = 33.3 + 55 \cos 2090t + 27.6 \cos 4180t - 13.8 \cos 8360t$ volts.
A sketch of this equation is shown in Fig. E-2.3.

Example 7.3

GIVEN: The periodic current wave shown in Fig. E-3.1, whose period is 0.0002 seconds.
TO FIND: The Fourier series through the fourth harmonic.
SOLUTION: The average value is

$$A_0 = \frac{\frac{1}{2}(10)(2\pi)}{2\pi} = 5 \text{ amp.}$$

Examination of the wave about the A_0 axis for symmetry shows that there is no half-wave symmetry but that there is sine symmetry at either M or N (Fig. E-3.2).

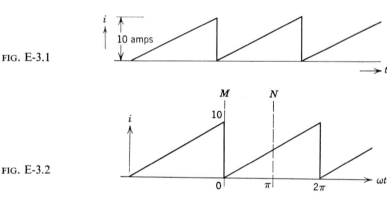

FIG. E-3.1

FIG. E-3.2

Let us use M as reference. There will be no cosine terms. For this origin, $f(x) = 10x/2\pi = 5x/\pi$, and

$$B_n = \frac{1}{\pi} \int_0^{2\pi} f(x) \sin nx \, dx$$

$$= \frac{1}{\pi} \int_0^{2\pi} \frac{5x}{\pi} \sin nx \, dx$$

$$= \frac{5}{\pi^2} \int_0^{2\pi} x \sin nx \, dx$$

Integrating by parts or from a table of integrals we obtain

$$B_n = \frac{5}{\pi^2} \left[-\frac{x \cos nx}{n} + \frac{\sin nx}{n^2} \right]_0^{2\pi}$$

The second term reduces to zero for all values of n. Then

$$B_n = -\frac{5(2\pi)}{\pi^2 n} = -\frac{10}{\pi n}.$$

Since the period is 0.0002 seconds,

$$\omega_1 = \frac{2\pi}{T} = \frac{2\pi}{0.0002} = 31{,}400 \text{ radians/sec,}$$

and the Fourier series may be written as

$$i = 5 - 3.18 \sin 31{,}400t - 1.59 \sin 62{,}800t$$
$$- 1.06 \sin 94{,}200t - 0.795 \sin 125{,}600t \text{ amp.}$$

7.3 DETERMINATION OF COEFFICIENTS BY NUMERICAL INTEGRATION

If the integrals used for determining the coefficients of the Fourier series are difficult to evaluate or if only graphical data are available, the coefficients may be determined by numerical integration. One method of doing this is to divide the interval from K_1 to $K_1 + 2\pi$ into m equal-increments, so $m \Delta x = 2\pi$ or $\Delta x = 2\pi/m$.

Then Equation 7.7 becomes

$$A_0 = \frac{1}{2\pi} \int_{K_1}^{K_1+2\pi} f(x)\, dx \simeq \frac{1}{2\pi} \sum_{k=1}^{m} [\text{avg}_k f(x)]\, \Delta x$$

$$\simeq \frac{1}{m} \sum_{k=1}^{m} \text{avg}_k f(x), \tag{7.10}$$

in which $\text{avg}_k f(x)$ is the average value of $f(x)$ in the kth increment. (It is often convenient to use the value at the mid-point of the increment as the average value.) In a similar way,

$$A_n = \frac{1}{\pi} \int_{K_1}^{K_1+2\pi} f(x) \cos nx\, dx \simeq \frac{2}{m} \sum_{k=1}^{m} \text{avg}_k[f(x) \cos nx] \tag{7.11}$$

and

$$B_n = \frac{1}{\pi} \int_{K_1}^{K_1+2\pi} f(x) \sin nx\, dx \simeq \frac{2}{m} \sum_{k=1}^{m} \text{avg}_k[f(x) \sin nx]. \tag{7.12}$$

Example 7.4

GIVEN: The waveform of Example 7.3.
TO FIND: The Fourier series by numerical integration.
SOLUTION: There is no standard as to the number of increments to use. This is normally determined by the accuracy required (the more increments the greater the accuracy) and the nature of the waveform to be

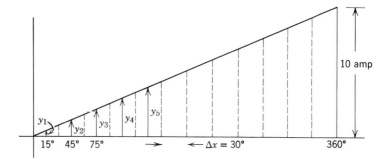

FIG. E-4.1

analyzed. Let us try 12 increments for our example. For $m = 12$, the increments are 30°. Fig. E-4.1 shows the waveform divided into 12 increments of 30° and the ordinate heights at the midincrement points. Normally this method would be used for waveshapes for which it is not possible to write convenient mathematical expressions. The ordinate heights y_1, y_2, y_3, etc. are usually measured directly from oscillographs of the waveform. This wave has been selected for its simplicity and for the possibility of checking the accuracy of the solution with the analytic method of Example 7.3. To keep our work orderly we establish Table E-4.1.

Table E-4.1

x Degrees	y	$\sin x$	Product $y \sin x$	$2x$ Degrees	$\sin 2x$	Product $y \sin 2x$
15	0.417	0.259	+0.108	30	+0.500	+0.210
45	1.25	0.707	+0.884	90	+1.000	+1.25
75	2.08	0.966	+2.01	150	+0.500	+1.04
105	2.92	0.966	+2.82	210	−0.500	−1.46
135	3.75	0.707	+2.65	270	−1.000	−3.75
165	4.58	0.259	+1.18	330	−0.500	−2.29
195	5.42	−0.259	−1.40	390	+0.500	2.71
225	6.25	−0.707	−4.42	450	+1.00	6.25
255	7.08	−0.966	−6.84	510	+0.500	3.54
285	7.92	−0.966	−7.65	570	−0.500	−3.96
315	8.75	−0.707	−6.19	630	−1.000	−8.75
345	9.58	−0.259	−2.48	690	−0.500	−4.79
Total	60.00		−19.33			−10.00

$$A_0 = \frac{60.0}{12} = 5.0, \quad B_1 = \frac{2(-19.33)}{12} = -3.22, \quad B_2 = \frac{2(-10.00)}{12} = -1.67$$

The first three nonzero terms in the series are $i = 5 - 3.22 \sin 31{,}400t - 1.67 \sin 62{,}800t$ amp. This equation may be compared with the analytic solution of Example 7.3.

7.4 RMS VALUES AND AVERAGE POWER OF NONSINUSOIDAL PERIODIC WAVES

In Chapter 5 the root-mean-square value of a periodic function, i, was given as

$$I_{rms} = \sqrt{\frac{1}{T}\int_{K}^{K+T} i^2\, dt}. \tag{7.13}$$

If i may be expressed as a Fourier series,

$$i = I_0 + I_1 \cos(\omega_1 t + \phi_1) + I_2 \cos(2\omega_1 t + \phi_2)$$
$$+ I_3 \cos(3\omega_1 t + \phi_3)\ldots,$$

$$I_{rms}^2 = \text{avg}(i^2) = \frac{1}{T}\int_K^{K+T} i^2\, dt = \frac{1}{2\pi}\int_{K_1}^{K_1+2\pi} i^2\, d(\omega_1 t). \tag{7.14}$$

$$= \text{avg}\{[I_0 + I_1\cos(\omega_1 t + \phi_1) + I_2\cos(2\omega_1 t + \phi_2)\ldots] \times$$
$$[I_0 + I_1\cos(\omega_1 t + \phi_1) + I_2\cos(2\omega_1 t + \phi_2)\ldots]\}$$

We perform the indicated operation in steps, the first step being to multiply the Fourier series by the first term of the series, I_0, and taking the average value of this product. The use of the orthogonality properties results in I_0^2 for this step. We then multiply the Fourier series by the second term of the series, $I_1 \cos(\omega_1 t + \phi_1)$ and find the average value of this product to be $I_1^2/2$. Continuation of this process results in

$$I_{rms}^2 = I_0^2 + \frac{I_1^2}{2} + \frac{I_2^2}{2} + \frac{I_3^2}{2}\ldots,$$

or

$$I_{rms} = \sqrt{I_0^2 + \frac{I_1^2}{2} + \frac{I_2^2}{2} + \frac{I_3^2}{2}\ldots} \tag{7.15}$$

If we expressed this result in terms of the rms value of each sinusoid,

$$I_{rms} = \sqrt{I_0^2 + I_{1rms}^2 + I_{2rms}^2 + I_{3rms}^2\ldots} \tag{7.16}$$

Similar expressions apply for voltages written as Fourier series.

In considering power flow, let

$$i = I_0 + I_1 \cos(\omega_1 t + \phi_1) + I_2 \cos(2\omega_1 t + \phi_2)$$
$$+ I_3 \cos(3\omega_1 t + \phi_3)\ldots, \tag{7.17}$$

and

$$v = V_0 + V_1 \cos(\omega_1 t + \phi_1') + V_2 \cos(2\omega_1 t + \phi_2')$$
$$+ V_3 \cos(3\omega_1 t + \phi_3')\ldots. \tag{7.18}$$

Then $p = vi$, and

$$\text{average } p = P = \frac{1}{T}\int_K^{K+T} vi\, dt = \frac{1}{2\pi}\int_{K_1}^{K_1+2\pi} vi\, d(\omega_1 t). \tag{7.19}$$

The use of the same technique as in obtaining I_{rms} results in

$$P = V_0 I_0 + \frac{V_1 I_1}{2} \cos(\phi_1' - \phi_1) + \frac{V_2 I_2}{2} \cos(\phi_2' - \phi_2) \ldots \quad (7.20)$$

The student should prove Equation 7.20.

It is indicated in Appendix D that a D'Arsonval instrument may be used to measure average values of voltage or current. It is also indicated that an electrodynamometer type of instrument may be used to measure rms value of voltage or current or it may be used to measure the average value of the product (voltage across the voltage coil times the current through the current coil). In the latter case this instrument may be connected to read average power.

The accurate measurement of any physical quantity requires engineering ability of high order. All instruments have inherent limitations and all need to be calibrated. The introduction of an instrument into a physical system changes the system from what it was without the instrument.

It is assumed that the student is concurrently taking a laboratory course in which he is being introduced to the problems of measurement. For our purposes in this text we shall assume that the instruments are ideal unless otherwise specified. This means that an ammeter, or current coil of a wattmeter, is assumed to have zero impedance and that a voltmeter or voltage coil of a wattmeter is assumed to have zero admittance.

Example 7.5

GIVEN: The circuit shown in Fig. E-5.1.

TO FIND: The steady-state current i_{ss}, the reading of the D'Arsonval instrument A_1, the reading of the instruments A_2 and V which are electrodynamometer instruments calibrated to read rms values, and the reading of the wattmeter P which is an electrodynamometer instrument calibrated to read average power.

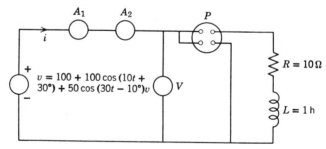

FIG. E-5.1

SOLUTION: We shall use the principle of superposition. For the d-c component,

$$I_{\text{dc}} = \tfrac{100}{10} = 10 \text{ amp.}$$

For the fundamental,
$$\omega_1 L = (10)(1) = 10 \ \Omega$$
$$\hat{Z}_{(1)} = 10 + j10 = 14.14e^{+j45°}$$

and
$$\hat{I}_{(1)} = \frac{100e^{j30°}}{14.14e^{+j45°}} = 7.07e^{-j15°},$$

For the third harmonic,
$$3\omega_1 L = (30)(1) = 30 \ \Omega,$$
$$\hat{Z}_{(3)} = 10 + j30 = 31.6e^{j71.6°},$$

and
$$\hat{I}_{(3)} = \frac{50e^{-j10°}}{31.6e^{+j71.6}} = 1.58e^{-j81.6°}.$$

Then
$$i = I_{\text{dc}} + \text{Re}(\hat{I}_{(1)}e^{j10t}) + \text{Re}(\hat{I}_{(3)}e^{j30t}).$$
$$= 10 + 7.07 \cos(10t - 15°) + 1.58 \cos(30t - 81.6°) \text{amp}$$

The ammeter, A_1, reads the average value of i or 10 amperes.

$$I_{\text{(rms)}} = \sqrt{10^2 + \frac{7.07^2 + 1.58^2}{2}} = \sqrt{126.3} = 11.2 \text{ amp.}$$

$$V_{\text{(rms)}} = \sqrt{100^2 + \frac{100^2 + 50^2}{2}} = 128 \text{ volts.}$$

Thus the ammeter, A_2, reads 11.2 amperes and the voltmeter, V, reads 128 volts.

$$P = V_{\text{dc}} I_{\text{dc}} + \frac{V_1 I_1}{2} \cos(\phi_{v_1} - \phi_{i_1}) + \frac{V_3 I_3}{2} \cos(\phi_{v_3} - \phi_{i_3})$$
$$= (10)(100) + \frac{(100)(7.07)}{2} \cos(30° + 15°)$$
$$+ \frac{(50)(1.58)}{2} \cos(-10° + 81.6°)$$
$$= 1000 + 250 + 12.5 = 1263 \text{ watts.}$$

Thus, the wattmeter P reads 1263 watts (actually we would probably not be able to read the instrument accurately to four places, and we should realize that the instrument may be limited to less accuracy than this).

Example 7.6

GIVEN: The periodic current wave shown in Fig. E-6.1.

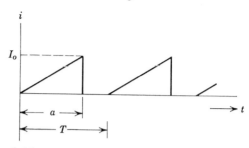

FIG. E-6.1

TO FIND: The rms value of this current.

SOLUTION: It is possible to determine the Fourier series of this wave and then evaluate the rms value from Equation 7.19. It is far easier to start with the defining expression for rms values:

$I_{rms}^2 = \text{avg}(i^2)$

i^2 as a function of t is shown in Fig. E-6.2.

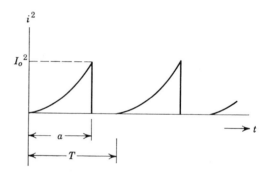

FIG. E-6.2

Since the function i may be expressed during the first period by $i = I_0 t/a$ for $0 < t < a$, and $i = 0$ for $T > t > a$.

$$I_{rms}^2 = \frac{1}{T}\int_0^a \frac{I_0^2 t^2}{a^2} \, dt = \frac{I_0^2 t^3}{T3a^2}\bigg|_0^a = \frac{I_0^2 a}{3T}$$

or

$$I_{rms} = \sqrt{\frac{a}{3T}} I_0$$

7.5 FREQUENCY SPECTRUM OF PERIODIC RECTANGULAR PULSES

The variation of harmonic content with changes in the shape of the wave may be studied for simple waveshapes. Consider the periodic wave consisting of rectangular pulses as shown in Fig. 7.4. We shall determine the expression for the Fourier series of this wave and study the variation of the harmonic content as the ratio a/T is varied. The

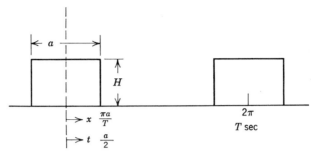

FIG. 7.4 Periodic wave of rectangular pulses.

choice of the origin as shown takes advantage of cosine symmetry, requiring $B_n = 0$.

$$A_0 = \frac{Ha}{T} \tag{7.21}$$

$$A_n = \frac{1}{\pi} \int_{-\pi a/T}^{\pi a/T} H \cos nx \, dx, \tag{7.22}$$

with $n = 1, 2, 3,$ etc.

$$A_n = \frac{2H}{\pi} \frac{\sin(n\pi a/T)}{n} = \frac{2Ha}{T} \left[\frac{\sin(n\pi a/T)}{n\pi a/T} \right] \tag{7.23}$$

The reason for writing Equation 7.23 as shown is that the function, $(\sin x)/x$, is well known and may be sketched as shown in Fig. 7.5. The reason for showing negative values of x will be given in the next section.

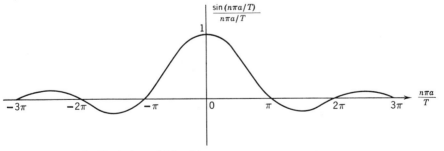

FIG. 7.5 Variation of $(\sin x)/x$ versus x.

As we vary the ratio, a/T, let us also vary H so that Ha/T is a constant $= A_0$. Then

$$A_n = 2A_0 \left[\frac{\sin(n\pi a/T)}{n\pi a/T} \right]. \tag{7.24}$$

Nonsinusoidal Periodic Waves—Steady State Response 349

If the ratio of a/T is $\frac{1}{2}$,

$$A_n = 2A_0 \left[\frac{\sin(n\pi/2)}{n\pi/2} \right],$$

and the magnitudes and signs of the cosine coefficients may be taken from a diagram such as Fig. 7.6a. The even harmonics are zero because of half-wave symmetry.

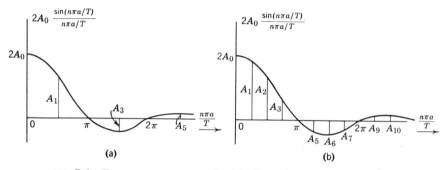

FIG. 7.6 Frequency spectrums for (a) $a/T = \frac{1}{2}$, and (b) $a/T = \frac{1}{4}$.

If the ratio of $a/T = \frac{1}{4}$,

$$A_n = 2A_0 \frac{\sin(n\pi/4)}{n\pi/4},$$

and the magnitude and signs of the cosine coefficients may be taken from a diagram such as Fig. 7.6b.

From this illustration it is clear that as the ratio a/T is decreased, the number of harmonics of relatively large magnitude increases. The question of significance of these harmonics must be settled in each individual case since the network functions are functions of frequency.

Let us consider T being held constant and a being made smaller, H being made larger so that aH is constant. The fundamental frequency remains constant and the number of harmonics of large magnitude increases. Consider only a finite number of harmonics, the first 1000 harmonics. As a approaches 0, H approaches infinity, and the amplitude of each of the 1000 harmonics approaches the same value, $2A_0$.

Suppose we consider a and H to be held constant, letting T become larger. The fundamental frequency decreases as T is increased. As T becomes very large the number of harmonics in a given portion of the curve of Fig. 7.6 increase. This curve may then be thought of as a continuous frequency spectrum instead of a discrete frequency spectrum. The amplitudes of the harmonics decrease as T is increased. In the limit, as the continuous frequency spectrum is approached, the amplitude of any particular frequency approaches zero!

7.6 EXPONENTIAL FORM OF FOURIER SERIES

Fourier series, which have been expressed in terms of sinusoidal time functions, may be expressed in exponential form by use of Euler's formula. Since $e^{j\phi} = \cos\phi + j\sin\phi$, $e^{-j\phi} = \cos\phi - j\sin\phi$, and thus

$$\cos\phi = \frac{e^{j\phi} + e^{-j\phi}}{2}, \tag{7.25}$$

and

$$\sin\phi = \frac{e^{j\phi} - e^{-j\phi}}{2j} \tag{7.26}$$

The Fourier series, in terms of sinusoids, is

$$f(x) = A_0 + \sum_{n=1}^{\infty}(A_n \cos nx + B_n \sin nx), \quad n = 1, 2, 3, \ldots \tag{7.27}$$

Then

$$f(x) = A_0 + \sum_{n=1}^{\infty}\left[\frac{A_n(e^{jnx} + e^{-jnx})}{2} + \frac{B_n(e^{jnx} - e^{-jnx})}{2j}\right]$$

$$= A_0 + \sum_{n=1}^{\infty}\left[\left(\frac{A_n - jB_n}{2}\right)e^{jnx} + \left(\frac{A_n + jB_n}{2}\right)e^{-jnx}\right]$$

$$= A_0 + \sum_{n=1}^{\infty}\left(\frac{A_n - jB_n}{2}\right)e^{jnx} + \sum_{n=1}^{\infty}\left(\frac{A_n + jB_n}{2}\right)e^{-jnx},$$

$$n = 1, 2, 3, \ldots \tag{7.28}$$

For the last summation of Equation 7.28 consider changing from positive values of n to negative values of n. Refer to the equations for A_n and B_n:

$$A_n = \frac{1}{\pi}\int_{K_1}^{K_1+2\pi} f(x) \cos nx\, dx, \tag{7.29}$$

and

$$B_n = \frac{1}{\pi}\int_{K_1}^{K_1+2\pi} f(x) \sin nx\, dx. \tag{7.30}$$

A change from positive values of n to negative values of n does not affect the magnitudes of A_n or B_n but does change the sign of B_n. Then Equation 7.28 may be written

$$f(x) = A_0 + \sum_{n=1}^{\infty}\left(\frac{A_n - jB_n}{2}\right)e^{jnx} + \sum_{n=-1}^{-\infty}\left(\frac{A_n - jB_n}{2}\right)e^{jnx}$$

$$= \sum_{-\infty}^{\infty}\left(\frac{A_n - jB_n}{2}\right)e^{jnx} = \sum_{-\infty}^{\infty}\hat{C}_n e^{jnx}, \tag{7.31}$$

in which $\hat{C}_n = (A_n - jB_n)/2$, and specifically $\hat{C}_0 = A_0$. $\hat{C}_n$ may be evaluated by use of Equations 7.29 and 7.30.

$$\hat{C}_n = \frac{A_n - jB_n}{2} = \frac{1}{2\pi} \int_{K_1}^{K_1+2\pi} [f(x) \cos nx - jf(x) \sin nx] \, dx$$

$$= \frac{1}{2\pi} \int_{K_1}^{K_1+2\pi} f(x) e^{-jnx} \, dx \qquad (7.32)$$

Let us use Equation 7.32 to determine $\hat{C}_n$ for the rectangular wave of Fig. 7.4.

$$\hat{C}_n = \frac{1}{2\pi} \int_{-\pi a/T}^{\pi a/T} H e^{-jnx} \, dx = \frac{H}{2\pi} \frac{e^{-jnx}}{n(-j)} \bigg|_{-\pi a/T}^{\pi a/T}$$

$$= \frac{H}{\pi n} \sin\left(\frac{n\pi a}{T}\right) = \frac{Ha}{T} \left[\frac{\sin(n\pi a/T)}{n\pi a/T}\right] \qquad (7.33)$$

$\hat{C}_n$ is real, thus $B_n = 0$ and $\hat{C}_n = A_n/2$. $f(x)$ may be determined by substitution into Equation 7.31. The exponential form is more concise than the trigonometric form; however, it requires the use of angular frequencies from $-\infty$ to ∞. It is this exponential form of the Fourier series that will be related in Chapter 9 to the Laplace transform of a periodic function.

7.7 SUMMARY

Nonsinusoidal periodic functions which are physically realizable may be expressed in a Fourier series. The Fourier series may be expressed in either trigonometric or exponential form. The coefficients of the terms of the series may be determined by analytical or numerical integration. For linear systems, superposition applies and the steady-state response to a periodic function may be obtained by adding the steady state responses to the individual frequencies.

FURTHER READING

Most introductory circuit texts include a discussion of Fourier series. Good discussions are in Clement and Johnson's *Electrical Engineering Science*, Chapter 16, McGraw-Hill, New York, 1960, and in Brenner and Javid's *Analysis of Electrical Circuits*, Chapter 19, McGraw-Hill, New York, 1959. For a discussion of the Gibbs phenonenon, see E. A. Guillemin's *The Mathematics of Circuit Analysis*, Chapter 7, M.I.T. Press, Cambridge, 1949.

PROBLEMS

7.1 Prove the orthogonality properties of the trigonometric functions.

7.2 Prove Equation 7.8. First determine the expression for A_1 by multiplying Equation 7.3 by cos x, and integrating the resulting product with respect to x over one period. Then generalize your result for A_n.

7.3 Prove Equation 7.9, using a technique similar to that outlined in problem 7.2.

7.4 In Example 7.1 show that the same result is obtained for A_n if the limits of integration are from 0 to 2π, instead of from $-\pi/2$ to $3\pi/2$.

7.5 The square wave of Fig. P-7.5 is like that of Fig. E-1.2 in Example 7.1 except that the origin is now chosen so that the wave has sine symmetry. Solve for B_n and then write the series as a function of x'. Show that this equation corresponds to that of Equation 3 of Example 7.1 if the relationship $x' = x + \pi/2$ is used.

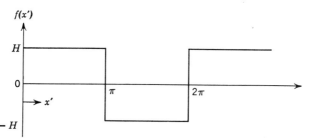

FIG. P-7.5

7.6 For each of the waveforms shown in Fig. P-7.6, indicate the period and the points you would choose as $x = 0$ in order to minimize the work of analysis. If some terms are zero because of symmetry, state the symmetry and give the terms that are zero.

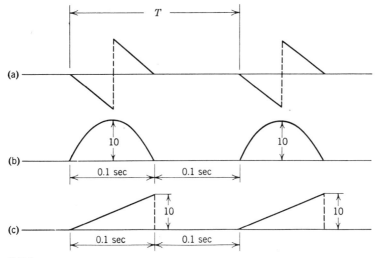

FIG. P-7.6

7.7 For a wave having cosine symmetry it is claimed that

$$\int_0^\pi f(x) \cos nx \, dx = \int_\pi^{2\pi} f(x) \cos nx \, dx,$$

and therefore only half the period needs to be integrated or,

$$A_n = \frac{2}{\pi} \int_0^\pi f(x) \cos nx \, dx$$

Prove that this is true.

7.8 For a wave having cosine symmetry and half-wave symmetry and $A_0 = 0$, it is claimed that

$$\int_0^{\pi/2} f(x) \cos nx \, dx = \int_{\pi/2}^\pi f(x) \cos nx \, dx.$$

If this result is combined with the results of problem 7.7,

$$A_n = \frac{4}{\pi} \int_0^{\pi/2} f(x) \cos nx \, dx.$$

Prove whether or not this is true.

7.9 A wave does not have half-wave symmetry but an origin can be chosen so that the wave displays cosine symmetry. Deduce whether or not an origin can be chosen so that the wave displays sine symmetry. Give your reasoning clearly.

7.10 For the wave shown in Fig. P-7.10, choose an origin which will minimize the number of terms in the series, and determine the first four nonzero terms of this series. Then write the series as a function of time, and make a sketch showing each of these terms and their resultant as a function of time.

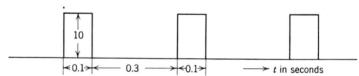

FIG. P-7.10

7.11 The wave shown in part b of problem 7.6 is one-half of a sinusoidal wave of amplitude 10. Showing the point you use as origin, determine the expression for the coefficients of the Fourier series. Evaluate the first four nonzero terms of the series and make a sketch showing each of these terms and their resultant as a function of time.

7.12 The wave shown in Fig. P-7.12 is a rectified (full-wave) sinusoidal wave. Indicate the period and the point you are using as origin. Evaluate the first four nonzero terms of the Fourier series and sketch these terms and their resultant as a function of time.

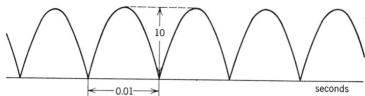

FIG. P-7.12

Partial answer: $|A_2| = 0.848$.

7.13 For the wave shown in Fig. P-7.13, choose an origin and determine the general expressions for the coefficients of the Fourier series. Evaluate the first four nonzero terms and make a sketch showing each of these terms and their resultant as a function of time.

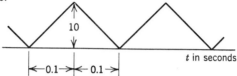

FIG. P-7.13

7.14 For the wave shown in Fig. P-7.14, choose an origin and evaluate the first four nonzero terms of the Fourier series.

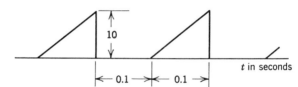

FIG. P-7.14

7.15 For the wave of Example 7.2, evaluate A_1 by numerical integration using 12 increments per period. Repeat, using 24 increments per period.

7.16 *a* For the wave of Problem 7.12, evaluate A_1 by numerical integration using 12 increments per period. Compare with the analytic solution.
b Repeat, using 24 increments per period.

7.17 Determine the average and rms values of the waves in Fig. P-7.17.

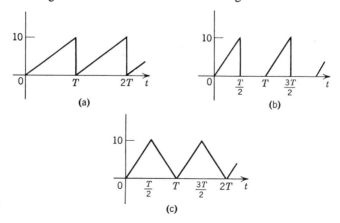

FIG. P-7.17

Nonsinusoidal Periodic Waves—Steady State Response 355

7.18 Determine the average and rms values of the waves in Fig. P-7.18.

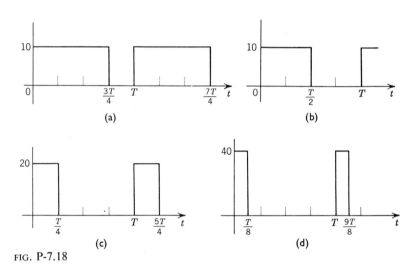

FIG. P-7.18

7.19 Prove Equation 7.20, showing your reasoning in detail.

7.20 For the circuit of Fig. P-7.20, determine i_{ss}, the reading of the D'Arsonval voltmeter, V_1, the rms readings of the ammeter A and voltmeter V_2 and the reading of the wattmeter P.

Partial answer: The ammeter reads 0.976 amp rms, $P = 9.48$ watts.

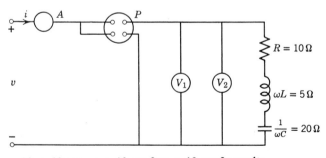

FIG. P-7.20 $v = 10 + 10 \cos \omega t + 10 \cos 2\omega t + 10 \cos 3\omega t$ volts.

356

7.21 For the circuit of Fig. P-7.21, determine i_{ss}, the rms readings of the ammeter A and voltmeter V and the reading of the wattmeter P.

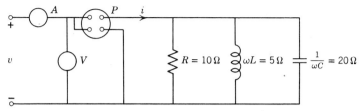

FIG. P-7.21 $v = 10 \cos \omega t + 10 \cos 2\omega t + 10 \cos 3\omega t$ volts.

7.22 For the circuit of Fig. P-7.22 assume $v = 10 + 10 \cos \omega t + 5 \sin 2\omega t$ volts. Determine the rms readings of the ammeters A_1, A_2, and A_3 and voltmeter V. Also determine the reading of the wattmeter, P.

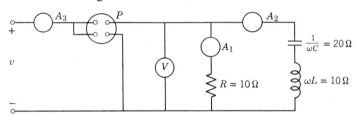

FIG. P-7.22

Partial answer: Ammeter A_3 reads 1.5 amp rms, $P = 16.2$ watts.

7.23 Same as problem 7.22, except use $v = 10 + 10 \cos(\omega t + 30°) + 5 \sin \omega t$ volts.

7.24 D'Arsonval instruments are sometimes used with rectifiers to measure alternating voltages or currents. A rectifier is an element which ideally has zero impedance for one direction of current and zero admittance for the other direction of current. The connection which allows the use of the instrument is shown in Fig. P-7.24.

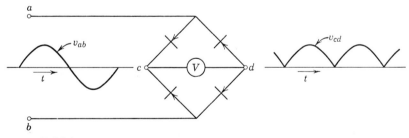

FIG. P-7.24

It is customary to calibrate the instrument with a sinusoidal source and a standard rms instrument, marking the scale of the D'Arsonval instrument in rms values. Thus if $v_{ab} = 141 \sin \omega t$ the instrument scale would be marked

Nonsinusoidal Periodic Waves—Steady State Response   357

100 volts. (The student should realize that if the instrument has been calibrated and scale marked with d-c source, it would give a reading for the fully rectified wave of $141(2/\pi) = 90$ volts.)

With this introduction let us study the error that is possible if this instrument is used on a nonsinusoidal wave. Specifically assume $v_{ab} = 141(\sin \omega t + 0.2 \sin 3 \omega t)$.
a Calculate the rms value of V_{ab}.
b Calculate the reading of the instrument.
Partial answer: (b) 106.7 volts.

7.25 Repeat problem 7.24, except use $v_{ab} = 141(\sin \omega t + 0.2 \cos 3 \omega t)$.
The student should carefully sketch this wave and observe that the resultant wave does not go through zero at $t = 0$ but at some other time which can be determined by "trial and error."

7.26 Repeat problem 7.24, except use $v_{ab} = 141(\sin \omega t + 0.2 \cos 2 \omega t)$.
The student should carefully sketch this wave and note that the two portions of the rectified wave are of different duration.

7.27 The triangular voltage wave of Fig. P-7.27a is across the capacitance of Fig. P-7.27b. Determine the rms value of the current i.

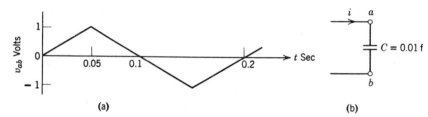

FIG. P-7.27

7.28 An amplitude modulated voltage may be written $v = (V_c + V_m \cos \omega_m t) \cos \omega_c t$ in which ω_c is the carrier angular frequency, ω_m is the modulating angular frequency, and $(V_m/V_c)(100)$ is the percentage of modulation. Assume ω_m and ω_c are integers.
a Write v as a Fourier series.
b Determine the rms value of v if $V_c = 100$ volts and the percentage of modulation is (1) zero, and (2) 100%.

7.29 The current of Fig. P-7.29a is in steady-state conditions through the circuit of Fig. P-7.29b. The average value of v_{cd} is zero.
a Sketch v_{ab}, v_{bc}, v_{cd}, and v_{ad} as functions of time, labeling values at key points for each curve.
b Determine the rms values for i, v_{ab}, v_{bc}, v_{cd}.
Partial answer: $I_{rms} = 5.77$ amp, $V_{cd_{rms}} = 9.1$ volts.

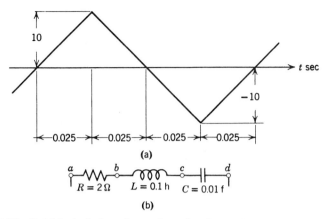

FIG. P-7.29

7.30 The circuit of Fig. P-7.30a is designed to reduce the alternating components of the rectified 60-cps input voltage shown in Fig. P-7.30b.

a Calculate the amplitudes of the d-c component, of the fundamental component v_{bc_1} and of the second harmonic component v_{bc_2} for the voltage v_{bc}. The voltage v_{ac} has the Fourier coefficients: $A_0 = 63.7$, $A_1 = 42.4$, and $A_2 = 8.5$.

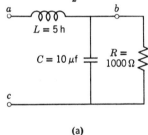
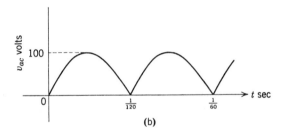

FIG. P-7.30

Partial answer: $V_{bc_1} = 1.53$ volts.

b Determine the ratios,

$$\frac{V_{bc_1}}{V_{bc_{dc}}} \quad \text{and} \quad \frac{V_{bc_2}}{V_{bc_{dc}}}.$$

Answer: $\dfrac{V_{bc_1}}{V_{bc_{dc}}} = 0.0243$, $\dfrac{V_{bc_2}}{V_{bc_{dc}}} = 0.00119$.

7.31 For the circuit shown in Fig. P-7.31, the switch is closed at $t = 0$. Determine

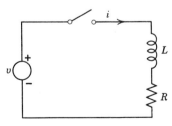

FIG. P-7.31

Nonsinusoidal Periodic Waves—Steady State Response 359

the complete solution for i. Use either the classical or the Laplace transform method. $v = 10 + 10 \cos 2t$ volts. $R = 2 \, \Omega$, $L = 1$ henry.

7.32 For the circuit shown in Fig. P-7.32 the switch is closed at $t = 0$. $v_{ab}(0-) = 5$ volts. Determine the complete solution for i. Use either classical or Laplace transform methods. $v = 10 + 10 \cos 2t$ volts, $R = 2$ ohms, $C = 1$ f.

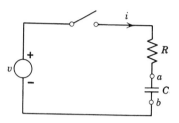

FIG. P-7.32

7.33 For the periodic pulse of Fig. P-7.33, consider aH and T to be held constant, but that the ratio a/T be varied from $\frac{1}{2}$ to 1. How does this variation affect the frequency content of the wave? Show clearly, using sketches similar to those of Fig. 7.6.

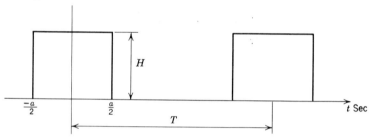

FIG. P-7.33

7.34 For the periodic pulse of Fig. P-7.33, assume aH and T are held constant. Let $A_0 = aH/T$.
   a For $a/T = \frac{1}{5}$, determine, in terms of A_0, the amplitude of the fundamental and each successive harmonic until the first zero amplitude is obtained.
   b Repeat for $(a/T) = \frac{1}{10}$.

7.35 a For the periodic pulse of Fig. P-7.33, assume $a = 0.1$ second, $H = 10$ and $T = 0.5$ second. Determine the amplitude of the d-c term, the fundamental, and each successive harmonic until the first zero amplitude is obtained.
   b Repeat except use $T = 1.0$ second.

7.36 For the square wave of Example 7.1, write the series in exponential form. Show that this is identical with the trigonometric solution obtained in Example 7.1.

8 Three-Phase Systems

Low-frequency (60 cps) electromagnetic power is usually generated by means of rotating machines called synchronous generators. The rotor of this machine consists of an electromagnet with the d-c current supplied through slip-rings from a separate source. The windings on the stator are placed in axial slots which are distributed around the machine. As the rotor turns, voltages are induced in the circuits of the stator because the flux linkages of these circuits change with time. By proper design of the rotor and by suitable location of the stator windings the voltage generated can become very nearly a sinusoidal function of time.

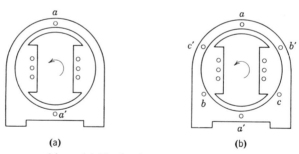

FIG. 8.1 Synchronous machines. (*a*) Single-phase generator. (*b*) Three-phase generator.

A two-pole machine must rotate at 3600 revolutions per minute to generate a frequency of 60 cycles per second, a four-pole machine must rotate at 1800 revolutions per minute for this purpose. In Fig. 8.1*a* is shown a two-pole machine with a winding, a-a', shown as a concentrated coil rather than distributed over a portion of the stator surface. This is a single-phase generator. If three identical windings, a-a', b-b', and c-c', as shown in Fig. 8.1*b*, are distributed properly over the stator surface, the machine has three sources of power in which the generated voltages differ from each other by 120°. This is called a three-phase generator. For the generator shown the voltage, $v_{bb'}$, would lag the voltage, $v_{aa'}$, by 120° and the voltage $v_{cc'}$ would lag the voltage $v_{bb'}$ by 120°. The sequence could be written $v_{aa'}, v_{bb'}, v_{cc'}, v_{aa'}$, etc., or simply abc. If the rotation of the machine were reversed, the sequence would be acb.

For an n-phase system the voltages would differ by integral values of $360°/n$, with the exception of the two-phase system in which there are two sources differing in phase angle by 90°, and which is thus arbitrarily defined as being one half of a four-phase system.

Polyphase machines and polyphase systems have many advantages over single-phase machines and systems. Some of these will become apparent as we consider certain aspects of three-phase systems. We limit our text discussion to three-phase systems since these are the most common and since the same principles apply to all polyphase systems.

8.1 BALANCED LOADS

The three windings of the generator of Fig. 8.1b are connected in one of two ways as shown in Fig. 8.2.

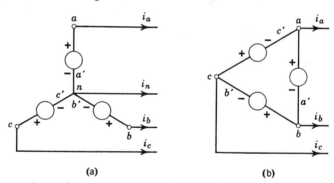

FIG. 8.2 Connections of a three-phase machine. (a) Wye (star) connection. (b) Delta (mesh) connection.

For the wye connection, terminals a', b', and c' are connected together and then called the common or "neutral" connection. Here we label this terminal n. If the currents i_a, i_b, and i_c are "balanced," meaning that they have the same magnitude but differ in phase by 120°, then $i_n = 0$ and there is no need for a lead to be connected to this terminal.

In the delta connection, terminal a' is connected to b, terminal b' is connected to c, and then c' is connected to a, the last connection being possible without resulting current flow for balanced voltages, since $v_{ac'}$ is zero before the connection is made.

362

The student may wonder why each winding is represented only by an ideal voltage source since there is surely mutual coupling between the various phases. Also there should be resistances and self-inductances to be concerned about. These are good questions and cannot be adequately answered without detailed study of these machines. Let us realize that in practice, not only is the load usually balanced but the rotor field current is automatically adjusted to maintain the system voltage at a fairly constant value so that our assumption of idealized voltage sources is a fairly good one at this time.

Since we have idealized the source, we may picture it as connected either in wye or delta; we shall consistently show it as a wye connection in order that we may analyze problems in which the neutral current is not zero.

The loads on three-phase generators may consist of either balanced three-phase motors or single-phase loads of various types. For a power system the single-phase loads are distributed among the various phases so that the net load is usually well balanced.

8.1.1 Voltages and Currents for Balanced Loads (Phasor Diagrams)

For balanced voltages and balanced loads a three-phase circuit may be treated as three separate single-phase circuits and the nature of the solution may be indicated on phasor diagrams.

For the balanced wye load a connection diagram is shown in Fig. 8.3a. The fact that the voltage $\hat{V}_{dn}$ is zero may readily be determined from Kirchhoff's current law about node d.

$$\hat{V}_{dn}\left[\frac{1}{\hat{Z}} + \frac{1}{\hat{Z}} + \frac{1}{\hat{Z}}\right] = \frac{\hat{V}_{an}}{\hat{Z}} + \frac{\hat{V}_{bn}}{\hat{Z}} + \frac{\hat{V}_{cn}}{\hat{Z}}. \tag{8.1}$$

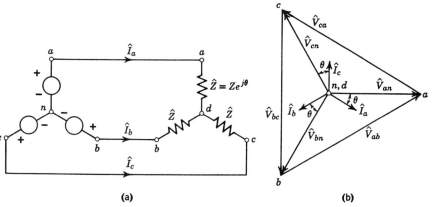

FIG. 8.3 Connection and phasor diagram for balanced wye load. (a) Connection diagram. (b) Phasor diagram.

Since the impedances in each phase are equal and since $\hat{V}_{an} + \hat{V}_{bn} + \hat{V}_{cn} = 0$, $\hat{V}_{dn} = 0$. Thus the terminals n and d may be connected together to indicate that the three single-phase circuits may be treated separately. $\hat{I}_a = \hat{V}_{ad}/\hat{Z} = \hat{V}_{an}/\hat{Z}$ or the current $\hat{I}_a$ lags $\hat{V}_{an}$ by the angle θ. Similarly $\hat{I}_b$ lags $\hat{V}_{bn}$ by the angle θ and $\hat{I}_c$ lags $\hat{V}_{cn}$ by the angle θ. In the phasor diagram of Fig. 8.3b, $\hat{V}_{an}$ has been selected as the reference phasor and the sequence is abc, $\hat{V}_{bn}$ lagging $\hat{V}_{an}$ by 120°, $\hat{V}_{cn}$ lagging $\hat{V}_{bn}$ by 120°. As mentioned in Chapter 5, page 230, it is permissible to associate the first subscript of the voltage with the head of the directed line segment and the second subscript with the tail of the directed line segment. We do this in Fig. 8.3b and in all figures in this chapter for this technique is helpful in polyphase analysis. The line-to-line voltages, $\hat{V}_{ab}$, $\hat{V}_{bc}$, and $\hat{V}_{ca}$ may be sketched directly as shown. For the wye connection the phasor diagram of Fig. 8.4 shows that the magnitude of a line voltage, V_{ab}, is related to the magnitude of a phase voltage, V_{an}, by $V_{ab} = 2V_{an} \sin 60° = \sqrt{3} V_{an}$.

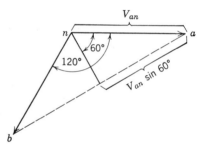

FIG. 8.4 Phase and line voltage for wye connection.

Let V_l refer to the amplitude of a line voltage and V_ϕ refer to the amplitude of a phase voltage. Then, for the wye connection

$$V_l = \sqrt{3} V_\phi. \tag{8.2}$$

The connection diagram shows that the line current, I_l, must equal the phase current, I_ϕ, or

$$I_l = I_\phi. \tag{8.3}$$

For the balanced delta load a connection diagram is shown in Fig. 8.5a. For this connection,

$$V_l = V_\phi, \tag{8.4}$$

while

$$I_l = \sqrt{3} I_\phi, \tag{8.5}$$

since the line current is the difference of two phase currents which are equal in magnitude but differ in phase angle by 120°.

In the delta connection, double subscripts are used with the phase currents for convenience. The convention is that a current such as i_{ab} is positive when in the direction from a to b through the load. Do not associate the subscripts with the ends of the phasor $\hat{I}_{ab}$. The current $\hat{I}_{ab}$ lags the voltage $\hat{V}_{ab}$ by the angle θ. The line current can be sketched in readily from Kirchhoff's current equations $\hat{I}_a = \hat{I}_{ab} - \hat{I}_{ca}$, $\hat{I}_b = \hat{I}_{bc} - \hat{I}_{ab}$, and $\hat{I}_c = \hat{I}_{ca} - \hat{I}_{bc}$.

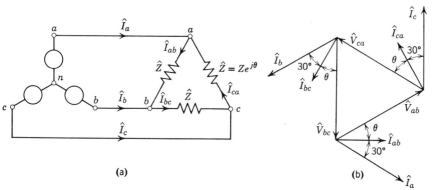

FIG. 8.5 Connection and phasor diagram for balanced delta load. (*a*) Connection diagram. (*b*) Phasor diagram.

The student should observe that the line voltage phasors of Fig. 8.5*b* are sketched identically to those of Fig. 8.3*b*. Note that line current phasors have identical angles in these two figures.

Example 8.1

GIVEN: The balanced 60-cps three-phase system shown in Fig. E-1.1 with sequence *acb*. $v_{an} = 100 \cos 377t$ volts, $v_{cn} = 100 \cos(377t - 120°)$ volts, and $v_{bn} = 100 \cos(377t + 120°)$ volts.

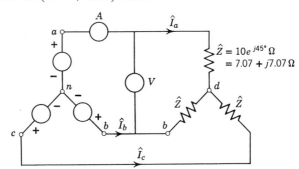

FIG. E-1.1

TO FIND: (*a*) The line currents $\hat{I}_a$, $\hat{I}_b$, and $\hat{I}_c$, the average three-phase power, and the readings of the ammeter A and voltmeter V which read rms values.

(b) Same as (a) except that the same three impedances are connected in delta instead of wye.

SOLUTION: (a) Since $\hat{V}_{dn} = 0$ for this balanced condition,

$$\hat{I}_a = \frac{\hat{V}_{ad}}{\hat{Z}} = \frac{\hat{V}_{an}}{\hat{Z}} = \frac{100e^{j0°}}{10e^{j45°}} = 10e^{-j45°} \text{ amp.} \tag{1}$$

The phasor diagram showing phase voltages and currents can now be sketched as shown in Fig. E-1.2. $\hat{I}_a$ lags $\hat{V}_{an}$ by 45°. Furthermore $\hat{I}_b$ lags $\hat{V}_{bn}$ by 45° and $\hat{I}_c$ lags $\hat{V}_{cn}$ by 45°.

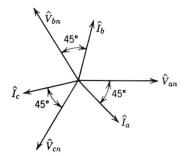

FIG. E-1.2

Thus,

$$\hat{I}_b = 10e^{j75°} \tag{2}$$

and

$$\hat{I}_c = 10e^{-j165°} \tag{3}$$

The average power per phase, P_ϕ, is

$$P_\phi = \frac{V_\phi I_\phi}{2} \cos \theta = \frac{100(10) \cos 45°}{2} = 354 \text{ watts,} \tag{4}$$

or

$$P_\phi = \frac{I_\phi^2 R}{2} = \frac{10^2(7.07)}{2} = 354 \text{ watts.} \tag{5}$$

The average three-phase power, $P_{3\phi}$, is

$$P_{3\phi} = 3P_\phi = 3(354) = 1062 \text{ watts.} \tag{6}$$

Since the currents and voltages are sinusoidal, the ammeter A reads

$$\frac{I_a}{\sqrt{2}} = \frac{10}{\sqrt{2}} = 7.07 \text{ amp,} \tag{7}$$

and the voltmeter V reads

$$\frac{V_{ab}}{\sqrt{2}} = \frac{\sqrt{3}\,100}{\sqrt{2}} = 122 \text{ volts.} \tag{8}$$

(b) For the delta-connected load, our connection diagram is shown in Fig. E-1.3.

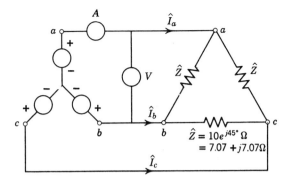

FIG. E-1.3

The amplitude of the line voltages is $\sqrt{3}(100)$ or 173 volts. These voltages are shown in Fig. E-1.4, the appropriate angles being determined from Fig. E-1.2. The current

$$\hat{I}_{ab} = \frac{\hat{V}_{ab}}{\hat{Z}} = \frac{173e^{-j30°}}{10e^{j45°}} = 17.3e^{-j75°} \text{ amp.}$$

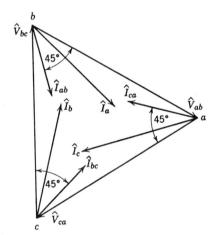

FIG. E-1.4

The current $\hat{I}_{bc}$ lags $\hat{V}_{bc}$ by 45°, is sketched as shown, and may be written

$$\hat{I}_{bc} = 17.3e^{j45°} \text{ amp.}$$

The current $\hat{I}_{ca}$ lags $\hat{V}_{ca}$ by 45° and may thus be written

$$\hat{I}_{ca} = 17.3e^{-j195°} = 17.3e^{j165°} \text{ amp.}$$

The line currents may be sketched in the phasor diagram. Since $\hat{I}_a = \hat{I}_{ab} - \hat{I}_{ca}$, it will lead $\hat{I}_{ab}$ by 30° and will have the magnitude of $\sqrt{3}(I_{ab}) = \sqrt{3}(17.3) = 30$ amp. Thus

$$\hat{I}_a = 30e^{-j45°} \text{ amp.} \tag{9}$$

Similarly

$$\hat{I}_b = 30e^{j75°} \text{ amp,} \tag{10}$$

and

$$\hat{I}_c = 30e^{-j165°} \text{ amp.} \tag{11}$$

The average power per phase is

$$P_\phi = \frac{I_\phi^2 R}{2} = \frac{(17.3)^2 7.07}{2} = 1062 \text{ watts.} \tag{12}$$

$$P_{3\phi} = 3P_\phi = 3186 \text{ watts.} \tag{13}$$

The ammeter A reads

$$\frac{30}{\sqrt{2}} = 21.2 \text{ amp.} \tag{14}$$

The voltmeter V reads, as before, 122 volts. (15)

The effect of changing the connection of the three impedances from wye to delta was to increase the line currents by a factor of 3 and the average power by a factor of 3. The phase currents were increased by a factor of $\sqrt{3}$.

8.1.2 Power Under Balanced Load Conditions

One advantage of the balanced three-phase machine over the single-phase machine is that the power, and therefore the torque, is not a function of time. We shall prove this for the wye-connected generator of Fig. 8.3. The power delivered by the source voltage v_{an}, is

$$p_a = v_{an} i_a, \tag{8.6}$$

with $v_{an} = V \cos \omega t$, and $i_a = I \cos(\omega t - \theta)$. Then

$$p_a = VI \cos \omega t \cos(\omega t - \theta),$$

$$= \frac{VI}{2} [\cos \theta + \cos(2\omega t - \theta)]. \tag{8.7}$$

If this were the only output of our generator the instantaneous power has an average value, $(VI/2)\cos \theta$, but also has a term $(VI/2)\cos(2\omega t - \theta)$ which pulsates at double the frequency of voltage or current. The torque would also have this pulsation; this is one reason why single-phase machines are noisier than polyphase machines.

For the source voltage v_{bn} the power output is

$$p_b = v_{bn} i_b$$

$$= V \cos(\omega t - 120°) I \cos(\omega t - 120° - \theta),$$

$$= \frac{VI}{2} [\cos \theta + \cos(2\omega t - 240° - \theta)]. \tag{8.8}$$

For the source voltage v_{cn} the power output is

$$p_c = v_{cn}i_c,$$
$$= V\cos(\omega t - 240°)I\cos(\omega t - 240° - \theta),$$
$$= \frac{VI}{2}[\cos\theta + \cos(2\omega t - 120° - \theta)]. \tag{8.9}$$

The total instantaneous power is the sum of these and thus

$$p_a + p_b + p_c = \tfrac{3}{2}VI\cos\theta. \tag{8.10}$$

This shows that a balanced machine under steady-state sinusoidal conditions has a constant power flow and therefore constant torque. It also shows that average three-phase power is three times the average power per phase.

It is useful to review our definitions for power and to introduce some new definitions which are helpful in 60-cps systems. Figure 8.6 shows two conductors connecting two electrical systems. For the assumed

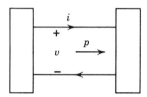

FIG. 8.6 Sketch for energy flow convention.

positive voltage and current conventions, the positive direction for energy flow is fixed and is to the right when $p = vi$ is positive. If v and i are each sinusoidal,

$$v = V\cos(\omega t + \alpha_v) = \text{Re}(\hat{V}e^{j\omega t}),$$
$$i = I\cos(\omega t + \alpha_i) = \text{Re}(\hat{I}e^{j\omega t}),$$

then the average power is

$$p_{\text{avg}} = (vi)_{\text{avg}} = P = \frac{VI}{2}\cos(\alpha_v - \alpha_i). \tag{8.11}$$

It is useful to consider a complex volt-ampere term, $\hat{V}\hat{I}^*/2$.

$$\frac{\hat{V}\hat{I}^*}{2} = \frac{VI}{2}e^{j(\alpha_v - \alpha_i)} = \frac{VI}{2}\cos(\alpha_v - \alpha_i)$$
$$+ j\frac{VI}{2}\sin(\alpha_v - \alpha_i),$$
$$= P + jQ. \tag{8.12}$$

The term, $VI/2$, with the units of volt amperes, is significant since the rating of equipment is given in such terms. The term $P = 1/2\,\text{Re}(\hat{V}\hat{I}^*)$, with the units of watts, is significant since it represents average rate of energy transfer. The term Q, which is equal to $VI/2 \sin(\alpha_v - \alpha_i)$, is helpful in circuit calculations and is given the units of reactive volt amperes or vars.

One application is with parallel circuits in which case the net average power is the algebraic sum of the individual average powers and the net reactive volt amperes is the algebraic sum of the individual reactive volt amperes. If $\hat{I} = \hat{I}_1 + \hat{I}_2$, then

$$\frac{\hat{V}\hat{I}^*}{2} = \frac{\hat{V}}{2}[\hat{I}_1^* + \hat{I}_2^*] = \frac{\hat{V}\hat{I}_1^*}{2} + \frac{\hat{V}\hat{I}_2^*}{2}$$

$$= P_1 + jQ_1 + P_2 + jQ_2$$
$$= P_1 + P_2 + j(Q_1 + Q_2). \tag{8.13}$$

A balanced three-phase system with balanced load consists of three single-phase systems in which the voltages and currents have the same amplitude but differ from system to system by 120° in phase. These three systems have the same P and Q. Thus the average power for a balanced three-phase system, $P_{3\phi}$, is three times the average power for one of the phases, P_ϕ.

$$P_{3\phi} = 3P_\phi = 3 \frac{V_\phi I_\phi}{2} \cos \angle_{\hat{I}_\phi}^{\hat{V}_\phi}. \tag{8.14}$$

Similarly

$$Q_{3\phi} = 3Q_\phi = 3 \frac{V_\phi I_\phi}{2} \sin \angle_{\hat{I}_\phi}^{\hat{V}_\phi}. \tag{8.15}$$

The volt-ampere rating of a three-phase machine means

$$\left(\frac{VI}{2}\right)_{3\phi} = 3\left(\frac{V_\phi I_\phi}{2}\right). \tag{8.16}$$

(Many texts use I_m and V_m to indicate the amplitudes of sinusoidal current and voltage while using I and V to indicate their rms values. Although this is convenient for steady-state sinusoidal calculations, it is not convenient in general.)

If rms values of voltage and current are given, then we need to recall that for sinusoidal quantities,

$$V = \sqrt{2}V_{\text{rms}}, I = \sqrt{2}I_{\text{rms}}$$

and Equation 8.14 could be written

$$P_{3\phi} = 3V_{\phi\text{rms}}I_{\phi\text{rms}} \cos \angle_{\hat{I}_\phi}^{\hat{V}_\phi}. \tag{8.17}$$

If the load is passive each phase may be represented by an impedance, $Ze^{j\theta}$. Equations 8.14 and 8.15 become

$$P_{3\phi} = \frac{3V_\phi I_\phi}{2} \cos \theta, \qquad (8.18)$$

and

$$Q_{3\phi} = \frac{3V_\phi I_\phi}{2} \sin \theta. \qquad (8.19)$$

Note that Q has been defined so that it is positive when θ is positive or when the load is inductive. A triangle showing the relationship between three phase volt amperes, power, and vars is shown in Fig. 8.7.

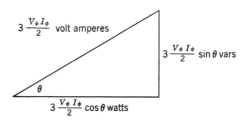

FIG. 8.7 Triangle for volt amperes, watts, and vars.

Example 8.2

GIVEN: Two balanced 3ϕ loads are in parallel on a system with line-to-line voltages of 400 volts rms. One load is an induction motor with an output of 10 horsepower (1 hp = 746 watts) operating at a lagging power factor of 0.8 and an efficiency of 85%. The other load is a 3ϕ resistive load of 5 kw.

TO FIND: The magnitude of the line current (rms).

SOLUTION: The power input to the motor is power out/eff. $P_{3\phi} = 746(10)/0.85 = 8760$ watts. The vars of the motor load are

$$Q_{3\phi} = P_{3\phi} \tan \cos^{-1}(0.8) = 8760(0.75) = 6570 \text{ vars}.$$

The net average power is then

$8760 + 5000 = 13{,}760$ watts.

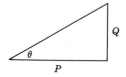

The net vars is 6570 vars and the three-phase volt amperes is

$$\left(\frac{VI}{2}\right)_{3\phi} = 3\left(\frac{V_\phi I_\phi}{2}\right) = \sqrt{P_{3\phi}^2 + Q_{3\phi}^2},$$

$$3\frac{V_\phi I_\phi}{2} = \sqrt{13{,}760^2 + 6570^2} = 15{,}200 \text{ volt amperes}.$$

It does not matter whether we consider the loads connected in delta or wye since for a wye connection the phase voltage is the line-to-line voltage divided by $\sqrt{3}$, $V_\phi = V_l/\sqrt{3}$ and the phase current is equal to the line current, $I_\phi = I_l$; while for the delta connection $V_\phi = V_l$ and $I_\phi = I_l/\sqrt{3}$. Thus, we may write

$$3\frac{V_\phi I_\phi}{2} = \frac{\sqrt{3}V_l I_l}{2} = \sqrt{3}V_{l_{rms}}I_{l_{rms}} = 15{,}200$$

or

$$I_{l_{rms}} = \frac{15{,}200}{\sqrt{3}V_{l_{rms}}} = \frac{15{,}200}{\sqrt{3}(400)} = 22 \text{ amp.}$$

8.2 UNBALANCED THREE-PHASE LOADS (BALANCED VOLTAGES)

With balanced voltages but unbalanced phase loads the technique will depend primarily on whether the loads are connected in delta or wye. If they are connected in delta the same technique as used for the balanced loads may be used except that now each phase current must be calculated individually and each line current then determined from the phase currents.

If the loads are connected in wye it is generally not desirable to convert to a delta because of the additional complex algebra. One should consider using mesh current or node voltage approaches. One should also consider using Thévenin's theorem. If there is a connection between the neutral of the generator and the common connection of the load and if the impedance of this branch is not zero, the nodal approach is usually best. If the impedance of this branch is zero, the phase currents can immediately be calculated since the phase voltages are determined.

Example 8.3

GIVEN: The unbalanced load shown in Fig. E-3.1 connected to a balanced, three-phase, 60-cycle, 208-volt (rms) line. The phase sequence is abc.

TO FIND: The line currents $\hat{I}_a$, $\hat{I}_b$, and $\hat{I}_c$, and the phase voltages $\hat{V}_{ad}$, $\hat{V}_{bd}$, and $\hat{V}_{cd}$, using $\hat{V}_{ab}$ as reference.

SOLUTION: We shall first determine the phase voltages of the generator. A funicular diagram of the line voltages is shown in Fig. E-3.2. $V_l = \sqrt{2}V_{l_{rms}} = 294$ volts. From the diagram,

$$\hat{V}_{an} = \frac{294}{\sqrt{3}}e^{-j30°} = 170e^{-j30°} = 147 - j85 \text{ volts,}$$

$$\hat{V}_{bn} = 170e^{-j150°} = -147 - j85 \text{ volts,}$$

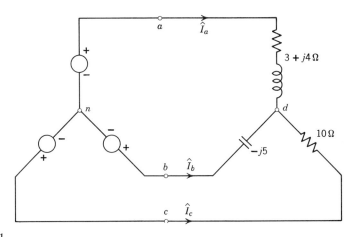

FIG. E-3.1

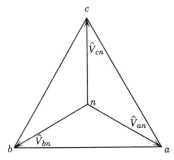

FIG. E-3.2

and

$$\hat{V}_{cn} = 170e^{+j90°} = 0 + j170 \text{ volts.}$$

Since our load is unbalanced the voltage between d and n will undoubtedly not be zero as in previous balanced systems. As a consequence the load phase voltages are not equal to the generator phase voltages and our approach to the solution must be altered.

Several approaches are possible but let us try the node-voltage method. As there are only two nodes, only one equation is required:

$$(\hat{Y}_{ad} + \hat{Y}_{bd} + \hat{Y}_{cd})\hat{V}_{dn} = \hat{V}_{an}\hat{Y}_{ad} + \hat{V}_{bn}\hat{Y}_{bd} + \hat{V}_{cn}\hat{Y}_{cd} \qquad (1)$$

where

$$(\hat{Y}_{ad} + \hat{Y}_{bd} + \hat{Y}_{cd}) = \frac{3 - j4}{25} + \frac{0 + j5}{25} + \frac{1}{10},$$

$$= 0.12 - j0.16 + j0.2 + 0.1,$$

$$= 0.22 + j0.04 \text{ mhos.}$$

Three-Phase Systems 373

Then
$$\hat{V}_{dn} = \frac{(170e^{-j30°})(0.12 - j.16) + (170e^{-j150°})(j0.2) + (170e^{j90°})(0.1)}{0.22 + j0.04},$$

$$= \frac{4.04 - j33.9 + 17 - j29.4 + j17}{0.22 + j0.04} = \frac{21.0 - j46.3}{0.22 + j0.04},$$

$$= \frac{2.78 - j10.8}{0.05} = 55.6 - j220 \text{ volts}. \tag{2}$$

$\hat{V}_{dn}$ is now located on the phasor diagram, Fig. E-3.3. We next calculate $\hat{V}_{ad}$, $\hat{V}_{bd}$, and $\hat{V}_{cd}$, the phase voltages of the load using the phasor

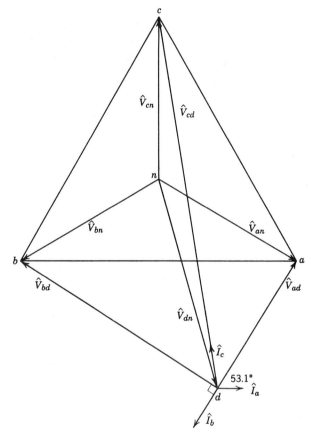

FIG. E-3.3

diagram as a guide. From Kirchhoff's voltage law,

$$\hat{V}_{ad} = \hat{V}_{an} - \hat{V}_{dn}$$
$$= 147 - j85 - 55.6 + j220 = 91.4 + j135 \text{ volts}, \tag{3}$$
$$\hat{V}_{bd} = \hat{V}_{bn} - \hat{V}_{dn}$$
$$= -147 - j85 - 55.6 + j220 = -203 + j135 \text{ volts}, \tag{4}$$

and

$$\hat{V}_{cd} = \hat{V}_{cn} - \hat{V}_{dn}$$
$$= 0 + j170 - 55.6 + j220 = -55.6 + j390 \text{ volts.} \quad (5)$$

The currents, then, are

$$\hat{I}_a = \frac{\hat{V}_{ad}}{\hat{Z}_{ad}} = \frac{91.4 + j135}{3 + j4} = 32.6 + j1.6 \text{ amp,} \quad (6)$$

$$\hat{I}_b = \frac{\hat{V}_{bd}}{\hat{Z}_{bd}} = \frac{-203 + j135}{-j5} = -27.0 - j40.6 \text{ amp,} \quad (7)$$

and

$$\hat{I}_c = \frac{\hat{V}_{cd}}{\hat{Z}_{cd}} = \frac{-55.6 + j390}{10} = -5.6 + j39.0 \text{ amp.} \quad (8)$$

As a check the sum of the currents ($\hat{I}_a + \hat{I}_b + \hat{I}_c$) must be zero.

Example 8.4

GIVEN: The circuit shown in Fig. E-4.1, which is the circuit for a device used to determine the phase sequence of a three-phase system.

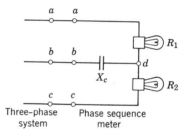

FIG. E-4.1

TO FIND: An analysis of the performance of this instrument. R_1 and R_2 are two lamps of approximately equal resistance, and for convenience it will be assumed the reactance $1/\omega C$ of the capacitor is very near the resistance of the lamps. The leads a, b, c, of the instrument are attached to the terminals of the line, the phase sequence of which is to be determined. One of the two lamps, R_1 or R_2, will burn more brightly depending on the phase sequence of the system.

To formulate our problem, let us assume that the phase sequence of the system is known, and is abc, and that we are to determine which lamp will be brightest for this sequence, and then are to label the instrument for future use. The brightness of a lamp is dependent on the current through it, so we must determine which lamp will have the greater current. This can be done by applying Thévenin's theorem to the circuit of our device as follows. Let us first remove lamp R_1 from the circuit and find Thévenin's voltage across the resulting open circuit, $\hat{V}_T = \hat{V}_{ad}$. This can best be done by use of a phasor diagram as shown in Fig. E-4.2. Since we selected $R = 1/\omega C$ the current $\hat{I}_1$ flowing in

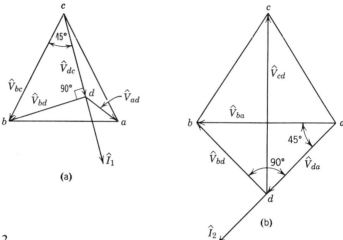

FIG. E-4.2

the remaining loop, bdc (see Fig. E-4.3), will lead the voltage $\hat{V}_{bc}$ by 45°, as shown in Fig. E-4.2a. The voltage $\hat{V}_{bd}$ must lag this current by 90°, the voltage $\hat{V}_{dc}$ must be in phase with this current, and $\hat{V}_{bd} + \hat{V}_{dc} = \hat{V}_{bc}$. Then $\hat{V}_{ad}$ may be drawn as shown in Fig. E-4.2a. This is the open circuit or Thévenin's voltage V_T.

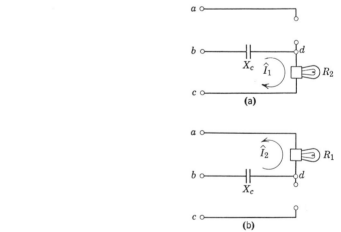

FIG. E-4.3

In Fig. E-4.3b we show the circuit diagram drawn for our circuit with lamp R_1 replaced and lamp R_2 removed. Fig. E-4.2b shows the phasor diagram for this circuit. $\hat{V}_{cd}$ is the open circuit or Thévenin's voltage for this circuit.

376

We can see then that the open circuit voltage $\hat{V}_{cd}$ is much greater than the open circuit voltage $\hat{V}_{ad}$.

Thévenin's impedance $\hat{Z}_T$ is now calculated for both circuits of Fig. E-4.3. Since we chose $1/\omega C \simeq R_1 \simeq R_2 = R$, $\hat{Z}_T$ is the same for both circuits, and is

$$\hat{Z}_T = \frac{\left(\dfrac{1}{j\omega C}\right)(R)}{R + \dfrac{1}{j\omega C}} = \frac{(-j1)(R)}{1 - j1} = \left(\frac{1 - j1}{2}\right) R.$$

The current through the respective lamps is then

$$\hat{I}_{R1} = \frac{\hat{V}_{ad}}{\hat{Z}_T + R_1},$$

and

$$\hat{I}_{R2} = \frac{\hat{V}_{cd}}{\hat{Z}_T + R_2},$$

and since $V_{cd} > V_{ad}$ and the denominators of both expressions are identical,

$$I_{R2} > I_{R1}$$

and lamp R_2 burns brightest. We can now label our instrument, putting a-b-c sequence next to lamp R_2 and a-c-b next to lamp R_1, and our instrument is ready for use.

8.3 POWER MEASUREMENTS IN THREE-PHASE SYSTEMS

The instrument used for power measurements in three-phase systems is generally the electrodynamometer type of instrument described in Appendix D. We assume that the current coil of this instrument has zero impedance, that the voltage coil has zero admittance, and that the reading is the average value of the product (voltage across the voltage coil times the current through the current coil.) The voltages and currents need not be sinusoidal.

If such an instrument is connected to a two-wire system it will read the average power being transmitted. If this instrument is connected to a system having more than two wires there may be some correlation between the reading of the instrument and the power or there may not be such correlation. The instrument is not capable of "seeing" the system, all we can expect is that its reading will be the average value of the product mentioned above.

8.3.1 The Two-Wattmeter Method for Measuring Average Power Transmitted Via Three Conductors

The measurement of average power in a three-conductor system may be performed by connecting three wattmeters as shown in Fig. 8.8. We shall show that the impedances $\hat{Z}_a$, $\hat{Z}_b$, and $\hat{Z}_c$ need not be identical nor need the line voltages be balanced. If the point d of the load were readily available, then the common connection terminal, o, of the wattmeters' voltage coils could be connected to point d and wattmeter P_c would read the average power dissipated in $\hat{Z}_c$, etc. However, the point d is often not available and a connection can be made as shown.

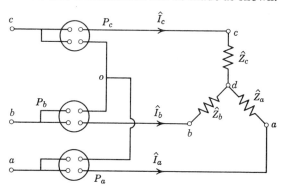

FIG. 8.8 Three wattmeters measuring power for three-conductor system.

For this connection the average power read by wattmeter P_a is $\mathrm{avg}(v_{ao}i_a)$, the average power read by wattmeter P_b is $\mathrm{avg}(v_{bo}i_b)$, and the average power read by wattmeter P_c is $\mathrm{avg}(v_{co}i_c)$. The total algebraic sum of the wattmeter readings is

$$P_a + P_b + P_c = \mathrm{avg}(v_{ao}i_a) + \mathrm{avg}(v_{bo}i_b) + \mathrm{avg}(v_{co}i_c). \tag{8.20}$$

The expression for the total instantaneous power supplied to the load is

$$p(\text{total}) = v_{ad}i_a + v_{bd}i_b + v_{cd}i_c. \tag{8.21}$$

Since $v_{ad} = v_{ao} + v_{od}$, $v_{bd} = v_{bo} + v_{od}$ and $v_{cd} = v_{co} + v_{od}$, Equation 8.21 may be written

$$p(\text{total}) = v_{ao}i_a + v_{bo}i_b + v_{co}i_c + v_{od}(i_a + i_b + i_c),$$

and since $i_a + i_b + i_c = 0$,

$$P(\text{total}) = \mathrm{avg}(v_{ao}i_a) + \mathrm{avg}(v_{bo}i_b) + \mathrm{avg}(v_{co}i_c). \tag{8.22}$$

Comparison of Equations 8.20 and 8.22 shows that the algebraic sum of the three wattmeters is equal to the average power supplied via the three conductors. This is independent of the voltage between terminals o and d. The terminal o can therefore be connected to line b.

This means that wattmeter, P_b, reads zero and may be removed resulting in the connection shown in Fig. 8.9.

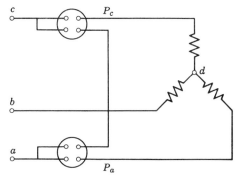

FIG. 8.9 Two-wattmeter connection for three-conductor line.

The above proof can be extended to show that average power flowing via n conductors may be measured by $n - 1$ wattmeters.

Note that the reading of any one wattmeter may not be associated with any particular portion of the load. In fact one reading may need to be subtracted from the other to obtain the total average power.

Example 8.5

GIVEN: The balanced, 60-cycle, three-phase system shown in Fig. E-5.1. The line-to-line voltages are balanced with amplitude of 100 volts. The phase sequence is abc.

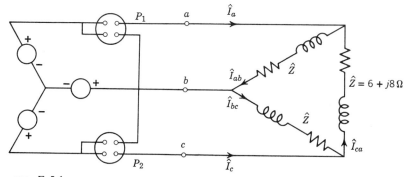

FIG. E-5.1

TO FIND: The reading of each wattmeter and the three-phase power.

SOLUTION: The phasor diagram of voltages is sketched as shown in Fig. E-5.2a. We choose $\hat{V}_{ab}$ as the reference phasor. The currents $\hat{I}_{ab}$, $\hat{I}_{bc}$, and $\hat{I}_{ca}$ are also sketched since we note that $\hat{Z} = 6 + j8 = 10e^{j53.1°}$ ohms and thus know that each phase current lags each phase voltage by 53.1°.

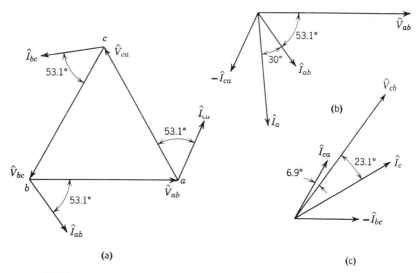

FIG. E-5.2

The amplitude of each phase current is $I_\phi = V_\phi/Z_\phi = 100/10 = 10$ amperes and the amplitude of each line current in this case is $\sqrt{3}I_\phi = 17.3$ amperes.

Wattmeter P_1 reads

$$\frac{V_{ab}I_a}{2} \cos\sphericalangle\genfrac{}{}{0pt}{}{\hat{V}_{ab}}{\hat{I}_a},$$

in which $\hat{I}_a = \hat{I}_{ab} - \hat{I}_{ca}$. The sketch in Fig. E-5.2b shows that the angle between $\hat{I}_a$ and $\hat{V}_{ab}$ is 83.1°. So

$$P_1 = \frac{100(17.3)}{2} \cos 83.1° = 104 \text{ watts}.$$

In a similar way $\hat{I}_c = \hat{I}_{ca} - \hat{I}_{bc}$ is found to lag $\hat{V}_{cb}$ by 23.1° as shown in Fig. E-5.2c. Thus

$$P_2 = \frac{V_{cb}I_c}{2} \cos\sphericalangle\genfrac{}{}{0pt}{}{\hat{V}_{cb}}{\hat{I}_c}$$

$$= \frac{100(17.3)}{2} \cos 23.1° = 795 \text{ watts}.$$

The average power per phase is

$$P_\phi = \frac{I_\phi^2 R_\phi}{2} = \frac{10^2(6)}{2} = 300 \text{ watts}$$

and

$$P_{3\phi} = 3P_\phi = 900 \text{ watts}.$$

We note that the two readings of the wattmeter should be added to obtain the total average power. This is not always the case. It is easy to show, for balanced loads, that the readings will need to be subtracted from each other for power factors less than 0.5. In this example, if we kept the average power constant but decreased the power factor to 0.5 lagging, P_1 would read zero and P_2 would read 900 watts. If the power factor were decreased further, P_1 would try to read backward; we would need to change the connections of either the current coil or the voltage coil to obtain a reading up-scale, and then this reading would need to be subtracted from the reading of P_2 to obtain the average three-phase power.

From an analysis standpoint it is usually possible to have a consistent use of voltage, current, and power conventions so that one knows the direction of average power. Assume that a wattmeter is connected with its current coil in series with a current, i_a, and its voltage coil connected from line a to terminal d. Then if the average value of $v_{ad}i_a$ is positive, average power is in the assumed positive direction of i_a at the current coil; if the average value of $v_{ad}i_a$ is negative, average power is opposite to the direction of i_a at the current coil. By assuming all line currents positive toward the load and then using average power expressions such as

$$\frac{V_{ad}I_a}{2} \cos\left(\frac{\hat{V}_{ad}}{\hat{I}_a}\right)$$

we assure ourselves that if this expression is positive the average power is in the direction toward the load. This method will apply whether or not loads are balanced.

In actual measurement work, wattmeters usually have polarity markings to assist one in determining whether or not the readings should be added or subtracted. In case of doubt, the experimenter should always perform some test to remove the doubt. Such a test would often involve changing the nature of the load.

8.4 HARMONICS IN THREE-PHASE SYSTEMS

We consider here the effect of odd harmonics which may be present in the voltage sources of a three-phase machine. These sources have half-wave symmetry because of the symmetrical construction of the machine. Consider a balanced three-phase source with sequence abc in which

$$v_{an} = V_1 \cos(\omega t + \alpha_1) + V_3 \cos(3\omega t + \alpha_3)$$
$$+ V_5 \cos(5\omega t + \alpha_5) + V_7 \cos(7\omega t + \alpha_7). \quad (8.23)$$

Then

$$v_{bn} = V_1 \cos(\omega t + \alpha_1 - 120°) + V_3 \cos[3(\omega t - 120°) + \alpha_3]$$
$$+ V_5 \cos[5(\omega t - 120°) + \alpha_5] + V_7 \cos[7(\omega t - 120°) + \alpha_7],$$
$$= V_1 \cos(\omega t + \alpha_1 - 120°) + V_3 \cos(3\omega t + \alpha_3)$$
$$+ V_5 \cos(5\omega t + \alpha_5 - 240°) + V_7 \cos(7\omega t + \alpha_7 - 120°). \quad (8.24)$$

Similarly

$$v_{cn} = V_1 \cos(\omega t + \alpha_1 - 240°) + V_3 \cos(3\omega t + \alpha_3)$$
$$+ V_5 \cos(5\omega t + \alpha_5 - 120°) + V_7 \cos(7\omega t + \alpha_7 - 240°). \quad (8.25)$$

These equations show that for a fundamental quantity having the sequence abc, the third harmonics are in phase, the fifth harmonics have the sequence acb and the seventh harmonics have the sequence abc. This phenomenon is continued for higher harmonics. The student should show that there are no third-harmonic components in the line voltages v_{ab}, v_{bc}, and v_{ca}. For the balanced circuit of Fig. 8.10 we must treat each harmonic separately, and may solve for the line currents on a

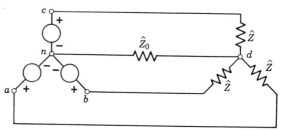

FIG. 8.10 Wye connection with harmonics.

phase basis. Note that for the third harmonics the voltages are in phase and if $\hat{Z}_0$ were infinite (no neutral connection), there would be no currents. If $\hat{Z}_0$ is present, note that the circuit is as shown in Fig. 8.11a, the equivalent circuit for which is shown in Fig. 8.11b. In these figures, $\hat{Z}$ and $\hat{Z}_0$ are the impedances calculated for the third harmonic.

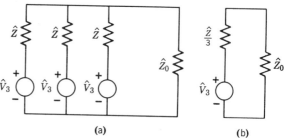

FIG. 8.11 Circuit of Fig. 8.10 for third harmonic.

Example 8.6

GIVEN: The circuit of Fig. E-6.1 in which the line-to-line voltages are balanced and have an rms value of 150 volts. The generator phase voltages are balanced, have an rms value of 100 volts, and contain only a third harmonic in addition to the fundamental frequency. The sequence of the fundamental is abc, $1/\omega C = 8$ ohms at the fundamental frequency.

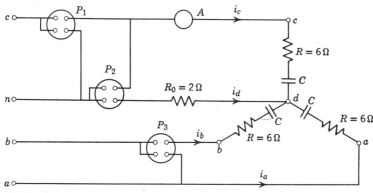

FIG. E-6.1

TO FIND: The reading of the ammeter A which reads rms values and the readings of the wattmeters P_1, P_2, and P_3.

SOLUTION: We note from Equations 8.23, to 8.25 that the third harmonics cancel in line-to-line voltages v_{ab}, v_{bc}, or v_{ca}. Thus the rms value of 150 volts is all of fundamental frequency and the rms value of the fundamental component of the phase voltage is $150/\sqrt{3}$ volts. Since

$$V_{\phi\text{rms}} = \sqrt{V^2_{1\text{rms}} + V^2_{3\text{rms}}},$$

then

$$V_{3\text{rms}} = \sqrt{100^2 - \left(\frac{150}{\sqrt{3}}\right)^2} = 50 \text{ volts}.$$

The current in the neutral connection which has only the third harmonic component can be calculated by referring to Fig. 8.11 and noting that in this case,

$$\hat{Z}_0 + \frac{\hat{Z}}{3} = 2 + \frac{6 - j(8/3)}{3} = 4 - j.89 \text{ ohms}.$$

Thus the current in the neutral connection has the rms magnitude of $50/\sqrt{4^2 + 0.89^2} = 12.2$ amp and this current leads the third harmonic of phase voltage by $\tan^{-1} 0.89/4 = 12.5°$.

Each phase current has a third harmonic with one third the magnitude of the current in the neutral connection. Also each phase current has a fundamental component with rms magnitude of $150/\sqrt{3}Z_1 = 150/\sqrt{3}(10) = 8.66$ amp, the current leading its respective phase voltage by $\tan^{-1} 8/6 = 53.1°$.

The ammeter A reads the rms value of the line or phase current. In this case

$$A = \sqrt{I_{1\text{rms}}^2 + I_{3\text{rms}}^2},$$
$$= \sqrt{8.66^2 + (12.2/3)^2} = 9.53 \text{ amp}.$$

Wattmeter P_1 is so connected that it reads

$$P_1 = V_{1\text{rms}} I_{1\text{rms}} \cos\sphericalangle{\hat{V}_1 \atop \hat{I}_1} + V_{3\text{rms}} I_{3\text{rms}} \cos\sphericalangle{\hat{V}_3 \atop \hat{I}_3}$$

$$= \frac{150}{\sqrt{3}} 8.66(0.6) + 50\left(\frac{12.2}{3}\right) \cos 12.5°$$

$$= 450 + 198 = 648 \text{ watts}.$$

Wattmeter P_2 is so connected that there is no fundamental component of current in the current coil. So

$$P_2 = V_{3\text{rms}}(3I_{3\text{rms}}) \cos\sphericalangle{\hat{V}_3 \atop \hat{I}_3}$$

$$= 50(12.2) \cos 12.5° = 595 \text{ watts}.$$

Wattmeter P_3 has no third harmonic voltage impressed on the

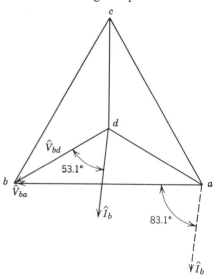

FIG. E-6.2

voltage coil, so

$$P_3 = V_{ba_{rms}} I_{b1_{rms}} \cos \underset{\hat{I}_{b1}}{\overset{\hat{V}_{ba_1}}{\angle}}$$
$$= 150(8.66) \cos 83.1°$$
$$= 159 \text{ watts}$$

the angle of 83.1° being obtained from Fig. E-6.2.

8.5 SUMMARY

Three-phase systems are studied under sinusoidal steady state only. The usefulness of phasor diagrams is clearly demonstrated. The third, ninth, fifteenth, etc. harmonics of a balanced three-phase system are in phase and thus must be treated in a special manner. In determining the reading of instruments one must concentrate on the instrument's connections and capabilities.

FURTHER READING

Nearly all introductory circuit texts have some material on three-phase systems. The student may wish to consult Brenner and Javid's *Analysis of Electrical Circuits*, Chapter 17, McGraw-Hill, New York, 1959, or Clement and Johnson's *Electrical Engineering Science*, Chapter 18, McGraw-Hill, New York, 1960.

PROBLEMS

8.1 Three phase is normally not supplied to a residence. It is common practice to supply three wires, one of the wires being grounded. The rms voltage between the other two wires is 240 volts. The rms voltage between either of these wires and the grounded wire is 120 volts. The 240 volts is usually used for water heaters, clothes dryers, and electric stoves; 120 volts is used for other applications. Discuss the advantages and disadvantages of this method in comparison with other methods that are possible.

8.2 *a* Draw a schematic diagram for a star-connected four-phase generator. Assuming the phase voltages are balanced, find the magnitude and phase angles of the line-to-line voltages if $v_{an} = 100 \cos 377t$ volts and the phase sequence is *abcd*. Draw a phasor diagram of the system.

Answer: $\hat{V}_{ab} = 141.4e^{j45°}$, $\hat{V}_{bc} = 141.4e^{-j45°}$
$\hat{V}_{cd} = 141.4e^{-j135°}$, $\hat{V}_{da} = 141.4e^{+j135°}$

b Show the schematic of the four-phase generator mesh-connected. What is the relationship between line-to-line and phase currents for this system?

8.3 What is the magnitude of the phase voltage of a six-phase star-connected generator if the line-to-line voltage is 100 volts?

Answer: 100 volts.

8.4 What is the magnitude of the line-to-line voltage of a 12-phase star-connected generator if the phase voltage is 100 volts.

Answer: 51.7 volts.

8.5 Three impedances, $\hat{Z}_1 = \hat{Z}_2 = \hat{Z}_3 = 8 - j6$ ohms, are connected in wye to a wye-connected balanced three-phase generator. $v_{an} = 100 \cos 377t$ volts, and the generator phase sequence is abc.
  a Determine the line currents $\hat{I}_a$, $\hat{I}_b$, and $\hat{I}_c$. Assume these positive into load.
  b Find the readings of rms ammeters connected in each line and rms voltmeters connected between lines a and b and lines b and c.
  c Show a completely labeled (magnitudes and angles) phasor diagram showing all voltages and currents.
  d Calculate the average three-phase power.
  e Find the sum of the three currents $\hat{I}_a + \hat{I}_b + \hat{I}_c$.

Partial answer: a, $\hat{I}_a = 10e^{+j36.9°}$, $\hat{I}_b = 10e^{-j83.1°}$.
  b, 7.07 amp, 122.0 volts.
  d, $P = 1200$ watts.

8.6 Repeat problem 8.5 with the three impedances connected in delta.

Partial answer: a, $\hat{I}_a = 30e^{j36.9°}$.
  b, 21.2 amp.
  d, $P = 3600$ watts.

8.7 Three impedances, each $5 + j8.66$ ohms, are connected in delta across a balanced three-phase power supply as shown in Fig. P-8.7. The sequence is acb. Assume $v_{ab} = 100 \cos \omega t$ volts.
  a Make a phasor diagram showing all voltages and currents.
  b Determine all currents in exponential form.
  c Determine the average three phase power.

Partial answer: b, $\hat{I}_a = 17.32e^{-j30°}$ amp.
  c, $P_{3\phi} = 750$ watts.

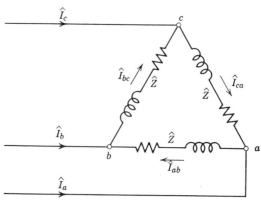

FIG. P-8.7

8.8 Figure P-8.8 shows two balanced wye-connected loads in parallel. The amplitude of the **phase** (line-to-neutral) voltage of the wye-connected generator supplying the system is 100 volts, and the phase sequence is abc.

a Find the phase currents $\hat{I}_{ad}$, $\hat{I}_{bd}$, $\hat{I}_{cd}$, $\hat{I}_{ae}$, $\hat{I}_{be}$, and $\hat{I}_{ce}$ for each load. (Use voltage $\hat{V}_{an} = 100e^{j0°}$ as reference.)

b Find the line currents $\hat{I}_a$, $\hat{I}_b$, and $\hat{I}_c$.

c Find the line-to-line voltages $\hat{V}_{ab}$, $\hat{V}_{bc}$, and $\hat{V}_{ca}$.

d Using a scale of $1'' = 50$ volts and $1'' = 5$ amperes carefully construct a phasor diagram showing all of the quantities in parts a, b, and c.

e What do each of the meters shown read? (All meters are calibrated to give rms values.)

f What is the total average power and power factor of each load?

g What is the total average power and power factor of the combined system?

h What effect would reversing the phase sequence of the generator have on the meter readings?

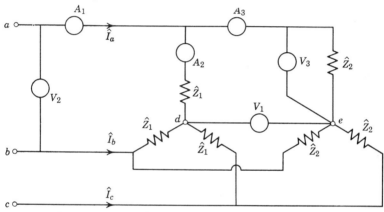

FIG. P-8.8 $\hat{Z}_1 = 20e^{j60°}\,\Omega$; $\hat{Z}_2 = 20e^{j0°}\,\Omega$.

Partial answer: *a*, $\hat{I}_{ad} = 5e^{-j60°}$ amp, $\hat{I}_{be} = 5e^{-j120°}$ amp.
 b, $\hat{I}_a = 7.5 - j4.34$ amp $= 8.66e^{-j30°}$ amp.
 c, $\hat{V}_{bc} = 173.2e^{-j90°}$ volts.
 e, $A_1 = 6.13$ amp, $V_1 = 0$ volts.
 g, $P = 1125$ watts, power factor $= 0.866$.

8.9 Repeat problem 8.8 but with $\hat{Z}_2 = 10e^{-j90°}\,\Omega$.

Partial answer: $A_2 = 3.53$ amp.
 $P = 375$ watts.
 $A_3 = 7.07$ amp.

8.10 Repeat problem 8.8 but with $\hat{Z}_1 = 20e^{+j30°}\,\Omega$ and $\hat{Z}_2 = 10e^{-j45°}\,\Omega$, and phase sequence acb.

Partial answer: $\hat{I}_b = 12.3e^{j141.9°}$ amp.
 $P = 1710$ watts.

8.11 Repeat problem 8.8 but with $\hat{Z}_1 = 20e^{j0°}\,\Omega$ and $\hat{Z}_2 = 20e^{-j90°}\,\Omega$.

8.12 In problem 8.8 show that $\hat{I}_{ad} + \hat{I}_{ac}$ is equal to the phase voltage $\hat{V}_{an}$ divided by $(\hat{Z}_1 \hat{Z}_2)/(\hat{Z}_1 + \hat{Z}_2)$.

8.13 In problem 8.8 reconnect the wye impedances, $\hat{Z}_1$ and $\hat{Z}_2$, to form two balanced delta-connected loads in parallel. (Omit voltmeters V_1 and V_3.)
 a Calculate the phase currents for each load. (Assume the voltage $\hat{V}_{an}$ is specified as $\hat{V}_{an} = 100e^{j0°}$.)
 b Calculate the line currents $\hat{I}_a$, $\hat{I}_b$, and $\hat{I}_c$.
 Partial answer: $\hat{I}_a = 26.1e^{-j30°}$ amp.
 c Construct a phasor diagram showing each phase current for each load, the currents $\hat{I}_a$, $\hat{I}_b$, and $\hat{I}_c$, and the voltages $\hat{V}_{ab}$, $\hat{V}_{bc}$, and $\hat{V}_{ca}$. (Voltage $\hat{V}_{an}$ is specified as reference.)
 d Find the total average power to each load.
 e Find the total average power and the power factor for the system.

8.14 Figure P-8.14 shows a balanced wye load in parallel with a balanced delta load. The generator supplying this system is a balanced three-phase generator, having phase sequence acb, and line-to-line voltage of 300 volts amplitude,

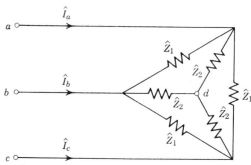

FIG. P-8.14 $\hat{Z}_1 = 30e^{j60°}\,\Omega;\ \hat{Z}_2 = 17.3e^{-j90°}\,\Omega$.

 a Use line-to-line voltage $\hat{V}_{ab}$ as reference and draw a phasor diagram showing $\hat{V}_{ab}$, $\hat{V}_{bc}$, $\hat{V}_{ca}$, $\hat{V}_{ad}$, $\hat{V}_{bd}$, $\hat{V}_{cd}$, $\hat{I}_{ad}$, $\hat{I}_{ab}$, and $\hat{I}_{ac}$ and $\hat{I}_a$, $\hat{I}_b$, and $\hat{I}_c$.
 b Calculate the total power and the power factor of the system.
 Answer: $P = 2250$ watts, power factor $= 0.866$.

8.15 Repeat problem 8.14 for: $\hat{Z}_1 = 30e^{-j90°}\,\Omega;\ \hat{Z}_2 = 17.3e^{+j60°}\,\Omega$.
Partial answer: $\hat{I}_a = 10e^{+j90°}$ amp, $P = 1300$ watts, power factor $= 0.5$.

8.16 Repeat problem 8.14 for: $\hat{Z}_1 = 60e^{j45°}\,\Omega;\ \hat{Z}_2 = 10e^{-j65°}\,\Omega$
Partial answer: $\hat{I}_a = 16.5e^{+j65.3°}$ amp, $P = 3500$ watts, power factor $= 0.816$.

8.17 Show that the power output of a four-phase balanced machine under steady-state sinusoidal conditions is not a function of time.

8.18 The following balanced 3φ loads are connected in parallel across a system having 220 volts rms between lines.
  1 An induction motor having an output of 20 horsepower with an efficiency of 85% and a lagging power factor of 0.8.
  2 A unity power factor load of 8 kw.
  3 A 10-kva load having a power factor of 0.8 leading.
  Determine the rms value of the line current and the power factor for the combined loads.
  Partial answer: rms line current is 89.9 amperes, power factor is 0.98.

8.19 A balanced three-phase, 60-cps system with 220 volts (rms) line to line has two balanced loads connected in parallel. One load is an induction motor having an output of 20 horsepower with an efficiency of 85% and a lagging power factor of 0.8. The other load is a unity power factor load of 8 kilowatts.
  a What is the rms value of the line current?
  b It is desired to improve the power factor to unity by placing three capacitors in delta across the lines. What should be the rating of each capacitor in microfarads?
  c For part *b* what should the microfarad rating be for each capacitor if they were connected in wye instead of delta?
  Answer: *a*, 74.8 amperes; *b*, 240 μf; and *c*, 720 μf.

8.20 In Fig. P-8.20 the power supply is balanced, each line-to-line voltage having an amplitude of 100 volts. The sequence is *abc*. Use $\hat{V}_{ab}$ as the reference phasor.

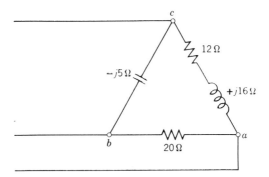

FIG. P-8.20

  a Determine all currents.
  Partial answer: $\hat{I}_{ab} = 5e^{j0°}$ amp, $\hat{I}_{bc} = 20e^{-j30°}$ amp, and $\hat{I}_{ca} = 5e^{+j66.9°}$ amp.
  b Determine the total average power.
  Answer: $P_T = 400$ watts.

8.21 Repeat problem 8.20 for the circuit shown in Fig. P-8.21, the power supply being the same.
  Partial answer: $\hat{I}_a = 17.32e^{j0°}$ amp, $P = 683$ watts.

Three-Phase Systems 389

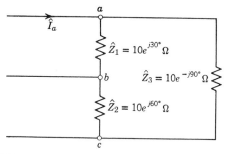

FIG. P-8.21

8.22 For the circuit shown in Fig. P-8.22 the amplitude of the line-to-line voltage is 100 volts, the sequence is *abc*.
   a Use $\hat{V}_{ab}$ as reference phasor and calculate $\hat{I}_a$, $\hat{I}_b$, and $\hat{I}_c$.
   b Repeat part *a* for sequence *acb*.

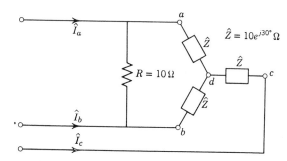

FIG. P-8.22

8.23 Repeat problem 8.22, but replace the 10-ohm resistance between lines *a* and *b* with a capacitance reactance of 10 Ω.
   Partial answer: $\hat{I}_a = 5.77e^{+j60°}$ amp, $\hat{I}_b = 11.5e^{-j120°}$ amp.

8.24 In Fig. P-8.24 the phase sequence is *abc*.
   $v_{an} = 100 \cos(377t - 30°)$ volts.
   a With the switch **closed** find
   $\hat{I}_a$, $\hat{I}_b$, $\hat{I}_c$ and $\hat{I}_n$.

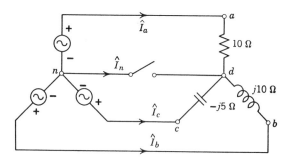

FIG. P-8.24

   Partial answer: $\hat{I}_n = 16.3 - j3.66$ amp.
   b With the switch **open** find $\hat{I}_a$, $\hat{I}_b$, $\hat{I}_c$, and $\hat{V}_{dn}$ using the mesh-current method.
   Partial answer: $\hat{V}_{dn} = -63.4 + j100$ volts, $\hat{I}_a = 21.2e^{-j45°}$ amp.

c With the switch open use Thévenin's theorem to find the current $\hat{I}_a$.
d Repeat part b using the node-voltage method.

8.25 In Fig. P-8.25 the sequence is *abc* and the amplitude of the line-to-line voltage is 100 volts.
  a Determine all currents using $\hat{V}_{ab}$ as reference phasor.
  Partial answer: $\hat{I}_0 = 3.96e^{+j151°}$ amp, $\hat{I}_c = 6.94e^{+j104.5}$ amp.
  b Repeat part *a* for sequence *acb*.

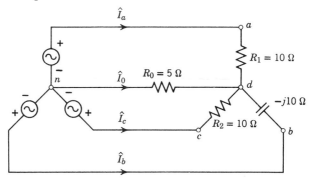

FIG. P-8.25

8.26 In the **two**-phase system with the sequence indicator connected as shown in Fig. P-8.26, lamp *B* is brighter. What is the phase sequence? Prove. Assume *R* of lamp $\simeq 1/\omega C$.

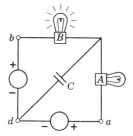

FIG. P-8.26

8.27 For the circuit of Fig. P-8.27 the amplitude of the line-to-line voltages is

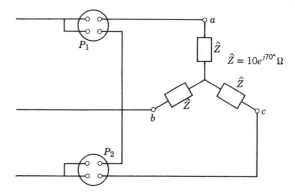

FIG. P-8.27

173 volts. Determine the readings of the wattmeters P_1 and P_2 for both sequence *abc* and sequence *acb*. Should the readings be added or subtracted so that the algebraic sum is equal to the average three-phase power?

Partial answer: $P_1 = -150$ watts, $P_2 = 663$ watts, for sequence *abc*.

8.28 This is a balanced three-phase system (Fig. P-8.28); each load impedance, $\hat{Z}$, is $5 + j8.66$ ohms and each line-to-line voltage is 100 volts rms. For the phase-sequence indicator assume that the resistance of each lamp is equal to the reactance of the capacitor. Determine the reading of the wattmeters P_1 and P_2. Show that the sum of the wattmeter readings equals the sum of the I^2R losses.

Partial answer: Sequence is *acb*; wattmeter P_2 reads 1500 watts.

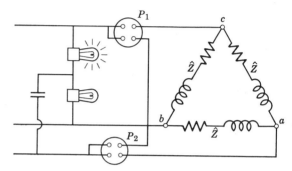

FIG. P-8.28

8.29 Two wattmeters, P_a and P_c, are connected to read the power of a balanced three-phase load. P_a, with current coil in line *a* and voltage coil from *a* to *b*, reads 1200 watts. P_c, with current coil in line *c* and voltage coil from *c* to *b*, reads 400 watts. A balanced three-phase load, purely capacitive, is now placed in parallel with the original load, and the wattmeter readings change, but do not reverse direction. P_a now reads 1000 watts, P_c reads 600 watts.
a Was the original load inductive or capacitive? Explain.
b What is the phase sequence? Explain. Answer: *acb*.

8.30 Power plants need to measure vars as well as watts. One way in which this is done is to use an electrodynamometer instrument, connecting the current coil in series with one line, and the voltage coil across the other two lines (Fig. P-8.30). Show that the reading will be proportional to vars for balanced loads. If the instrument has been calibrated as a wattmeter, by what factor should the reading be multiplied to obtain three-phase vars?

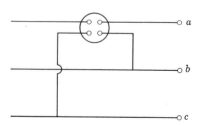

FIG. P-8.30

8.31 For Fig. P-8.31, $\hat{V}_{an} = 100e^{j0°}$ volts, $\hat{V}_{bn} = 100e^{-j120°}$ volts, and $\hat{V}_{cn} = 100e^{j120°}$ volts.

Determine:
a The voltage $\hat{V}_{dn}$.
b The current through the inductance.
c The reading of the wattmeter, P_a.

Partial answer: a, $\hat{V}_{dn} = 25 - j75$ volts.
c, $P_a = 1770$ watts.

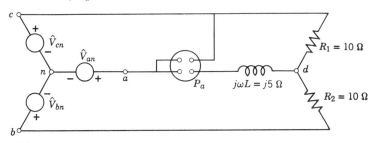

FIG. P-8.31

8.32 For the circuit shown in Fig. P-8.32 the line-to-line voltages are balanced three-phase. The phase sequence is *abc*. Voltmeter *V* reads 300 volts rms. The resistance of each lamp is equal to $1/\omega C$.
a Determine which lamp is brighter. (Prove your answer.)
b Calculate $\hat{I}_a$, using $\hat{V}_{ab}$ as the reference phasor.
c Determine the reading of wattmeter *P*.

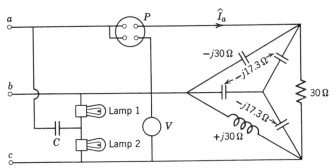

FIG. P-8.32

8.33 For the circuit shown in Fig. P-8.33 the **line-to-line voltages are balanced** three-phase. The sequence is *abc*. Voltmeter *V* reads 100 volts rms. $R = X_L = 10^6$ Ω. $\hat{Z}_1 = \hat{Z}_2 = \hat{Z}_3 = 10e^{j60°}$ Ω.
a Determine the reading of voltmeter V_1.
  Answer: $V_1 = 136.6$ volts.
b Find $\hat{I}_a$ (rms) using $\hat{V}_{ab}$ as reference phasor.
  Answer: $7.32e^{-j90°}$ amp.
c Determine the reading of wattmeter *P*.
  Answer: $P = 633$ watts.

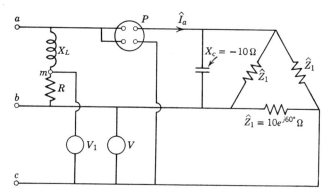

FIG. P-8.33

8.34 The phase voltages (Fig. P-8.34) are balanced and have only the third harmonic in addition to the fundamental frequency. The rms phase voltage is 100 volts; the rms line-to-line voltage is 150 volts. At the fundamental frequency, $\omega L = 6$ ohms. $R = 8$ ohms.

a Determine the phase sequence.

Answer: *abc*.

b Determine the readings of the wattmeters P_1 and P_2.

Answer: $P_1 = 155$ watts, $P_2 = 510$ watts.

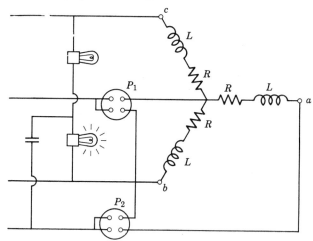

FIG. P-8.34

8.35 For the circuit shown in Fig. P-8.35 the generator voltages are balanced, three-phase. $v_{an} = 141.4 \cos 100t + 70.7 \cos 300t$ volts. The phase sequence of the fundamental is *abc*. All meters are ideal meters and read rms values.

a With the **switch closed**, determine the readings of ammeters A_0 and A_1.

b With the switch open, ammeter A_0 reads 5.6 amperes. What is the rms magnitude of the fundamental current through this meter?

Answer: 2.52 amp.

c With the **switch open**, what is the total average **third-harmonic power**?

Answer: 250 watts.

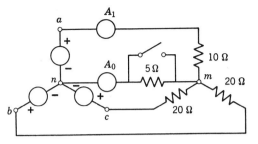

FIG. P-8.35

8.36 The generator voltages are balanced (Fig. P-8.36). Each voltage has only a fundamental and a third-harmonic frequency. Voltmeter V_1 reads 100 volts, V_2 reads 150 volts. For the fundamental the sequence is abc. The reactance of each capacitor, C, is -10 ohms at the frequency of the fundamental.

a Determine the readings of the ammeters I_a and I_n.

Answer: $I_a = 17.32$ amp, $I_n = 45$ amp.

b Determine the readings of each of the three wattmeters. Show your reasoning clearly.

Answer: $P_1 = 0$, $P_2 = 0$, $P_3 = 650$ watts.

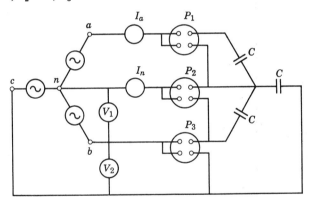

FIG. P-8.36

8.37 In Fig. P-8.37 the phase voltages and line voltages are balanced. Phase voltage is 126.8 volts rms and line-to-line voltage is 173 volts rms. Only the

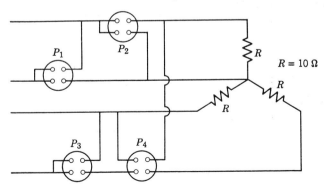

FIG. P-8.37

fundamental and third harmonic are present. Determine each wattmeter reading.

Partial answer: $P_1 = 1820$ watts, $P_3 = 1500$ watts.

8.38 In Fig. P-8.38 line-to-neutral rms voltages are balanced and equal to 141 volts. Line-to-line rms voltages are balanced and equal to 173 volts. V reads 236 volts. $R_1 = 10$ ohms. $R_0 = 5$ ohms. Only the fundamental and third harmonic voltages are present. $R_2 = 10^6$ ohms.

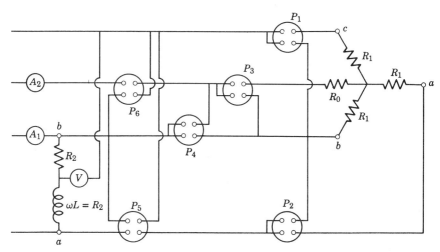

FIG. P-8.38

a Determine the phase sequence.

Answer: *abc*.

b Determine the readings of the ammeters A_1 and A_2 which are calibrated in rms values.

Answer: $A_1 = 10.75$ amp, $A_2 = 12$ amp.

c Determine the readings of the wattmeters P_1, P_2, P_3, P_4, P_5, and P_6.

Partial answer: $P_1 = 1500$ watts, $P_3 = 1200$ watts, and $P_4 = 1400$ watts.

d Now assume the three resistors, R_1, are replaced by three inductances with $\omega L = 10$ ohms, and then answer parts *b* and *c*.

Partial answer: $P_1 = 866$ watts, $A_1 = 10.41$ amp, and $P_5 = 1730$ watts.

8.39 In the circuit shown in Fig. P-8.39 the generator voltages are balanced three-phase and have fundamental and third-harmonic voltages only. The phase sequence is *abc*. Voltmeter V_1 reads 206 volts rms. Voltmeter V_2 reads 346 volts rms. Find the readings of ammeter A, voltmeter V_3, and wattmeters P_1 and P_2.

Partial answer: $P_1 = 6500$ watts, $V_3 = \sqrt{222^2 + 10^2}$ volts.

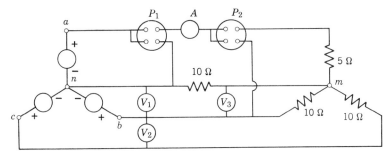

FIG. P-8.39

8.40 For the circuit shown in Fig. P-8.40 the generator voltages are balanced three-phase. $v_{an} = 141.4 \cos 100t + 70.7 \cos 300t$ volts. The phase-sequence of the fundamental is abc. All meters are ideal meters and read rms value.
a With the switch open, find the readings of V_1, V_2, V_3, and A_1.

Answer: $V_1 = 111.6$ volts, $V_3 = 50$ volts, and $A_1 = 10$ amp.
b With the switch closed, find the readings of V_3, A_0, and A_1.

Answer: $A_0 = 10.6$ amp, $V_3 = 35.4$ volts, and $A_1 = 10.55$ amp.

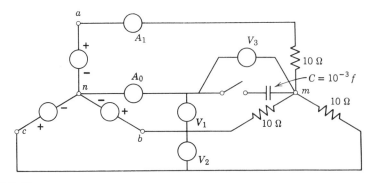

FIG. P-8.40

8.41 For the circuit shown in Fig. P-8.41 the generator is balanced three-phase with sequence abc. $v_{an} = 100 \cos 300t + 60 \cos 900t$ volts. At fundamental frequency $\hat{Z}_1 = 17.32e^{-j15°}$ ohms, $\hat{Z}_2 = 17.32e^{j15°}$ ohms, and $\hat{Z}_3 = 17.05e^{j0°}$ ohms.
a Find the readings of all instruments shown.
Partial answer: A_0 reads 4.24 amp, A_1 reads 14.2 amp, $P_1 = 1500$ watts, and $P_3 = 180$ watts.
b Calculate the total power.

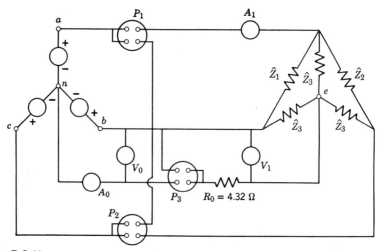

FIG. P-8.41

8.42 A balanced three-phase wye-connected generator has phase sequence acb.
$v_{an} = 200 \cos \omega t + 100 \cos(3\omega t + 30°) + 50 \cos(5\omega t - 50°)$ volts.
  a Write the expressions for the line-to-line voltages v_{ab}, v_{bc}, and v_{ca}.
  b Calculate the reading of an rms voltmeter connected across lines a and b;. across any phase (line-to-neutral).
  c If the generator were delta-connected, what would an rms voltmeter connected between lines a and b read?

8.43 Figure P-8.43 shows a delta-connected balanced 3ϕ generator. Consider that the generator phase voltages have fundamental and third-harmonic voltages. Voltmeter V_e reads 94.9 volts and voltmeter V_f reads 90.0 volts when the switch shown is open. What does voltmeter V_g read with the switch open? With the switch closed?

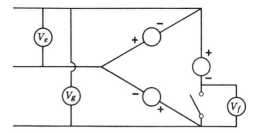

FIG. P-8.43

8.44 The phase voltages of the balanced three-phase generator shown in Fig. P-8.44 contain fundamental and third-harmonic voltages. The wye-connected load is balanced and purely resistive. Ammeter A_0 reads 15 amp rms. Voltmeter V_1 reads 520 volts rms. Wattmeter P_0 reads 1500 watts. Find the readings of the instruments V_0, V_2, A_1, P_1, and P_2. What is the total power?
Partial answer: $P_1 = 16{,}875$ watts, $V_0 = 316$ volts.

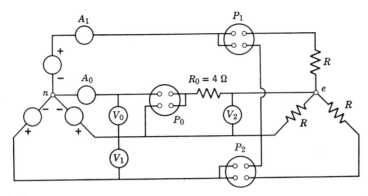

FIG. P-8.44

9 Pulses, Impulses, Dependent Sources

Thus far, the sources have been usually limited to sinusoidal functions and the order of the differential equations have been usually limited to second order. With these limitations it was possible to present the classical and Laplace transform methods in a fairly concise form and to consider a great number of problems of practical interest. It was also possible to show how the phasor domain may be obtained from either the classical or the Laplace transform approach.

We have thus far avoided pulse sources of the type mentioned in chapter 1, such as 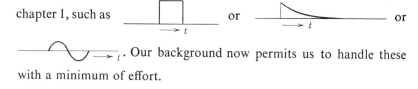. Our background now permits us to handle these with a minimum of effort.

9.1 UNIT STEP FUNCTION, GATE FUNCTION, AND SHIFTED TIME FUNCTION

In order to write the expressions for pulses in proper mathematical form, it is convenient to define the unit step function, $u(t)$:

$$u(t) = \begin{cases} 0 & \text{for } t < 0 \\ 1 & \text{for } t > 0 \end{cases} \tag{9.1}$$

It is also convenient to define a delayed unit step function $u(t - a)$:

$$u(t - a) = \begin{cases} 0 & \text{for } t < a \\ 1 & \text{for } t > a \end{cases} \tag{9.2}$$

These are sketched in Figs. 9.1a and 9.1b. In Fig. 9.1c is shown the "gate" function, $u(t - a_1) - u(t - a_2)$, which is defined by

$$u(t - a_1) - u(t - a_2) = \begin{cases} 0 & \text{for } t < a_1 \\ 1 & \text{for } a_1 < t < a_2 \\ 0 & \text{for } t > a_2. \end{cases} \quad (9.3)$$

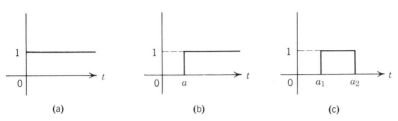

FIG. 9.1 Unit step and gate function. (a) $u(t)$. (b) $u(t - a)$. (c) $u(t - a_1) - u(t - a_2)$.

We need to be careful in the nomenclature that we use with unit step functions and to keep in mind that multiplying a function of time by the delayed unit step function, $u(t - a)$, gives a result which is zero until $t = a$ and then is equal to the function. Multiplication of a function of time by a gate function, $u(t - a_1) - u(t - a_2)$, gives a result which is zero except for the interval $a_1 < t < a_2$, during which it is equal to the function of time.

The shifted time function, $f(t - a)u(t - a)$ is often the most useful way in which to express part or all of a driving or response function. The reason for this is that if $x(t)u(t)$ is the response to the driving function $f(t)u(t)$, then $x(t - a)u(t - a)$ is the response to the driving function $f(t - a)u(t - a)$. Also the Laplace transform of $f(t - a)u(t - a)$ can readily be determined if the Laplace transform of $f(t)u(t)$ is known.

Figure 9.2 illustrates the writing of modified ramp functions and shifted time functions. The shifted time function may be obtained algebraically from the form using the gate function, or may be obtained directly by considering the function as a summation of shifted time functions. We demonstrate with the function of Fig. 9.2b. From an algebraic point of view,

$$f(t) = 5(t - 1)[u(t - 1) - u(t - 3)]$$
$$= 5(t - 1)u(t - 1) - 5(t - 1)u(t - 3). \quad (9.4)$$

The first term of Equation 9.4 is a shifted time function, the second term would be a shifted time function if it were $5(t - 3)u(t - 3)$ instead of $5(t - 1)u(t - 3)$. This change is possible by adding and

Pulses, Impulses, Dependent Sources 401

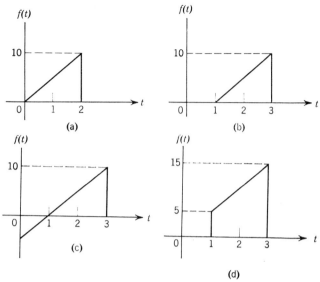

FIG. 9.2 Writing of time functions using gate functions and shifted time functions.
(a) $f(t) = 5t[u(t) - u(t - 2)] = 5tu(t) - 5(t - 2)u(t - 2) - 10u(t - 2)$.
(b) $f(t) = 5(t - 1)[u(t - 1) - u(t - 3)] = 5(t - 1)u(t - 1) - 5(t - 3)u(t - 3) - 10u(t - 3)$.
(c) $f(t) = 5(t - 1)[u(t) - u(t - 3)] = -5u(t) + 5tu(t) - 5(t - 3)u(t - 3) - 10u(t - 3)$.
(d) $f(t) = 5t[u(t - 1) - u(t - 3)] = 5(t - 1)u(t - 1) + 5u(t - 1) - 5(t - 3)u(t - 3) - 15u(t - 3)$.

subtracting the quantity $10u(t - 3)$. Thus

$$f(t) = 5(t - 1)u(t - 1) - 5(t - 1)u(t - 3)$$
$$+ 10u(t - 3) - 10u(t - 3) = 5(t - 1)u(t - 1)$$
$$- 5(t - 3)u(t - 3) - 10u(t - 3). \qquad (9.5)$$

Figure 9.3 shows how this same function of Fig. 9.2b may be considered as a graphical summation of three separate time functions, each time function continuing indefinitely after it has started. This graphical

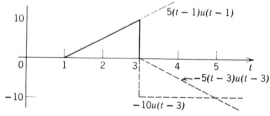

FIG. 9.3 The function of Fig. 9.2b shown as the sum of three separate functions.

method of summing shifted time functions is sometimes easier than the method previously described.

9.2 THE RESPONSE OF CIRCUITS TO PULSE DRIVING FUNCTIONS

With the application of pulses to circuits we are faced with the solution of differential equations of the form:

$$\frac{d^2x}{dt^2} + a_1\frac{dx}{dt} + a_0x = f_1(t)u(t)$$

$$+ f_2(t - a_2)u(t - a_2) + f_3(t - a_3)u(t - a_3). \quad (9.6)$$

In the classical method there are several procedures for solution. One procedure is to obtain the solution for each interval of time by the method given in Chapter 3, the initial conditions for each interval being obtained from the solution for the previous interval. A better procedure is to realize that the principle of superposition allows us to consider the net response as the sum of the responses to each source acting independently. Each initial condition is a source. Thus, the complete response to $f_1(t)u(t)$ **and** the initial conditions may be thought of as the sum of the complete response to $f_1(t)u(t)$ with zero initial conditions plus the complete responses to each initial condition with $f_1(t)u(t)$ and the other initial conditions made zero. To this group of responses we can simply add the complete responses to such driving functions as $f_2(t - a_2)u(t - a_2)$ with zero initial conditions at $t = a_2$. Our solution may therefore be written

$$x = x_1(t)u(t) + x_2(t - a_2)u(t - a_2) + x_3(t - a_3)u(t - a_3), \quad (9.7)$$

in which $x_1(t)u(t)$ is the solution to the driving function $f_1(t)u(t)$ with the proper initial conditions; $x_2(t - a_2)u(t - a_2)$ is the solution to the driving function, $f_2(t - a_2)u(t - a_2)$, with **zero stored energy at $t = a_2$**; and $x_3(t - a_3)u(t - a_3)$ is the solution to the driving function, $f_3(t - a_3)u(t - a_3)$ with **zero stored energy at $t = a_3$**. Zero stored energy means zero currents through inductances and zero voltages across capacitances.

In the Laplace transform method we shall need to determine the transforms of additional functions. We will develop now the transforms of the step functions and the shifted time function; additional transforms will be developed later as needed.

The Laplace transform of the unit step function, $u(t)$, is

$$\mathcal{L}u(t) = \int_0^\infty e^{-st}\,dt = \frac{1}{s}. \quad (9.8)$$

Since the limits of integration are from 0 to ∞, $u(t)$ has exactly the same transform as the number one. For the same reason $f(t)u(t)$ has the same transform as $f(t)$.

Also

$$\mathscr{L}u(t-a) = \int_0^\infty u(t-a)e^{-st}\,dt = \int_a^\infty e^{-st}\,dt = \frac{e^{-st}}{-s}\bigg|_a^\infty = \frac{e^{-as}}{s}. \quad (9.9)$$

The Laplace transform of the delayed step function is the transform of the step function multiplied by the term e^{-as}.

The Laplace transform of the shifted time function $f(t-a)u(t-a)$ can be obtained readily if we know the transform of $f(t)u(t)$, Fig. 9.4 shows that, by definition,

$$F(s) = \mathscr{L}[f(t)u(t)] = \mathscr{L}[f(t')u(t')] = \int_0^\infty f(t')e^{-st'}\,dt'.$$

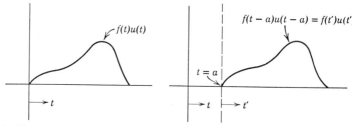

FIG. 9.4 Sketches showing $f(t)u(t)$ and $f(t-a)u(t-a)$. $a > 0$. Thus $\mathscr{L}[f(t)u(t)] = \mathscr{L}[f(t')u(t')]$.

Then, since $t' = t - a$,

$$F(s) = \int_a^\infty f(t-a)e^{-s(t-a)}\,dt,$$

the lower limit becoming a because at $t' = 0$, $t = a$. We can change the lower limit back to 0 by introducing $u(t-a)$; so

$$F(s) = e^{as}\int_0^\infty f(t-a)u(t-a)e^{-st}\,dt,$$
$$= e^{as}\mathscr{L}[f(t-a)u(t-a)]. \quad (9.10)$$

Finally

$$\mathscr{L}[f(t-a)u(t-a)] = e^{-as}F(s). \quad (9.11)$$

This is a very important transform since knowledge of a particular transform, for instance, that $\mathscr{L}[\cos \omega t\, u(t)] = s/(s^2 + \omega^2)$, permits us to write immediately that

$$\mathscr{L}[\cos \omega(t-a)u(t-a)] = \frac{se^{-as}}{s^2 + \omega^2}.$$

Example 9.1

GIVEN: The circuit of Fig. E-1.1 with $v_{cd}(0+) = 10$ volts.

TO FIND: The current i in literal form by both the classical and Laplace transform methods. Then evaluate i for the constants given.

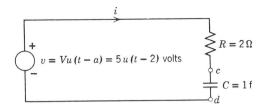

FIG. E-1.1

SOLUTION: (a) *Classical method.* Because superposition applies, the response i can be considered the sum of the response to the initial voltage on the capacitor and the response to the voltage source $Vu(t-a)$. Thus we are obtaining independently the solutions to the circuits of Figs. E-1.2a and E-1.2b and then finding $i = i' + i''$. For the circuit of Fig. E-1.2a, for $t > 0$,

$$Ri' + \frac{1}{C}\int_0^t i'\, dt + v_{cd}(0+) = 0, \tag{1}$$

or

$$R\frac{di'}{dt} + \frac{i'}{C} = 0,$$

which has the solution, $i' = Ae^{-t/RC}$ for $t > 0$. The constant A is evaluated as $-[v_{cd}(0+)]/R$ from Equation 1, so

$$i' = -\frac{v_{cd}(0+)}{R} e^{-t/RC} u(t), \tag{2}$$

the symbol $u(t)$ being introduced to indicate clearly that i' is zero for $t < 0$ and is equal to $v_{cd}(0+)e^{-t/RC}/R$ for $t > 0$.

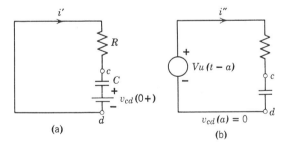

FIG. E-1.2

For the circuit of Fig. E-1.2b we need the response, i'', to the delayed time function, $Vu(t-a)$, with all other sources zero. Since the current response to $Vu(t)$ would be $(V/R)e^{-t/RC}u(t)$, we may write, "by inspection," that the response, i'', to $Vu(t-a)$ is

$$i'' = \frac{V}{R} e^{-(t-a)/RC} u(t-a). \tag{3}$$

Then the actual solution by the classical method is

$$i = i' + i'' = -\frac{v_{cd}(0+)}{R} e^{-t/RC} u(t) + \frac{V}{R} e^{-(t-a)/RC} u(t-a). \quad (4)$$

The insertion of numbers in Equation 4 results in

$$i = -5e^{-0.5t} + 2.5e^{-0.5(t-2)} u(t-2) \text{ amperes}. \quad (5)$$

A sketch of i vs. t is shown in Fig. E-1.3. Note the discontinuity at $t = 2$ seconds.

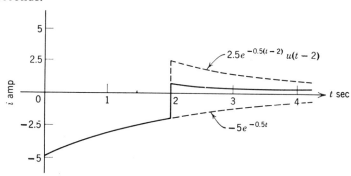

FIG. E-1.3

The same results could have been obtained by considering two separate intervals of time, $0 < t < a$ and $t > a$, the initial conditions for $t > a$ being obtained from the solution for the previous interval. However, the method we demonstrated is superior in that it involves less algebra.

(b) *Laplace transform method.* The transform circuit is sketched in Fig. E-1.4, the transform of $u(t-a)$ being e^{-as}/s. Kirchhoff's voltage equation gives

$$\left(R + \frac{1}{Cs}\right) I(s) = -\frac{v_{cd}(0+)}{s} + \frac{Ve^{-as}}{s}, \quad (6)$$

or

$$I(s) = \frac{-v_{cd}(0+) + Ve^{-as}}{R(s + 1/RC)},$$

$$= \frac{-v_{cd}(0+)}{R(s + 1/RC)} + \frac{Ve^{-as}}{R(s + 1/RC)}. \quad (7)$$

Then

$$i = \frac{-v_{cd}(0+)}{R} e^{-t/RC} u(t) + \frac{V}{R} e^{-(t-a)/RC} u(t-a), \quad (8)$$

which is the same solution as obtained by the classical method. It is important to realize that in obtaining the inverse transform of Equation 7, the term e^{-as} is interpreted as a time delay factor.

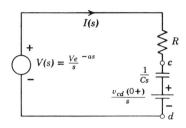

FIG. E-1.4

Example 9.2

GIVEN: The circuit of Fig. E-2.1a with the voltage driving function of Fig. E-2.1b.

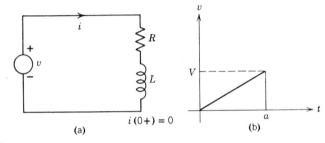

FIG. E-2.1.

TO FIND: The current i by both classical and Laplace transform methods.

(a) *Classical method.* The voltage driving function may be written as

$$v = \frac{Vt}{a} u(t) - \frac{V(t-a)}{a} u(t-a) - Vu(t-a). \tag{1}$$

We shall first obtain the solution for i to the driving function $(Vt/a)u(t)$ with zero initial conditions, then we shall write down the solution for $-[V(t-a)/a]u(t-a)$ "by inspection." In determining the solution for $-Vu(t-a)$ we shall recall the solution for the driving function $Vu(t)$.

For the driving function, $(Vt/a)\,u(t)$, Kirchhoff's voltage equation is

$$L\frac{di}{dt} + Ri = \frac{Vt}{a}, \quad t > 0. \tag{2}$$

Our driving function, which increases linearly with time, is known as a "ramp" function. The method of attack is the same as that used for sinusoidal driving functions in Section 3.12; we shall perform such operations on Equation 2 as will result in a homogeneous equation. Differentiation of Equation 2 twice results in

$$\frac{d^2}{dt^2}\left(L\frac{di}{dt} + Ri\right) = 0,$$

which has the characteristic equation

$$m^2(Lm + R) = 0. \tag{3}$$

The solution to Equation 2 may be written

$$i = K_1 + K_2 t + Ae^{-Rt/L}, \tag{4}$$

in which K_1 and K_2 are the constants for the source response and A is the constant of the free response. The constants K_1 and K_2 are evaluated by substituting the source response into Equation 2. We obtain

$$K_2 = \frac{V}{Ra},$$

and

$$K_1 = -\frac{LV}{R^2 a}.$$

The application of the initial condition, $i(0+) = 0$, to Equation 4 permits us to show that

$$A = -K_1 = \frac{LV}{R^2 a},$$

and therefore the solution to the driving function, $Vtu(t)/a$, is

$$i = \left[-\frac{LV}{R^2 a} + \frac{Vt}{Ra} + \frac{LV}{R^2 a} e^{-Rt/L} \right] u(t). \tag{5}$$

"By inspection," the solution to the driving function,

$$-\frac{V(t-a)}{a} u(t-a),$$

is

$$i = \left[\frac{LV}{R^2 a} - \frac{V(t-a)}{Ra} - \frac{LV}{R^2 a} e^{-R(t-a)/L} \right] u(t-a). \tag{6}$$

In order to obtain the response to the driving function $-Vu(t-a)$, we obtain the response to the driving function $Vu(t)$ with $i(0+) = 0$. This response can readily be shown to be

$$i = \frac{V}{R}(1 - e^{-Rt/L})u(t),$$

and thus the response to $-Vu(t-a)$ is

$$i = -\frac{V}{R}[1 - e^{-R(t-a)/L}]u(t-a). \tag{7}$$

The actual response to the driving function of Equation 1 is the sum of Equations 5 to 7 or

$$i = \left[-\frac{LV}{R^2 a} + \frac{Vt}{Ra} + \frac{LV}{R^2 a} e^{-Rt/L}\right] u(t)$$

$$+ \left[\frac{LV}{R^2 a} - \frac{V(t-a)}{Ra} - \frac{LV}{R^2 a} e^{-R(t-a)/L}\right] u(t-a)$$

$$- \frac{V}{R}[1 - e^{-R(t-a)/L}] u(t-a). \tag{8}$$

(b) Laplace transform method. In the transform domain the following equation must be satisfied.

$$(Ls + R)I(s) = \mathcal{L}\left[\frac{Vt}{a} u(t) - \frac{V(t-a)}{a} u(t-a) - Vu(t-a)\right]. \tag{9}$$

We need the transform of $f(t) = tu(t)$. This may be obtained from the Table of Transforms in Chapter 3 as the transform of te^{-at} for $a = 0$; however, it may be better to develop the transform.

$$\mathcal{L}[tu(t)] = \int_0^\infty t e^{-st} \, dt.$$

Integrate by parts, letting $t = u$, $e^{-st} \, dt = dv$. Then $dt = du$ and $-(e^{-st}/s) = v$, and

$$\mathcal{L}[tu(t)] = -\frac{te^{-st}}{s}\bigg|_0^\infty + \frac{1}{s}\int_0^\infty e^{-st} \, dt$$

$$= 0 + \frac{e^{-st}}{-s^2}\bigg|_0^\infty = \frac{1}{s^2}. \tag{10}$$

The Laplace transform of $(Vt/a)u(t)$ is V/as^2. The Laplace transform of the shifted time function, $[V(t-a)u(t-a)]/a$, is therefore Ve^{-as}/as^2. Equation 9 may be written

$$(Ls + R)I(s) = \frac{V}{as^2} - \frac{Ve^{-as}}{as^2} - \frac{Ve^{-as}}{s}, \tag{11}$$

or

$$I(s) = \frac{V(1 - e^{-as} - ase^{-as})}{Las^2(s + R/L)},$$

$$= \frac{V}{Las^2(s + R/L)} - \frac{Ve^{-as}}{Las^2(s + R/L)} - \frac{Ve^{-as}}{Ls(s + R/L)}. \tag{12}$$

The first term is expanded by partial fractions,

$$\frac{V}{Las^2(s + R/L)} = \frac{K_1}{s} + \frac{K_2}{s^2} + \frac{A_1}{s + R/L},$$

and the constants K_1, K_2, and A_1 are evaluated by the usual techniques to obtain:

$$K_2 = \frac{V}{Ra},$$

$$A_1 = \frac{LV}{R^2 a},$$

and

$$K_1 = -A_1 = -\frac{LV}{R^2 a}.$$

The last term of Equation 12 may be written

$$\left[\frac{V}{Ls(s + R/L)}\right] e^{-as},$$

and the quantity within the brackets may be expanded by partial fractions. The quantity e^{-as} then acts on each term of the expansion, being interpreted as a time delay factor in obtaining the inverse transform.

$$\frac{V}{Ls(s + R/L)} = \frac{K_3}{s} + \frac{A_2}{s + R/L},$$

from which $K_3 = V/R$, and $A_2 = -V/R$.

The inverse transform of Equation 12 may now be written as

$$i = \left[-\frac{LV}{R^2 a} + \frac{Vt}{Ra} + \frac{LV}{R^2 a} e^{-Rt/L}\right] u(t)$$
$$+ \left[\frac{LV}{R^2 a} - \frac{V(t-a)}{Ra} - \frac{LV}{R^2 a} e^{-R(t-a)/L}\right] u(t-a)$$
$$- \frac{V}{R}[1 - e^{-[R(t-a)]/L}] u(t-a). \tag{13}$$

Equation 13 is identical with Equation 8, the solution by the classical method.

9.3 IMPULSES

Impulses, in a mechanics sense, are high values of forces integrated over an interval of time that is short in comparison with other time intervals of interest. It is found that this impulse is essentially equal to the change in momentum of the masses on which these forces act. Examples would be the firing of a gun, the contact of two billiard balls or the striking of a tennis ball. Since the same force often acts on several bodies, the net changes in momentum must be equal, and we speak of the "conservation of momentum."

Similar phenomena occur in electromagnetism, in which large currents or voltages occur for relatively short lengths of time. If we idealize our circuit model we are apt to find infinite values of current or voltage appearing. Some illustrations will show this and permit us to define an impulse function.

Have you ever noticed the arcing that occurs when a switch is opened in an inductive d-c circuit? Has it occurred to you that the voltages to cause arcing are much higher than the d-c sources in the circuit? Perhaps you have been "shocked" to verify this experimentally.

A simple illustration will show why high voltages occur. In the circuit of Fig. 9.5 the switch is opened at $t = 0$ with a finite value for $i(0-)$. If we assume the switch is perfect and does change the current from $i(0-)$ to zero in zero time, then v_{bc} becomes infinitely large because

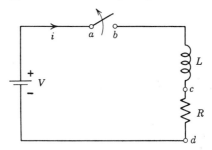

FIG. 9.5 The opening of a switch in an inductive circuit.

$v_{bc} = L(di/dt)$. Also the voltage across the switch would be infinitely large in order to satisfy Kirchhoff's voltage law. The realist will point out that infinite voltages will not exist, that capacitances have been neglected, (including the changing capacitance between switch blades as the switch is opened), and that there will probably be a breakdown of the dielectric about the switch terminals. The realist is right, but we do not have the needed information on capacitances and on dielectric breakdown, and we need some principles to guide us in obtaining approximate answers. A similar lack of information exists in impulse problems in mechanics.

Let us obtain a physical concept for a voltage impulse by assuming that the circuit of Fig. 9.5 is adequate for the short interval of time, ϵ, required for the switch to break the circuit. A voltage equation which holds for both positive and negative time is

$$V = v_{ab} + L\frac{di}{dt} + Ri. \tag{9.12}$$

Pulses, Impulses, Dependent Sources 411

A useful concept is obtained by integrating Equation 9.12 from $t = 0-$ to $t = 0 + \epsilon$. From a physical standpoint the voltages are finite quantities whose integrals exist.

$$\int_{0-}^{0+\epsilon} V\,dt = \int_{0-}^{0+\epsilon} v_{ab}\,dt + \int_{i(0-)}^{i(0+\epsilon)} L\,di + \int_{0-}^{0+\epsilon} Ri\,dt,$$

or

$$\int_{0-}^{0+\epsilon} V\,dt = \int_{0-}^{0+\epsilon} v_{ab}\,dt - Li(0-) + \int_{0-}^{0+\epsilon} Ri\,dt, \tag{9.13}$$

since $Li(0 + \epsilon)$ is zero from the statement of the problem.

If we now take the limit of Equation 9.13 as $\epsilon \to 0$, we obtain

$$0 = \int_{0-}^{0+} v_{ab}\,dt - Li(0-),$$

or

$$\int_{0-}^{0+} v_{ab}\,dt = Li(0-). \tag{9.14}$$

Although ϵ cannot physically be zero, inspection of Equation 9.13 reveals that the terms, $\int_{0-}^{0+\epsilon} V\,dt$ and $\int_{0-}^{0+\epsilon} Ri\,dt$, become smaller as ϵ becomes smaller and thus Equation 9.14 should be a good approximation if the switching time is relatively short. We observe that, at $t = 0$, v_{ab} is infinitely large; however, the integral, $\int_{0-}^{0+} v_{ab}\,dt$, is finite, in this case being equal to $Li(0-)$ volt seconds. It is customary to say that a voltage impulse having a magnitude of $Li(0-)$ volt seconds occurs at $t = 0$ and to write this symbolically as

$$v_{ab} = Li(0-)\delta(t). \tag{9.15}$$

The symbol $\delta(t)$ is called a unit impulse function and has the units of reciprocal seconds. Thus from a physical standpoint it is convenient to think of a unit impulse function as being zero for all time except at $t = 0$ where it is defined by the integral,

$$\int_{0-}^{0+} \delta(t)\,dt = 1. \tag{9.16}$$

In case someone questions the validity of such an integral we can retreat to a more secure position by stating that the interval over which we are integrating is not really zero, it is simply so small in comparison with other time intervals that we call it zero for convenience.

Although engineers have used the impulse function for about a hundred years, mathematicians have not considered it respectable until recently. This has occurred in a relatively new branch of mathematics called the theory of distributions.[1] Consider the integral,

$$\int_{-\infty}^{\infty} f(t)\,\delta(t)\,dt = f(0), \tag{9.17}$$

in which $f(t)$ is continuous at $t = 0$. The integral and the distribution function, $\delta(t)$, are both considered to be **defined by** the quantity $f(0)$ assigned to $f(t)$. The distribution function, $\delta(t)$, is then a peculiar function which if multiplied by a real function, $f(t)$, and the product integrated from $t = -\infty$ to $t = +\infty$ results in the value of $f(t)$ at $t = 0$. A similar expression applies to a delayed unit impulse function, $\delta(t - a)$.

$$\int_{-\infty}^{\infty} f(t)\,\delta(t - a)\,dt = f(a), \tag{9.18}$$

if $f(t)$ is continuous at $t = a$.

We note that Equation 9.17 [for $f(t) = 1$] gives the same results as Equation 9.16, and thus we can write

$$\int_{-\infty}^{\infty} \delta(t)\,dt = \int_{0-}^{0+} \delta(t)\,dt = 1. \tag{9.19}$$

The integral may also be interpreted as

$$\int_{-\infty}^{t} \delta(t)\,dt = u(t), \tag{9.20}$$

from which we might deduce that

$$\frac{du(t)}{dt} = \delta(t). \tag{9.21}$$

With regard to the Laplace transform of a unit impulse function, it is apparent that the lower limit of the integral must be considered as $0-$ if the impulse is to be included

$$\mathscr{L}[\delta(t)] = \int_{0-}^{\infty} \delta(t)e^{-st}\,dt = \int_{0-}^{0+} \delta(t)\,dt = 1. \tag{9.22}$$

We recall that the reason we have been using $0+$ values in the past was because we were not prepared to handle integrals of impulse functions. From now on we will use $0-$ values as initial conditions in the Laplace transform method since impulses, if they exist, will automatically be included in the solution.

In our original illustration $\int_{0-}^{0+} v_{ab}\,dt = Li(0-)$ and thus $v_{ab} = Li(0-)\delta(t)$. $Li(0-)$ has the units of volt seconds and is called the

---

[1] For a discussion of the impulse function as distribution see A. Papoulis, *The Fourier Integral and its Application*, Appendix I, McGraw-Hill, New York, 1962.

magnitude of the impulse function. In sketching an impulse function the magnitude is placed at the side of the arrow. Thus the function, $f(t) = 3\delta(t) - A\delta(t - a)$, would be sketched as shown in Fig. 9.6.

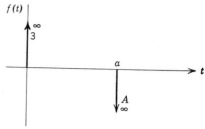

FIG. 9.6 Sketch of impulse functions. $f(t) = 3\delta(t) - A\delta(t - a)$.

9.3.1 Voltage Impulses in Inductive Circuits

We return now to the simple series RL circuit of the previous article and apply both classical and Laplace transform methods in solving for the voltage across the switch. The circuit is resketched in Fig. 9.7. The switch is opened at $t = 0$.

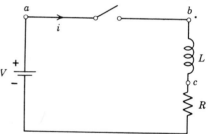

FIG. 9.7 The opening of a switch in an inductive circuit.

In the classical method we write a voltage equation which holds for all time.

$$V = v_{ab} + L\frac{di}{dt} + Ri. \tag{9.23}$$

It is convenient to consider time as consisting of three intervals, $t < 0$, $t = 0$, and $t > 0$. For $t < 0$ the switch is closed and $v_{ab} = 0$. At $t = 0$ a voltage impulse exists which is evaluated by integrating Equation 9.23 between $t = 0-$ and $t = 0+$ to give

$$\int_{0-}^{0+} v_{ab}\, dt = Li(0-). \quad \text{Thus at } t = 0, \, v_{ab} = Li(0-)\,\delta(t).$$

For $t > 0$, $i = 0$ and from Equation 9.23 we obtain $v_{ab} = Vu(t)$. These partial solutions are combined to give the total solution for v_{ab}.

$$v_{ab} = Li(0-)\,\delta(t) + Vu(t). \tag{9.24}$$

With regard to the polarity of the impulsive voltage across the switch, we observe from Equation 9.24 that a is plus with respect to b if $i(0-)$ is positive. Since Kirchhoff's voltage law must be obeyed, there is an impulsive voltage across L with b being negative with respect to c.

In the Laplace transform method we take the transform of Equation 9.23 which holds for all time. We use our revised definition of the transform,

$$\mathscr{L}f(t) = \int_{0-}^{\infty} f(t)e^{-st}\,dt = \int_{0-}^{0+} f(t)e^{-st}\,dt + \int_{0+}^{\infty} f(t)e^{-st}\,dt$$

$$= \int_{0-}^{0+} f(t)\,dt + \int_{0+}^{\infty} f(t)e^{-st}\,dt. \quad (9.25)$$

This involves only one change in our list of transforms; this change is for $\mathscr{L}[df(t)/dt]$.

$$\mathscr{L}\left[\frac{df(t)}{dt}\right] = \int_{0-}^{0+} df(t) + \int_{0+}^{\infty} \frac{df(t)}{dt} e^{-st}\,dt,$$

$$= f(0+) - f(0-) + sF(s) - f(0+),$$

$$= sF(s) - f(0-). \quad (9.26)$$

We can now write the transform of Equation 9.23 as

$$\frac{V}{s} = V_{ab}(s) + LsI(s) - Li(0-) + RI(s). \quad (9.27)$$

However, since $i = 0$ for $t > 0$, $I(s) = 0$, and we obtain

$$\frac{V}{s} = V_{ab}(s) - Li(0-),$$

or

$$V_{ab}(s) = Li(0-) + \frac{V}{s}. \quad (9.28)$$

The inverse Laplace transform of a constant is an impulse function, so

$$v_{ab} = Li(0-)\,\delta(t) + Vu(t), \quad (9.29)$$

which agrees with the solution by the classical method.

This simple example demonstrates that a voltage impulse may cause an instantaneous change in flux linkages. However, it is possible to have equal but opposite impulsive voltages in a circuit without having a change in the flux linkages of the circuit. This is illustrated by Fig. 9.8 in which the switch is opened at $t = 0$ with known nonzero values for $i_1(0-)$, $i_2(0-)$, and $i_3(0-)$.

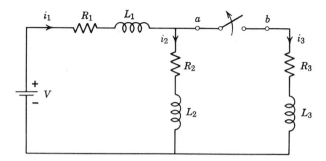

FIG. 9.8

Two independent voltage equations which hold for all time are

$$V = R_1 i_1 + L_1 \frac{di_1}{dt} + R_2 i_2 + L_2 \frac{di_2}{dt}, \quad (9.30)$$

and

$$V = R_1 i_1 + L_1 \frac{di_1}{dt} + v_{ab} + R_3 i_3 + L_3 \frac{di_3}{dt}. \quad (9.31)$$

Integration of Equation 9.30 between $t = 0-$ and $t = 0+$ yields

$$0 = \int_{i_1(0-)}^{i_1(0+)} L_1 \, di_1 + \int_{i_2(0-)}^{i_2(0+)} L_2 \, di_2,$$

or

$$L_1 i_1(0-) + L_2 i_2(0-) = L_1 i_1(0+) + L_2 i_2(0+). \quad (9.32)$$

We observe that Equation 9.32 represents a statement of the conservation of flux linkages for the loop around which the voltage equation was written. Furthermore, since $i_1(0+) = i_2(0+)$, this equation permits solving for this current value.

Integration of Equation 9.31 from $t = 0-$ to $t = 0+$ yields

$$L_1 i_1(0-) + L_3 i_3(0-) = L_1 i_1(0+) + L_3 i_3(0+) + \int_{0-}^{0+} v_{ab} \, dt. \quad (9.33)$$

Since $i_3(0+) = 0$, and since $i_1(0+)$ is known from Equation 9.32, the voltage impulse across the switch, $\int_{0-}^{0+} v_{ab} \, dt$, may be calculated. Thus we observe that flux linkages are not conserved for the loop around which Equation 9.31 is written because the switch acts as an instantaneous sink for flux linkages.

From this illustration we can deduce a principle which is known as the conservation of flux linkages and which may be stated as follows. "The net flux linkages of a circuit path cannot be changed instantaneously without the application of an external voltage impulse." This principle is analogous to the principle of conservation of momentum in mechanics.

The Laplace transform of Equation 9.30 is

$$\frac{V}{s} = R_1 I_1(s) + L_1 s I_1(s) - L_1 i_1(0-) + R_2 I_2(s) + L_2 s I_2(s) - L_2 i_2(0-).$$

(9.34)

Since $i_1 = i_2$ for $t > 0$ and since i_1 and i_2 are each finite at $t = 0$, $I_1(s) = I_2(s)$. Therefore Equation 9.34 may be written

$$\frac{V}{s} = (R_1 + R_2) I_1(s) + (L_1 + L_2) s I_1(s) - L_1 i_1(0-) - L_2 i_2(0-). \quad (9.35)$$

If we compare the last two terms in Equation 9.35 with Equation 9.32, we observe that the Laplace transform method has automatically applied the principle of conservation of flux linkages.

Example 9.3

GIVEN: The circuit of Fig. E-3.1, steady state existing with the switch closed. The switch is opened at $t = 0$. $R_1 = R_2 = 10\ \Omega$, $L_1 = 2$ henries, $L_2 = 3$ henries, $V = 100$ volts.

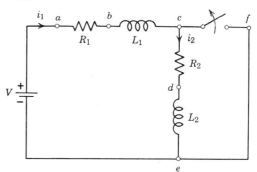

FIG. E-3.1.

TO FIND: The expressions for i_1 and v_{de} for $t > 0$ by (a) the classical method, and (b) the Laplace transform method.

SOLUTION: (a) *Classical method.* Two independent voltage equations which hold for all time are

$$V = R_1 i_1 + L_1 \frac{di_1}{dt} + R_2 i_2 + L_2 \frac{di_2}{dt}, \quad (1)$$

and

$$V = R_1 i_1 + L_1 \frac{di_1}{dt} + v_{ef}. \quad (2)$$

For the loop of Equation 1 there is no external voltage impulse; thus flux linkages are conserved, and the current at $t = 0+$ may be evaluated by integrating this equation from $t = 0-$ to $t = 0+$. This results in

$$L_1 i_1(0+) + L_2 i_2(0+) = L_1 i_1(0-) + L_2 i_2(0-). \quad (3)$$

Since $i_1(0-) = 10$ amp, $i_2(0-) = 0$ amp, and $i_2(0+) = i_1(0+)$,
$[2 + 3]i_1(0+) = 2(10)$,
or
$i_1(0+) = 4$ amperes. (4)

For $t > 0$, since $i_1 = i_2$, Equation 1 may be modified to be

$$(L_1 + L_2)\frac{di_1}{dt} + (R_1 + R_2)i_1 = V,$$

or

$$5\frac{di_1}{dt} + 20i_1 = 100. \qquad (5)$$

The steady-state response of i_1 is observed to be $V/(R_1 + R_2) = 5$; the free response has the form Ae^{mt} in which

$$m = -\frac{R_1 + R_2}{L_1 + L_2} = -4.$$

Thus

$$i_1 = 5 + Ae^{-4t}. \qquad (6)$$

Since $i_1(0+) = 4$, $A = -1$, and

$$i_1 = 5 - e^{-4t} \text{ amp.} \qquad t > 0. \qquad (7)$$

A sketch of i_1 versus time is shown in Fig. E-3.2. The discontinuity at $t = 0$ should indicate that impulses are present in our inductive circuit.

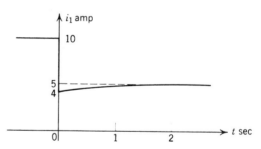

FIG. E-3.2

If we wish to express i_1 as a function of time for all t it is possible to do so with the use of unit step functions. If we assume $i_1 = 10$ amp for $-\infty < t < 0$, this part may be written as $10[1 - u(t)]$. Combining this with Equation 7 we obtain

$$i_1 = 10[1 - u(t)] + (5 - e^{-4t})u(t) \text{ amp.} \qquad (8)$$

To evaluate v_{de} for $t \gg 0$, we observe that

$$v_{de} = L_2 \frac{di_2}{dt}. \qquad (9)$$

At $t = 0$, i_2 changes immediately from zero to 4 amperes. For $t > 0$, $i_2 = i_1$ and v_{de} may be obtained from our solution for i_1. The voltage impulse at $t = 0$ may be evaluated by integrating Equation 9 from $t = 0-$ to $t = 0+$.

$$\int_{0-}^{0+} v_{de}\, dt = \int_{i_2(0-)}^{i_2(0+)} L_2\, di_2 = L_2 i_2(0+) - L_2 i_2(0-),$$

or

$$\int_{0-}^{0+} v_{de}\, dt = 3(4) = 12 \text{ volt seconds.} \tag{10}$$

The expression for v_{de} at $t = 0$ may be written

$$v_{de} = 12\delta(t) \text{ volts.} \tag{11}$$

For $t > 0$, $i_2 = i_1 = 5 - e^{-4t}$ amp from Equation 7, so

$$v_{de} = L_2 \frac{d}{dt}(5 - e^{-4t})$$

$$= 12 e^{-4t} \text{ volts.} \quad t > 0. \tag{12}$$

The complete expression for v_{de} is obtained by combining Equations 11 and 12:

$$v_{de} = 12\delta(t) + 12 e^{-4t} u(t) \text{ volts.} \tag{13}$$

(b) *Laplace transform method.* In the transform method the use of initial values at $t = 0-$ permits us to include impulses which occur at $t = 0$. Thus for this problem we show the circuit as in Fig. E-3.3 with the switch open.

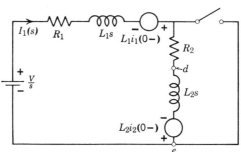

FIG. E-3.3

The transform voltage equation is

$$[(L_1 + L_2)s + R_1 + R_2]I_1(s) = \frac{V}{s} + L_1 i_1(0-) + L_2 i_2(0-). \tag{14}$$

The introduction of numbers gives

$$(5s + 20)I_1(s) = \frac{100}{s} + 20,$$

or

$$I_1(s) = \frac{4s + 20}{s(s + 4)} = \frac{A}{s} + \frac{B}{s + 4} \tag{15}$$

By Heaviside's expansion method, $A = 5$, and $B = -1$, and thus

$$i_1 = (5 - e^{-4t})u(t) \text{ amp.} \tag{16}$$

This solution is in accord with that of the classical method for $t > 0-$.

It is important to observe that transform voltages have the units of volt seconds or flux linkages and that in the transform voltage equations the principle of conservation of flux linkages is automatically invoked if initial conditions are those at $t = 0-$. The danger in this method is that one may not realize that discontinuities and impulses exist. In this particular calculation for i_1, one may not notice that $i_1(0+) \neq i_1(0-)$ and therefore may not realize that voltage impulses exist.

$$V_{de}(s) = L_2 s I_1(s) - L_2 i_2(0-)$$

$$= \frac{3s(4s + 20)}{s(s + 4)} = \frac{12s + 60}{s + 4}$$

$$= 12 + \frac{12}{s + 4}. \tag{14}$$

Then

$$v_{de} = 12\,\delta(t) + 12e^{-4t}u(t) \text{ volts,} \tag{15}$$

which agrees with the solution by the classical method.

9.3.2 Current Impulses in Capacitive Circuits

For an inductive circuit we have found that voltage impulses are associated with finite discontinuities in magnetic flux linkages. Such discontinuities and impulses occur because we have found it expedient to assume perfect switches and to assume that some capacitive effects were negligible. In an analogous manner, current impulses are associated with finite discontinuities in charge and occur because we assume that some resistive and inductive effects are negligible.

Let us consider first the circuit of Fig. 9.9 in which a voltage source, V, is to be connected to a capacitance, C, at $t = 0$. The inductance of the circuit is assumed to be negligibly small and the power dissipated in resistive elements is assumed to be negligibly small.

From a physical standpoint it is clear that at $t = 0$ the voltage across the capacitance must change from $v_{ab}(0-)$ to V, and that this will require a finite amount of charge, $\int_{0-}^{0+} i\,dt$, or a current impulse. One

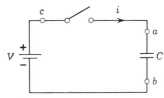

FIG. 9.9 Ideal voltage source applied to a capacitance.

approach is to write Kirchhoff's voltage law around the circuit for $t \gg 0-$.

$$-V + v_{ca} + \frac{1}{C}\int_{0-}^{t} i\, dt + v_{ab}(0-) = 0. \qquad (9.36)$$

Let us evaluate Equation 9.36 at $t = 0+$, for which $v_{ca} = 0$.

$$-V + \frac{1}{C}\int_{0-}^{0+} i\, dt + v_{ab}(0-) = 0,$$

or

$$\int_{0-}^{0+} i\, dt = C[V - v_{ab}(0-)]. \qquad (9.37)$$

Another approach is to write a current equation at the node a.

$$i = C\frac{dv_{ab}}{dt}. \qquad (9.38)$$

Integration of Equation 9.38 from $t = 0-$ to $t = 0+$ yields

$$\int_{0-}^{0+} i\, dt = \int_{v_{ab}(0-)}^{v_{ab}(0+)} C\, dv_{ab} = C[V - v_{ab}(0-)], \qquad (9.39)$$

since $v_{ab}(0+) = V$.

Thus the current impulse has a magnitude of $C[V - v_{ab}(0-)]$ and the expression for the current may be written

$$i = C[V - v_{ab}(0-)]\,\delta(t). \qquad (9.40)$$

In the Laplace transform method we may take the transform of Equation 9.36. This results in

$$-\frac{V}{s} + \frac{I(s)}{Cs} + \frac{v_{ab}(0-)}{s} = 0. \qquad (9.41)$$

We observe that $\mathcal{L}(v_{ca}) = 0$ since v_{ca} is finite at $t = 0$ and zero for $t > 0$. The transform circuit is shown in Fig. 9.10.

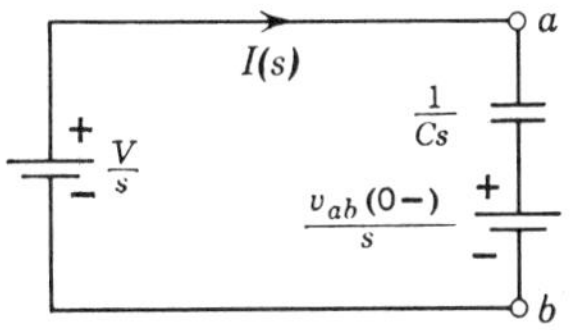

FIG. 9.10 Transform circuit for Fig. 9.9.

$I(s)$ may readily be determined from Equation 9.41.

$$I(s) = C[V - v_{ab}(0-)],$$

from which $i = C[V - v_{ab}(0-)]\,\delta(t)$, which agrees with the classical solution of Equation 9.40.

Figure 9.11 shows a circuit which demonstrates the principle known as conservation of charge. This principle states that the charge associated with a node or a group of nodes can not be changed instantaneously without the application of an external current impulse. (We have observed for the circuit of Fig. 9.9 in which the voltage source was applied to a capacitor that the charge at either node was immediately changed.) Now let us write a current equation about node b of Fig. 9.11.

$$C_1 \frac{dv_{ba}}{dt} + C_2 \frac{dv_{bc}}{dt} = 0. \tag{9.42}$$

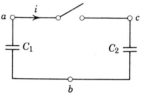

FIG. 9.11. Circuit with two charged capacitances.

Integration of this equation from $t = 0-$ to $t = 0+$ gives

$$\int_{v_{ba}(0-)}^{v_{ba}(0+)} C_1\, dv_{ba} + \int_{v_{bc}(0-)}^{v_{bc}(0+)} C_2\, dv_{bc} = 0,$$

or

$$C_1 v_{ba}(0+) + C_2 v_{bc}(0+) = C_1 v_{ba}(0-) + C_2 v_{bc}(0-). \tag{9.43}$$

Equation 9.43 states that the charge associated with node b at $t = 0+$ is the same as at $t = 0-$. This is logical since any charge which leaves the lower plate of C_1 must go to the lower plate of C_2 and vice versa. Since $v_{ba}(0+) = v_{bc}(0+)$, Equation 9.43 gives the expression for either of these.

$$v_{ba}(0+) = v_{bc}(0+) = \frac{C_1 v_{ba}(0-) + C_2 v_{bc}(0-)}{C_1 + C_2}. \tag{9.44}$$

An identical result would have been obtained by applying conservation of charge at the combined nodes a and c. However, conservation of charge does not apply to either of these nodes individually since an "external" current impulse occurs at $t = 0$.

The current impulse may be evaluated by recalling that $\Delta q = C \Delta v$, and thus

$$\int_{0-}^{0+} i\, dt = \Delta q = C_1[v_{ba}(0+) - v_{ba}(0-)]. \tag{9.45}$$

Substitution of Equation 9.44 into Equation 9.45 gives

$$\Delta q = \frac{C_1 C_2}{C_1 + C_2}[v_{bc}(0-) - v_{ba}(0-)], \tag{9.46}$$

and thus

$$i = \Delta q\, \delta(t) = \frac{C_1 C_2}{C_1 + C_2}[v_{bc}(0-) - v_{ba}(0-)]\, \delta(t) \tag{9.47}$$

Another approach to the circuit of Fig. 9.11 is to write a voltage equation which holds for $t > 0-$.

$$\underbrace{\frac{1}{C_2}\int_{0-}^{t} i\, dt + v_{cb}(0-)}_{v_{cb}} + \underbrace{\frac{1}{C_1}\int_{0-}^{t} i\, dt + v_{ba}(0-)}_{v_{ba}} + v_{ac} = 0. \tag{9.48}$$

Evaluating Equation 9.48 at $t = 0+$, with $v_{ac} = 0$, gives

$$\left(\frac{1}{C_1} + \frac{1}{C_2}\right)\int_{0-}^{0+} i\, dt = v_{bc}(0-) - v_{ba}(0-),$$

or

$$\Delta q = \int_{0-}^{0+} i\, dt = \frac{C_1 C_2}{C_1 + C_2}[v_{bc}(0-) - v_{ba}(0-)]. \tag{9.49}$$

This result agrees with that of Equation 9.46. $v_{ba}(0+)$ can be determined since $v_{ba}(0+) = v_{ba}(0-) + \Delta q/C_1$.

The Laplace transform method may also be used. The transform voltage equation for the circuit of Fig. 9.12 is

$$\left[\frac{1}{C_1 s} + \frac{1}{C_2 s}\right] I(s) = \frac{v_{bc}(0-)}{s} - \frac{v_{ba}(0-)}{s}. \tag{9.50}$$

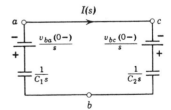

FIG. 9.12 Transform circuit for Fig. 9.11.

Then

$$I(s) = \frac{C_1 C_2}{C_1 + C_2}[v_{bc}(0-) - v_{ba}(0-)],$$

and

$$i = \frac{C_1 C_2}{C_1 + C_2} [v_{bc}(0-) - v_{bc}(0-)] \delta(t), \tag{9.51}$$

the same result as that of the classical method (see Equation 9.47).

The node method to determine the voltage $V_{ba}(s)$ would give the equation

$$V_{ba}(s)[C_1 s + C_2 s] = C_1 v_{ba}(0-) + C_2 v_{bc}(0-), \tag{9.52}$$

or

$$V_{ba}(s) = \frac{C_1 v_{ba}(0-) + C_2 v_{bc}(0-)}{(C_1 + C_2)s}$$

Then

$$v_{ba} = \frac{C_1 v_{ba}(0-) + C_2 v_{bc}(0-)}{C_1 + C_2} u(t). \tag{9.53}$$

The voltage v_{ba} is a constant for $t > 0$, in accord with the previous solutions. In Equation 9.52 the Laplace transform method has automatically applied the principle of conservation of charge since the right side of the equation is equal to the charge at the node b for $t = 0$.

Example 9.4

GIVEN: The circuit of Fig. E-4.1 with steady state existing. The switch is closed at $t = 0$. $V = 10$ volts, $R = 2$ ohms, $C_1 = 2$ f, $C_2 = 3$ f, $v_{cb}(0-) = -15$ volts.

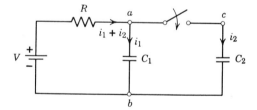

FIG. E-4.1

TO FIND: An outline of the solution for the current i_1 by classical methods and an actual solution by the Laplace transform method.

SOLUTION: Since $v_{ab}(0-) = V = 10$ volts, while $v_{cb}(0-) = -15$ volts, and since $v_{ab}(0+)$ must equal $v_{cb}(0+)$, it is clear that there are current impulses for C_1 and C_2 in order that their voltages may change instantaneously. This means that i_1 and i_2 are infinitely large at $t = 0$. One approach to a solution is to write Kirchhoff's current law about the nodes a and c.

$$C_1 \frac{dv_{ab}}{dt} + C_2 \frac{dv_{cb}}{dt} + \frac{v_{ab} - V}{R} = 0. \tag{1}$$

Integration of Equation 1 from $t = 0-$ to $t = 0+$ gives

$$C_1[v_{ab}(0+) - v_{ab}(0-)] + C_2[v_{cb}(0+) - v_{cb}(0-)] = 0. \tag{2}$$

This equation is an expression of the conservation of charge about nodes a and c.

Since $v_{ab}(0+) = v_{cb}(0+)$,

$$v_{ab}(0+) = \frac{C_1 v_{ab}(0-) + C_2 v_{cb}(0-)}{C_1 + C_2}. \tag{3}$$

Substitution of numbers results in

$$v_{ab}(0+) = \frac{2(10) + 3(-15)}{2 + 3} = -5 \text{ volts}. \tag{4}$$

Since the voltage across the capacitance C_1 of 2 farads changed instantaneously from 10 volts to -5 volts a current impulse of $-(15)(2)$ coulombs occurred or $i_1 = -30\,\delta(t)$ amp at $t = 0$. By similar argument $i_2 = 30\,\delta(t)$ amp at $t = 0$. The solution could be completed for $t > 0$ by standard methods; however, let us consider the loop method as an alternate approach. Two independent voltage equations, which hold for $t > 0-$, are

$$V = R(i_1 + i_2) + \frac{1}{C_1}\int_{0-}^{t} i_1\,dt + v_{ab}(0-) \tag{5}$$

and

$$V = R(i_1 + i_2) + \frac{1}{C_2}\int_{0-}^{t} i_2\,dt + v_{cb}(0-) + v_{ac}. \tag{6}$$

Since the voltages V, v_{ab}, v_{cb}, and v_{ac} are each finite, $R(i_1 + i_2)$ must be finite and thus $\int_{0-}^{0+}(i_1 + i_2)\,dt = 0$, or

$$\int_{0-}^{0+} i_1\,dt = -\int_{0-}^{0+} i_2\,dt. \tag{7}$$

A second relationship may be obtained by evaluating Equations 5 and 6 at $t = 0+$ or it may be written directly from the fact that $v_{ab}(0+) = v_{cb}(0+)$.

$$v_{ab}(0+) = \frac{1}{C_1}\int_{0-}^{0+} i_1\,dt + v_{ab}(0-) = \frac{1}{C_2}\int_{0-}^{0+} i_2\,dt + v_{cb}(0-). \tag{8}$$

Equations 7 and 8 permit evaluation of the current impulses at $t = 0$. Equations 5 and 6 may then be used to determine the solution for i_1 for $t > 0$.

In the Laplace transform method the equations are obtained by taking the transform of the time domain equations. From Equations 5 and 6,

$$\frac{V}{s} = R[I_1(s) + I_2(s)] + \frac{I_1(s)}{C_1 s} + \frac{v_{ab}(0-)}{s}, \tag{9}$$

$$\frac{V}{s} = R[I_1(s) + I_2(s)] + \frac{I_2(s)}{C_2 s} + \frac{v_{cb}(0-)}{s}. \tag{10}$$

The transform circuit may be sketched from the equations and is shown in Fig. E-4.2.

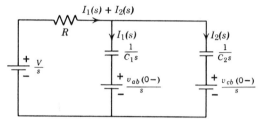

FIG. E-4.2

Substitution of numbers into Equations 9 and 10 and rearrangement of terms results in

$$\left[2 + \frac{1}{2s}\right]I_1(s) + 2I_2(s) = 0, \tag{11}$$

$$2I_1(s) + \left[2 + \frac{1}{3s}\right]I_2(s) = \frac{25}{s}. \tag{12}$$

Then

$$I_1(s) = \frac{\begin{vmatrix} 0 & 2 \\ \dfrac{25}{s} & 2 + \dfrac{1}{3s} \end{vmatrix}}{\begin{vmatrix} 2 + \dfrac{1}{2s} & 2 \\ 2 & 2 + \dfrac{1}{3s} \end{vmatrix}} = \frac{-30s}{s + \frac{1}{10}}$$

$$= -30 + \frac{3}{s + \frac{1}{10}}. \tag{13}$$

Finally

$$i_1 = -30\,\delta(t) + 3e^{-t/10}u(t) \text{ amp.} \tag{14}$$

It may be observed that the proper use of the transform method is dependent on a good understanding of the classical method.

Example 9.5

GIVEN: The circuit of Fig. E-5.1 with $v_{ab}(0-) = v_{bc}(0-) = v_{de}(0-) = 50$ volts.

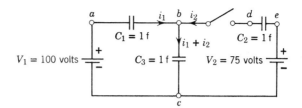

FIG. E-5.1

TO FIND: The current impulses and the voltages at $t = 0+$ by the classical method.

SOLUTION: Two voltage equations which hold at $t = 0+$ are

$$V_1 = \frac{1}{C_1}\int_{0-}^{0+} i_i\,dt + v_{ab}(0-) + \frac{1}{C_3}\int_{0-}^{0+}(i_1 + i_2)\,dt + v_{bc}(0-), \quad (1)$$

and

$$V_2 = \frac{1}{C_2}\int_{0-}^{0+} i_2\,dt + v_{ed}(0-) + \frac{1}{C_3}\int_{0-}^{0+}(i_1 + i_2)\,dt + v_{bc}(0-). \quad (2)$$

Let $\int_{0-}^{0+} i_1\,dt = \Delta q_1$ and $\int_{0-}^{0+} i_2\,dt = \Delta q_2$ and substitute numerical values to obtain

$$2\,\Delta q_1 + \Delta q_2 = 0 \quad (3)$$

$$\Delta q_1 + 2\,\Delta q_2 = 75. \quad (4)$$

Solving Equations 3 and 4, one obtains

$$\Delta q_1 = -25 \text{ coulombs,}$$

and

$$\Delta q_2 = 50 \text{ coulombs.}$$

These are current impulses; we can write the impulsive currents as

$$i_1 = \Delta q_1\,\delta(t) = -25\,\delta(t) \text{ amp,} \quad (5)$$

and

$$i_2 = \Delta q_2\,\delta(t) = 50\,\delta(t) \text{ amp.} \quad (6)$$

Then

$$v_{ab}(0+) = v_{ab}(0-) + \frac{\Delta q_1}{C_1} = 50 - 25 = 25 \text{ volts.} \quad (7)$$

$$v_{bc}(0+) = v_{bc}(0-) + \frac{(\Delta q_1 + \Delta q_2)}{C_3} = 50 + 25 = 75 \text{ volts.} \quad (8)$$

$$v_{de}(0+) = v_{de}(0-) - \frac{\Delta q_2}{C_2} = 50 - 50 = 0 \text{ volts.} \quad (9)$$

9.4 TIME RESPONSE TO SINGLE DRIVING FUNCTION (POLE-ZERO DIAGRAM)

One characteristic of the impulse function is that it has a simple transform. This is therefore a good time to review and integrate our knowledge of transform network functions and time response functions. For the purposes of this article we shall restrict our definition of a network function, $H(s)$, to one for which the driving function is always in the denominator.

$$H(s) = \frac{\text{Transform response function}}{\text{Transform driving function}} = \frac{R(s)}{D(s)}. \qquad (9.54)$$

This is not a serious restriction since the definition is in accord with the definition of a transfer function. It does restrict the definition of a driving-point immittance to a particular ratio; if the driving function is a transform voltage, the response function is a transform current and the network function is a transform admittance. Similarly, if the driving function is a transform current, the driving-point network function is a transform impedance.

The advantage in using the definition of Equation 9.54 is that the response function is always the product of the network function and the driving function.

$$R(s) = H(s)D(s). \qquad (9.55)$$

9.4.1 Response to Impulse and Step Driving Functions

In the form of Equation 9.55 it is easy to observe the changes in the transform response function as the driving function is changed. It is particularly helpful to start with an impulse driving function because its transform is not a function of s. Consider the driving function,

$$d(t) = B\,\delta(t), \qquad (9.56)$$

in which B is a constant having the units of volt-seconds or ampere-seconds. Then

$$D(s) = B,$$

and

$$R(s) = H(s)D(s) = BH(s), \qquad (9.57)$$

from which

$$\mathscr{L}^{-1}[R(s)] = r(t) = \mathscr{L}^{-1}[BH(s)] \qquad (9.58)$$

If the impulse function has a magnitude of one, $B = 1$ numerically, then $r(t)$ is known as the response to the unit impulse function or as the unit impulse response and has the units of volts or amperes. We distinguish carefully between two terms: (*1*) $\mathscr{L}^{-1}[BH(s)]$, for which $B = 1$ volt-second or 1 ampere-second, and which has the units of volts or amperes, and (*2*) $\mathscr{L}^{-1}[H(s)]$, which may be performed mathematically but which has no physical significance.

We recall from Chapter 6 that a transform network function can be written as a constant (scale factor) times the ratio of two polynomials in s. Each of the polynomials can be factored so let us assume for illustration that

$$H(s) = C \frac{(s - s_1)(s - s_2)}{(s - s_a)(s - s_b)}, \tag{9.59}$$

whose pole-zero diagram could be that of Fig. 9.13.

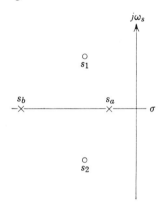

FIG. 9.13   Pole-zero diagram for $C \dfrac{(s - s_1)(s - s_2)}{(s - s_a)(s - s_b)}$.

From Equation 9.57, the transform response to the driving function, $B\,\delta(t)$, is

$$R(s) = BH(s) = BC \frac{(s - s_1)(s - s_2)}{(s - s_a)(s - s_b)},$$

$$= BC + \frac{A_1}{s - s_a} + \frac{A_2}{s - s_b} \qquad s_a \neq s_b \tag{9.60}$$

The term BC appears in Equation 9.60 because the numerator of $R(s)$ has the same degree as the denominator and it is therefore necessary to divide before expanding by partial fractions. The time response may now be written

$$r(t) = BC\,\delta(t) + A_1 e^{s_a t} + A_2 e^{s_b t}. \tag{9.61}$$

The term, $BC\,\delta(t)$, is the forced or source response, the terms, $A_1 e^{s_a t}$ and $A_2 e^{s_b t}$, are the free response. We observe that the coefficients in the exponents of the free response, s_a and s_b, are the poles of the network

Pulses, Impulses, Dependent Sources   429

function. Since we know that it is not possible for the free response of a passive network to increase indefinitely with time, it follows that the real parts of the poles, s_a and s_b, must be negative or zero. Thus the poles of a network function for a passive network must lie either in the left half of the s plane or on the imaginary axis.

For Equation 9.61 the coefficients A_1 and A_2 may be evaluated by Heaviside's expansion method. Thus

$$A_1 = BC \left. \frac{(s - s_1)(s - s_2)}{s - s_b} \right|_{s=s_a} = BC \frac{(s_a - s_1)(s_a - s_2)}{s_a - s_b}, \quad (9.62)$$

and

$$A_2 = BC \left. \frac{(s - s_1)(s - s_2)}{s - s_a} \right|_{s=s_b} = BC \frac{(s_b - s_1)(s_b - s_2)}{s_b - s_a} \quad (9.63)$$

We recall that in Chapter 6 we used pole-zero diagrams to determine the magnitude and phase of network functions at various values of ω. A similar method can be used to evaluate A_1 and A_2. For evaluating A_1 the terms $s_a - s_1$, $s_a - s_2$, and $s_a - s_b$ are shown in Fig. 9.14a as directed line segments. To evaluate A_2 in a similar manner the terms $s_b - s_1$, $s_b - s_2$, and $s_b - s_a$ are shown in Fig. 9.14b as directed line segments.

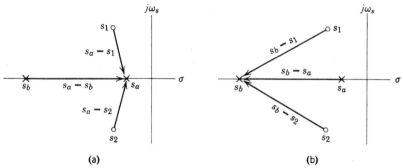

FIG. 9.14 Evaluation of coefficients of free response to impulse driving function. (a) Pole-zero diagram to evaluate A_1 in Equation 9.62. (b) Pole-zero diagram to evaluate A_2 in Equation 9.63.

Inspection of the pole-zero diagrams indicates that the magnitude of A_1 is directly proportional to the distances between the pole s_a and the zeros, s_1 and s_2, and is inversely proportional to the distance between the pole s_a and the pole s_b. The angle of A_1 is zero and the angle of A_2 is 180°, thus making the sign of A_2 negative. We would expect A_1 and A_2 to be real since the poles s_a and s_b have been shown as real.

If the poles s_a and s_b are complex conjugates instead of real, the coefficients A_1 and A_2 are complex conjugates. This can be observed graphically in Fig. 9.15 by noting that the magnitudes of the terms

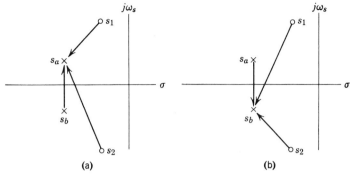

FIG. 9.15 Evaluation of coefficients of free response if poles are complex conjugates. (a) Pole-zero diagram to evaluate A_1 in Equation 9.62. (b) Pole-zero diagram to evaluate A_2 in Equation 9.63.

which determine A_1 and A_2 are equal and that the angle of A_2 is the negative of the angle of A_1.

Now let us replace the impulse driving function $B\,\delta(t)$, by a step driving function, $Bu(t)$, the transform network function being the same as before (Equation 9.59). The transform response function becomes

$$R(s) = H(s)D(s) = C\frac{(s - s_1)(s - s_2)}{(s - s_a)(s - s_b)}\left(\frac{B}{s}\right),$$

$$= BC\frac{(s - s_1)(s - s_2)}{s(s - s_a)(s - s_b)}. \qquad (9.64)$$

If either $s_1 = 0$ or $s_2 = 0$ the s terms in numerator and denominator may be cancelled. This would mean that the forced response would be zero. If $s_1 \neq 0$ and $s_2 \neq 0$ and if the poles are simple, Equation 9.64 may be expanded by partial fractions to give

$$R(s) = BC\frac{(s - s_1)(s - s_2)}{s(s - s_a)(s - s_b)} = \frac{K}{s} + \frac{A_3}{s - s_a} + \frac{A_4}{s - s_b}, \qquad (9.65)$$

the coefficient K being used to represent forced response and the subscripts 3 and 4 being employed with the coefficients of the free response since these coefficients will differ from those obtained with the impulse driving function.

The coefficients may be evaluated by means of a pole-zero diagram in a manner similar to that used for the impulse response. It is interesting to observe that the coefficient of the forced response, K, is

$$K = BC\frac{(s - s_1)(s - s_2)}{(s - s_a)(s - s_b)}\bigg|_{s=0}, \qquad (9.66)$$

and that the pole-zero diagram for evaluating K would appear as in Fig. 9.16. We may deduce from this illustration that if a step driving

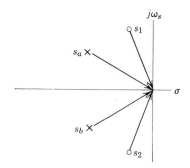

FIG. 9.16 Pole-zero diagram showing factors for evaluating K in Equation 9.66.

function $Bu(t)$ is applied to a network, the forced response function will be $Ku(t)$ in which

$$K = BH(s)|_{s=0}. \tag{9.67}$$

This deduction would not hold if one of the poles of $H(s)$ occurred at $s = 0$.

Example 9.6

GIVEN: The series RL circuit of Fig. E-6.1 with $i(0-) = 0$.

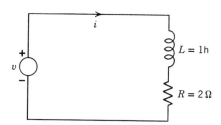

FIG. E-6.1

TO FIND: The current response with (a) $v = 10\,\delta(t)$ volts, and (b) $v = 10u(t)$ volts.

SOLUTION: (a)

$$Y(s) = \frac{1}{Ls + R} = \frac{1}{s+2}. \tag{1}$$

Since $V(s) = \mathscr{L}[10\,\delta(t)] = 10$,

$$I(s) = Y(s)V(s) = \frac{10}{s+2}, \tag{2}$$

and $i = 10e^{-2t}u(t)$ amperes. (3)

(b) Since
$$V(s) = \mathscr{L}[10u(t)] = \frac{10}{s},$$

$$I(s) = Y(s)V(s) = \frac{10}{s(s+2)} = \frac{K}{s} + \frac{A}{s+2}. \tag{4}$$

By the use of Heaviside's expansion method,
$$K = \frac{10}{s+2}\bigg|_{s=0} = 5, \quad \text{and} \quad A = \frac{10}{s}\bigg|_{s=-2} = -5.$$

Thus $i = 5u(t) - 5e^{-2t}u(t)$ amperes. $\tag{5}$

We observe that the transform network function is extremely important in determining the magnitude of the forced response and the nature of the free response.

9.4.2 Response to Exponential Driving Functions

If the driving function, $d(t) = Be^{s_g t}u(t)$, then
$$D(s) = \frac{B}{s - s_g}. \tag{9.68}$$

The subscript g is used to represent generator or source. We use the letter s_g deliberately in order to assist the student who may be reading other texts. Some writers, in order to rapidly enter an "s domain" of network functions, determine the steady-state response to an exponential driving function of the form $\hat{A}e^{st}$. The network functions are algebraically of the same form as transform network functions. However, the voltages and currents are not in the transform domain. The symbol s_g as used in this article corresponds to the symbol s used by these writers.

The transform response, using $H(s) = C\dfrac{(s - s_1)(s - s_2)}{(s - s_a)(s - s_b)}$, is

$$R(s) = H(s)D(s) = BC\frac{(s - s_1)(s - s_2)}{(s - s_a)(s - s_b)(s - s_g)},$$

$$= \frac{A_1}{s - s_a} + \frac{A_2}{s - s_b} + \frac{K}{s - s_g}. \tag{9.69}$$

A_1, A_2, and K may be evaluated in the same manner as for the step driving function. In fact we may consider the step driving function as a special case of the exponential driving function with $s_g = 0$.

Let us investigate the coefficient, K, of the forced response and determine how it depends on the nature of s_g. From Equation 9.69,

$$K = BC\frac{(s - s_1)(s - s_2)}{(s - s_a)(s - s_b)}\bigg|_{s=s_g} = BC\frac{(s_g - s_1)(s_g - s_2)}{(s_g - s_a)(s_g - s_b)}, \tag{9.70}$$

for $s_g \neq s_a$, $s_g \neq s_b$.

The magnitude of **K** is **BC** times the magnitude of the quantity $[(s_g - s_1)(s_g - s_2)]/[(s_g - s_a)(s_g - s_b)]$. Figure 9.17 shows how this may be evaluated for (a) s_g being negative real, (b) s_g being imaginary, and (c) s_g being complex.

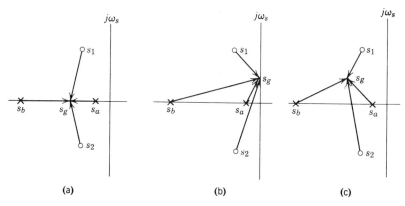

FIG. 9.17 Evaluation of $[(s_g - s_1)(s_g - s_2)]/[(s_g - s_a)(s_g - s_b)]$ for various values of s_g. (a) s_g is negative real. (b) s_g is imaginary. (c) s_g is complex.

Case (b), with s_g imaginary, would correspond to a complex driving function, $\hat{B}e^{j\omega t}$, such as was analyzed in Chapter 5, Section 5.9. We would recall that our actual driving function was $B\cos(\omega t + \phi)$, but that simplification was obtained by replacing this function by $\hat{B}e^{j\omega t}$. The coefficient K is complex so we indicate this with the caret, $\hat{K}$, and find the steady-state reponse as

$$r_{ss} = \text{Re}[\hat{K}e^{j\omega t}]. \tag{9.71}$$

Case (c), with s_g complex, would correspond to a complex driving function, $\hat{B}e^{(-\alpha+j\omega)t}$. This could be considered an extension of the technique of Section 5.9 in which we replaced the driving function, $Be^{-\alpha t}\cos(\omega t + \phi)$, by the function $\hat{B}e^{(-\alpha+j\omega)t}$. Again $\hat{K}$ is complex and we obtain the steady-state response as

$$r_{ss} = \text{Re}[\hat{K}e^{-(\alpha+j\omega)t}]. \tag{9.72}$$

Example 9.7

GIVEN: The series RL circuit of Fig. E-7.1, which is the same as that of Example 9.6. $i(0-) = 0$.

TO FIND: The current reponse with (a) $v = 10e^{-t}u(t)$ volts, (b) $v = 10\cos 2tu(t)$ volts, and (c) $v = 10e^{-t}\cos 2tu(t)$ volts.

SOLUTION: (a) Since

$$Y(s) = \frac{1}{Ls + R} = \frac{1}{s+2},$$

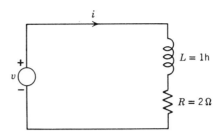

FIG. E-7.1

and

$$V(s) = \mathcal{L}[10e^{-t}] = \frac{10}{s+1},$$

$$I(s) = Y(s)V(s) = \frac{10}{(s+1)(s+2)}$$

$$= \frac{A}{s+2} + \frac{K}{s+1}. \tag{1}$$

By Heaviside's expansion method,

$$A = \frac{10}{s+1}\bigg|_{s=-2} = -10.$$

A can also be evaluated from Fig. E-7.2 using the scale factor of 10.

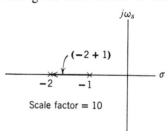

FIG. E-7.2

Similarly

$$K = \frac{10}{s+2}\bigg|_{s=-1} = 10$$

K can be evaluated from Fig. E-7.3 using the scale factor of 10.

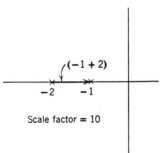

FIG. E-7.3

The time response is then

$$i = -10e^{-2t}u(t) + 10e^{-t}u(t) \text{ amp} \tag{3}$$

(b) For $v = 10 \cos 2t u(t)$, one approach is to replace $v = 10 \cos 2tu(t)$ by $\hat{v} = 10e^{j2t}u(t)$. Then $\hat{V}(s) = 10/(s - j2)$, and

$$\hat{I}(s) = Y(s)\hat{V}(s) = \frac{10}{(s+2)(s-j2)} = \frac{\hat{A}}{s+2} + \frac{\hat{K}}{s-j2}. \tag{4}$$

$$\hat{A} = \frac{10}{s-j2}\bigg|_{s=-2} = \frac{10}{-2-j2} = \frac{-5}{\sqrt{2}}e^{-j45°},$$

and

$$\hat{K} = \frac{10}{s+2}\bigg|_{s=j2} = \frac{5}{\sqrt{2}}e^{-j45°}.$$

Sketches which may be used to aid in evaluating $\hat{A}$ and $\hat{K}$ are shown in Figures E-7.4 and E-7.5.

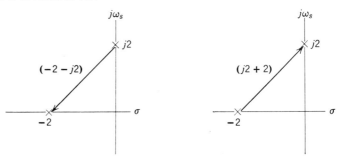

FIG. E-7.4   FIG. E-7.5

From the results obtained

$$\hat{i} = -\frac{5}{\sqrt{2}}e^{-j45°}e^{-2t}u(t) + \frac{5}{\sqrt{2}}e^{-j45°}e^{j2t}u(t) \tag{5}$$

and $i = Re(\hat{i}) = -2.5e^{-2t}u(t) + 2.5\sqrt{2}\cos(2t - 45°)u(t)$ amp.

The same result can be obtained by an alternate method. Since $V(s) = 10s/(s^2 + 4)$,

$$I(s) = Y(s)V(s) = \frac{10s}{(s+2)(s^2+4)} = \frac{A}{s+2} + \frac{\hat{K}_1}{s+j2} + \frac{\hat{K}_2}{s-j2} \tag{6}$$

In this approach, since $j2$ and $-j2$ are complex conjugate roots, $\hat{K}_1$ and $\hat{K}_2$ must be complex conjugates.

Then

$$A = \frac{10s}{(s+j2)(s-j2)}\bigg|_{s=-2} = -2.5$$

$$\hat{K}_1 = \frac{10s}{(s+2)(s-j2)}\bigg|_{s=-j2} = \frac{2.5}{\sqrt{2}}e^{j45°}$$

$$\hat{K}_2 = \hat{K}_1^* = \frac{2.5}{\sqrt{2}}e^{-j45°}$$

Sketches may be made to indicate the evaluation of A and $\hat{K}_1$; however, it is clear that this method has more algebra than the previous method. From Equation 6 and the values of the coefficients,

$$i = -2.5e^{-2t}u(t) + \frac{2.5}{\sqrt{2}}e^{j45°}e^{-j2t}u(t) + \frac{2.5}{\sqrt{2}}e^{-j45°}e^{j2t}u(t) \quad (7)$$

$$= -2.5e^{-2t}u(t) + 2.5\sqrt{2}\cos(2t - 45°)u(t) \text{ amp.} \quad (8)$$

(c) For $v = 10e^{-t}\cos 2t\, u(t)$, let us replace v with $\hat{v} = \hat{V}e^{s_g t}u(t) = 10e^{(-1+j2)t}u(t)$. Then

$$\hat{V}(s) = \frac{10}{s+1-j2},$$

and

$$\hat{I}(s) = Y(s)\hat{V}(s) = \frac{10}{(s+2)(s+1-j2)} = \frac{\hat{A}}{s+2} + \frac{\hat{K}}{s+1-j2}. \quad (9)$$

$$\hat{A} = \frac{10}{s+1-j2}\bigg|_{s=-2} = \frac{10}{-1-j2} = -\frac{10}{\sqrt{5}}e^{-j63.5°}$$

$$\hat{K} = \frac{10}{s+2}\bigg|_{s=-1+j2} = \frac{10}{1+j2} = \frac{10}{\sqrt{5}}e^{-j63.5°}$$

From these results

$$\hat{i} = -\frac{10}{\sqrt{5}}e^{-j63.5°}e^{-2t}u(t) + \frac{10}{\sqrt{5}}e^{-j63.5°}e^{(-1+j2)t}u(t), \quad (10)$$

and

$$i = \text{Re}(\hat{i}) = -2e^{-2t}u(t) + \frac{10}{\sqrt{5}}e^{-t}\cos(2t - 63.5°)u(t) \text{ amperes.} \quad (11)$$

It is interesting to note that for this case

$$Y(s_g) = Y(s)\big|_{s=s_g} = \frac{\hat{K}}{\hat{V}}.$$

For this complex frequency,

$$s_g = -1 + j2, \quad Y(s_g) = \frac{1}{s+2}\bigg|_{s=s_g} = \frac{1}{1+j2} = \frac{e^{-j63.5°}}{\sqrt{5}} \text{ mhos.}$$

Replacement of the real time function by the complex exponential time function is seen to be very effective in reducing the algebra required for solving parts b and c. We observe that if $\hat{v} = \hat{V}e^{s_g t}u(t)$, $\hat{V}(s) = \hat{V}/(s - s_g)$. Also, from the transform response $\hat{I}(s)$, the steady-state time response may be obtained since

$$\hat{i}_{ss} = \hat{I}e^{s_g t}, \text{ with } \hat{I} = \hat{I}(s)[s - s_g]\big|_{s=s_g}.$$

This means that

$$\frac{\hat{i}_{ss}}{\hat{v}} = \frac{\hat{I}}{\hat{V}} = \frac{\hat{I}(s)[s - s_g]}{\hat{V}(s)[s - s_g]}\bigg|_{s=s_g} = Y(s)\bigg|_{s=s_g} = Y(s_g). \quad (12)$$

We recall that in the phasor method for sinusoidal driving functions we replaced real sinusoidal time functions with complex exponential functions in which the coefficient of the exponent was $j\omega$. Now we are extending this method by letting the coefficient, s_g, be complex. Thus it is possible to define driving point immittance or transfer function at a complex frequency.

9.5 THE USE OF IMPULSE AND STEP FUNCTIONS TO APPROXIMATE A PULSE OF GENERAL SHAPE

Pulses may come in a variety of shapes, many of which are not easy to handle mathematically. However, we can always obtain approximate answers by dividing the pulse into a number of increments each of which may be considered a single rectangular pulse. This is similar to the method used to find the coefficients of Fourier series by numerical methods in Chapter 7.

Consider the pulse of Fig. 9.18a. As a first approximation we may replace this pulse with the two rectangular pulses shown in Fig. 9.18b. The pulse can then be expressed mathematically as

$$f(t) \cong A_1 u(t) + [A_2 - A_1] u(t - t_1) - A_2 u(t - t_2). \tag{9.73}$$

It may be more convenient to approximate a pulse by a group of impulse functions. In Fig. 9.18c we show two impulses, each of which is assumed to be equivalent to one of the rectangular pulses of Fig. 9.18b, and to occur at the mid point of the time interval of the rectangular pulse. The original pulse may then be approximated by

$$f(t) \cong B_1 \delta\left(t - \frac{t_1}{2}\right) + B_2 \delta\left(t - \frac{t_2 + t_1}{2}\right), \tag{9.74}$$

in which

$$B_1 = A_1 t_1,$$

and

$$B_2 = A_2(t_2 - t_1).$$

In replacing a given pulse by a number of rectangular pulses or impulses, how do we determine the number needed? This depends on the characteristics of the circuit we are dealing with. In case of doubt we make several calculations, increasing the number of pulses with each calculation, until we judge the resulting changes in the time response to be negligible.

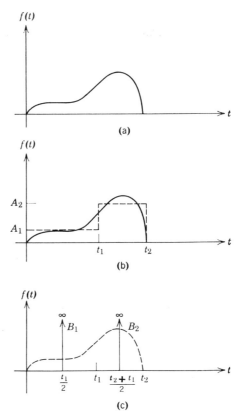

FIG. 9.18 Approximation of pulse by step functions and impulse functions. (*a*) Actual pulse. (*b*) Replacement of pulse by group of step functions. (*c*) Replacement of pulse by impulse functions.

9.6 THE LAPLACE TRANSFORM AND THE FOURIER TRANSFORM

The purpose of this section is to show the relationship that exists between the Laplace transform which we have been using for linear circuit analysis and another transform, the Fourier transform, which is useful in analyzing pulses from a frequency standpoint.

We have been using a Laplace transform with a range of integration from zero to infinity, the lower limit now being interpreted as $0-$ in order to include impulses at $t = 0$. This transform is written

$$F(s) = \int_{0-}^{\infty} f(t) e^{-st} \, dt, \tag{9.75}$$

the quantity, e^{-st}, being referred to as the kernel or kernel function of the transformation. This transform is sometimes called the unilateral

Pulses, Impulses, Dependent Sources 439

Laplace transform because the limits of the integral are from zero to infinity. The principal reason for its popularity is that σ, the real part of s, can be chosen large enough to ensure convergence of the integral for all functions we need in analysis of physical systems. Such functions include the step function, exponential functions, and periodic functions.

The French mathematician and astronomer, P. S. Laplace, introduced more transforms than the one we have been using. He used transforms with kernels of t^s as well as e^{-st} and also used transforms with a range of integration from $-\infty$ to ∞. This last form is known as the bilateral or double-sided Laplace transform and is usually written with the kernel e^{-st}.

$$F_{\text{II}}(s) = \int_{-\infty}^{\infty} f(t)e^{-st}\,dt, \tag{9.76}$$

in which the subscript II is used to indicate the bilateral transform.

This transform is not useful for such functions as periodic functions which are presumed to extend from a time of $-\infty$ to a time of $+\infty$ since the value of σ which serves to make the integral converge for positive values of t will cause the integral not to converge for negative values of t. However, there are many functions which will have a bilateral transform for $\sigma = 0$. Such functions include all pulses of finite area. If $\sigma = 0$, then we might as well replace the variable, $s = \sigma + j\omega_s$, by the variable $j\omega_s$. Equation 9.76 becomes

$$F_{\text{II}}(j\omega_s) = \int_{-\infty}^{\infty} f(t)e^{-j\omega_s t}\,dt, \tag{9.77}$$

in which form it is known as the bilateral Fourier transform or simply as the Fourier transform. The subscript II will be dropped since we shall not use the unilateral transform. The subscript s is retained for the variable, ω_s, to remind us that ω_s is a variable associated with the transform and need have no relation to the ω used for a time function such as $V\cos \omega t$. In fact we should realize that the transforms are not limited to dealing with time functions. We may write

$$F_{\text{II}}(s) = \int_{-\infty}^{\infty} f(x)e^{-sx}\,dx,$$

in which x may be any variable and in which the units of s or ω_s are the reciprocal of the units of x.

Since the Fourier transform may be considered a modified bilateral Laplace transform with $\sigma = 0$, it is relatively easy to determine the Fourier transform of simple pulses. Consider the pulse of Fig. 9.19. It may be written as a function of time,

$$f(t) = Hu(t) - Hu(t - a). \tag{9.78}$$

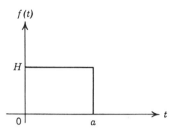

FIG. 9.19   A single rectangular pulse.

Since the pulse is zero for all time except $0 < t < a$, the unilateral and bilateral Laplace transforms are the same.

$$F(s) = F_{II}(s) = \frac{H}{s} - \frac{H}{s} e^{-as}. \tag{9.79}$$

The Fourier transform may be obtained in this case by simply substituting $j\omega_s$ for s. Thus

$$F(j\omega_s) = \frac{H}{j\omega_s} - \frac{H}{j\omega_s} e^{-j\omega_s a}. \tag{9.80}$$

The Fourier transform is easier to interpret since it is a function only of the real variable ω_s, while the Laplace transform is a function of the complex variable s. A further gain is obtained if the pulse has cosine or sine symmetry about its midpoint and if zero time is chosen for this midpoint. Fig. 9.20 shows the pulse of Fig. 9.19 with zero time chosen at the midpoint; the pulse shows cosine symmetry about this point.

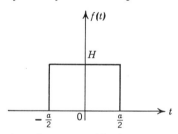

FIG. 9.20   The pulse of Fig. 9.19 with zero time chosen at midpoint.

The unilateral Laplace transform would now omit half the pulse. Either the bilateral Laplace transform or Fourier transform will include the entire pulse. We use the Fourier transform,

$$\begin{aligned} F(j\omega_s) &= \int_{-a/2}^{a/2} H e^{-j\omega_s t}\, dt = \frac{H e^{-j\omega_s t}}{-j\omega_s}\bigg|_{-a/2}^{a/2} \\ &= H\left[\frac{e^{-j\omega_s a/2} - e^{j\omega_s a/2}}{-j\omega_s}\right] = \frac{2H \sin(\omega_s a/2)}{\omega_s} \\ &= Ha\left[\frac{\sin(\omega_s a/2)}{\omega_s a/2}\right]. \end{aligned} \tag{9.81}$$

An alternate method of performing the integration is to use the expansion, $e^{-j\omega_s t} = \cos \omega_s t - j \sin \omega_s t$. Then

$$F(j\omega_s) = \int_{-a/2}^{a/2} H \cos \omega_s t\, dt - \int_{-a/2}^{a/2} jH \sin \omega_s t\, dt$$

$$= \frac{H \sin \omega_s t}{\omega_s} \bigg|_{-a/2}^{a/2} = \frac{2H \sin(\omega_s a/2)}{\omega_s}.$$

In this latter method the second term of the integration is zero because the pulse has cosine symmetry about the zero time reference. Thus for these cases the Fourier transform will be real. Similarly, if the pulse has sine symmetry about the zero time reference, the Fourier transform will be purely imaginary.

The expression of Equation 9.81 is shown graphically in Fig. 9.21.

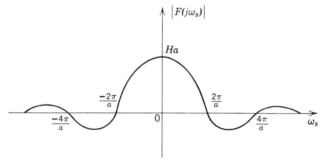

FIG. 9.21 Frequency spectrum of pulse in Fig. 9.20.

We observe that this single rectangular pulse in the time domain may be interpreted as having a continuous frequency spectrum with ω_s ranging from $-\infty$ to $+\infty$. However, we carefully distinguish between variations of the mathematical frequency, ω_s, and variations of the radian frequency, ω, in driving functions of the form $A \cos \omega t$. The first zeros of amplitude correspond to frequencies of $\pm 1/a$ cycles per second; the second zeros of amplitude correspond to frequencies of $\pm 2/a$ cycles per second. Thus, if the pulse width, a, were 10^{-6} seconds, the first zeros of amplitude would correspond to $\pm 10^6$ cycles per second. These positive frequencies are of significance physically since the phasor domain is related to the Fourier transform domain in a manner similar to its relation to the Laplace transform domain. For example, the phasor, network function, $\hat{H}$, may be obtained from the Fourier network function

$$\hat{H} = H(j\omega_s)\big|_{\omega_s = \omega}.$$

Also since $H(j\omega_s)$ is the ratio of the Fourier response transform, $R(j\omega_s)$, to the Fourier driving function transform, $D(j\omega_s)$, we observe that there is a direct relationship between *positive* values of ω_s and the ω of a sinusoidal driving function, $A \cos \omega t$.

If the pulse width, a, were made vanishingly small while the area of the pulse, Ha, were maintained constant, we would have an impulse function of $Ha\delta(t)$. The Fourier transform of this function is the constant, Ha. This means that for an impulse function, $Ha\delta(t)$, the frequency spectrum is a constant, Ha, as shown in Fig. 9.22.

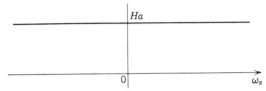

FIG. 9.22 Frequency spectrum of $Ha\,\delta(t)$.

Tables of Fourier transform pairs are available in the literature.[2] Also it is possible to obtain the time function from the inverse Fourier transform, which is written

$$f(t) = \frac{1}{2\pi} \int_{-\infty}^{\infty} F(j\omega_s) e^{j\omega_s t}\, d\omega_s. \tag{9.82}$$

This equation is the same as that for the inverse Laplace transform, Equation 3.23, with s replaced by $j\omega_s$. The integrals for the inverse transform are, unfortunately, more difficult to evaluate than the integrals of the direct transform.

Example 9.8

GIVEN: The pulse of Fig. E-8.1.

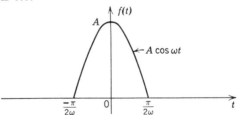

FIG. E-8.1

TO FIND: The Fourier transform of this function.

SOLUTION:

$$\begin{aligned}F(j\omega_s) &= \int_{-\infty}^{\infty} f(t)e^{-j\omega_s t}\, dt = \int_{-\pi/2\omega}^{\pi/2\omega} A\cos\omega t\, e^{-j\omega_s t}\, dt \\ &= \int_{-\pi/2\omega}^{\pi/2\omega} A\cos\omega t (\cos\omega_s t - j\sin\omega_s t)\, dt \\ &= \int_{-\pi/2\omega}^{\pi/2\omega} A\cos\omega t \cos\omega_s t\, dt, \end{aligned} \tag{1}$$

---

[2] See G. A. Campbell and R. M. Foster, *Fourier Integrals for Practical Applications*, Van Nostrand, New York, 1948.

the last step being true because the pulse has cosine symmetry. Since
$$\cos\alpha\cos\beta = \frac{\cos(\alpha+\beta)+\cos(\alpha-\beta)}{2},$$

$$\begin{aligned}
F(j\omega_s) &= \int_{-\pi/2\omega}^{\pi/2\omega} \frac{A}{2}[\cos(\omega+\omega_s)t + \cos(\omega-\omega_s)t]\,dt \\
&= \frac{A}{2}\left[\frac{\sin(\omega+\omega_s)t}{\omega+\omega_s} + \frac{\sin(\omega-\omega_s)t}{\omega-\omega_s}\right]\Big|_{-\pi/2\omega}^{\pi/2\omega} \\
&= A\left[\frac{\cos\left(\dfrac{\pi\omega_s}{2\omega}\right)}{\omega+\omega_s} + \frac{\cos\left(\dfrac{\pi\,\omega_s}{2\,\omega}\right)}{\omega-\omega_s}\right] \\
&= \frac{2A\omega\cos\left(\dfrac{\pi\,\omega_s}{2\,\omega}\right)}{\omega^2-\omega_s^2}.
\end{aligned} \qquad (2)$$

In order that we may observe the frequency characteristic of this function let $A = 1$ and $\omega = 1$. Then data may be determined for a sketch as shown in Fig. E-8.2. We observe that for this short pulse the maximum magnitude of $F(j\omega_s)$ does not occur at $\omega_s = \omega = 1$.

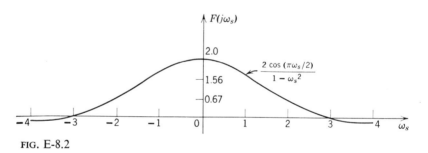

FIG. E-8.2

9.7 THE LAPLACE TRANSFORM AND PERIODIC FUNCTIONS

We may have observed that the unilateral Laplace transform of $\cos\omega t$ is identical to that of $\cos\omega t\,u(t)$. This is because the lower limit of the transform integral is zero and thus the transform ignores all values of $f(t)$ for negative values of t. If we consider $\cos\omega t$ to be a periodic function which exists for all time, then taking the transform of $\cos\omega t$ has the effect of making it equal to the semiperiodic function $\cos\omega t\,u(t)$. Thus for a driving function of $\cos\omega t$ the Laplace transform method will give a response which includes the transient terms that arise as if this driving function were introduced at $t = 0$. This is true for any periodic function.

The Laplace transform of a periodic function, or really of a semi-periodic function starting at $t = 0$, may be obtained by using the transform of the shifted time function, Equation 9.11. Let $\int_{0-}^{T} f(t)e^{-st}\,dt$ be the transform of the first period of a time function such as shown in Fig. 9.23a. Then $e^{-Ts}\int_{0-}^{T} f(t)e^{-st}\,dt$ is the transform of the second period shown in Fig. 9.23b. The transform of the semiperiodic function is an infinite series of such terms, or

$$F(s) = (1 + e^{-Ts} + e^{-2Ts} + e^{-3Ts} + \ldots)\int_{0-}^{T} f(t)e^{-st}\,dt$$

$$= \frac{1}{1 - e^{-Ts}}\int_{0-}^{T} f(t)e^{-st}\,dt. \tag{9.83}$$

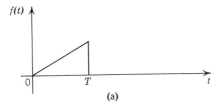

(a)

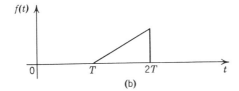

(b)

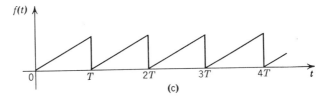

(c)

FIG. 9.23 A semiperiodic function as the sum of individual pulses. (a) First period. (b) Second period. (c) Semiperiodic function.

Because the infinite series, $1 + x + x^2 + x^3 \ldots$, can be written in the closed form $1/(1 - x)$, the Laplace transform of any semiperiodic function can be written as $1/(1 - e^{-Ts})$ times the transform of the first period.

Let us demonstrate the technique with the simple circuit of Fig. 9.24b, with the current source shown in Fig. 9.24a. It is desired to obtain the voltage v_{ab}.

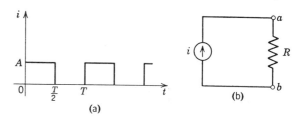

FIG. 9.24 Semiperiodic current source in resistive circuit. (a) Current source. (b) Circuit.

For the first period the current may be written,

$$i = Au(t) - Au(t - T/2),$$

and the Laplace transform for the semiperiodic wave is, therefore,

$$I(s) = \left[\frac{A}{s} - \frac{Ae^{-Ts/2}}{s}\right]\left(\frac{1}{1 - e^{-Ts}}\right). \tag{9.84}$$

The expression for $V_{ab}(s)$ is

$$V_{ab}(s) = RI(s) = \frac{RA}{s}\frac{(1 - e^{-Ts/2})}{(1 - e^{-Ts})} = \frac{RA}{s(1 + e^{-Ts/2})}. \tag{9.85}$$

There are two approaches for interpreting Equation (9.85). One approach is to recognize that $1/(1 + x) = 1 - x + x^2 - x^3 + \ldots$, and write

$$V_{ab}(s) = \frac{RA}{s}[1 - e^{-Ts/2} + e^{-Ts} - e^{-3Ts/2} + e^{-2Ts} \ldots],$$

from which

$$v_{ab} = RAu(t) - RAu(t - T/2) + RAu(t - T)$$
$$- RAu(t - 3T/2) + \ldots \tag{9.86}$$

We recognize Equation 9.86 as correctly representing the voltage across the resistance R.

Another approach is to expand the expression for $V_{ab}(s)$ in Equation 9.85 by partial fractions. In order to do this we must factor the quantity $(1 + e^{-Ts/2})$. The zeros of this quantity occur for $e^{-Ts/2} = -1$, which occur if $-Ts/2 = +jn\pi$, with $n = \pm 1, \pm 3, \pm 5$, etc. Thus the roots are $s_n = -j(2\pi n/T)$, with $n = \pm 1, \pm 3, \pm 5$, etc., and the term $(1 + e^{-Ts/2})$ may be written

$$1 + e^{-Ts/2} = \ldots \left(\right)\left(s - j\frac{6\pi}{T}\right)\left(s - j\frac{2\pi}{T}\right)\left(s + j\frac{2\pi}{T}\right)$$
$$\times \left(s + j\frac{6\pi}{T}\right)\left(\right) \ldots \tag{9.87}$$

Thus Equation 9.85 may be expanded,

$$V_{ab}(s) = \frac{RA}{s(1 + e^{-Ts/2})} = \frac{C_0}{s} + \sum_{n=-\infty}^{n=+\infty} \frac{\hat{C}_n}{s - s_n}, \qquad (9.88)$$

for n being odd.

The quantity

$$\left.\frac{(s - s_n)}{s(1 + e^{-Ts/2})}\right|_{s=s_n}$$

may be evaluated by differentiating both numerator and denominator before making the substitution. Then $\hat{C}_n = -2RA/s_n T = +RA/j\pi n$. We may then write $v_{ab}(t)$,

$$v_{ab}(t) = \frac{RA}{2} u(t) + \sum_{n=-\infty}^{\infty} \hat{C}_n e^{s_n t} u(t)$$

$$= \frac{RA}{2} u(t) + \sum_{n=-\infty}^{\infty} \frac{RA}{j\pi n} e^{j(2\pi n/T)t} u(t), \ n \text{ odd}. \qquad (9.90)$$

We may recognize the summation part of Equation 9.90 as the exponential form of the Fourier series for which even harmonics are zero. Let $\omega_1 = 2\pi/T$, and use the fact that $(e^{j\theta} - e^{-j\theta})/2j = \sin \theta$. Then

$$v_{ab}(t) = \frac{RA}{2} u(t) + \sum_{n=1}^{\infty} \left[\frac{2RA}{\pi n} \sin n\omega_1 t\right] u(t), \qquad (9.91)$$

for n being odd.

Equation 9.91 correctly expresses the voltage in terms of a Fourier series starting at $t = 0$. Thus there is a strong tie between the Laplace transform of repeated functions and the Fourier series.

Example 9.9

GIVEN: The circuit of Fig. E-9.1 with the voltage shown applied at $t = 0$. Initial conditions are zero.

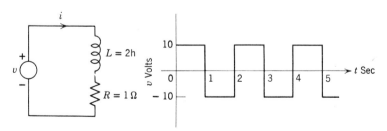

FIG. E-9.1

TO FIND: i as a function of time.

SOLUTION: v may be written

$$v = 10u(t) - 20u(t-1) + 20u(t-2) - 20u(t-3) + \ldots,$$

for which

$$V(s) = \frac{10}{s} - \frac{20e^{-s}}{s} + \frac{20e^{-2s}}{s} - \frac{20e^{-3s}}{s} + \ldots \tag{1}$$

Since

$$Y(s) = \frac{1}{Ls + R} = \frac{1}{2(s+0.5)},$$

$$I(s) = \frac{5}{s(s+0.5)} - \frac{10e^{-s}}{s(s+0.5)} + \frac{10e^{-2s}}{s(s+0.5)} - \frac{10e^{-3s}}{s(s+0.5)} + \ldots \tag{2}$$

The partial fraction expansion of $5/[s(s+0.5)]$ gives $A_1/s + A_2/(s+0.5)$ with $A_1 = 10$ and $A_2 = -10$. Thus the solution for the current i may be written

$$i = 10u(t) - 10e^{-0.5t}u(t) - 20u(t-1) + 20e^{-0.5(t-1)}u(t-1)$$
$$+ 20u(t-2) - 20e^{-0.5(t-2)}u(t-2) - 20u(t-3)$$
$$+ 20e^{-0.5(t-3)}u(t-3) + \ldots \text{ amp.} \tag{3}$$

Equation 3 correctly gives the complete solution. It would also have been possible to write the voltage for only the first period as

$$v = 10u(t) - 20u(t-1) + 10u(t-2),$$

from which

$$V(s) = \left[\frac{10}{s} - \frac{20e^{-s}}{s} + \frac{10e^{-2s}}{s}\right]\left[\frac{1}{1 - e^{-2s}}\right] \tag{4}$$

Then

$$I(s) = Y(s)V(s)$$
$$= \frac{5(1 - 2e^{-s} + e^{-2s})}{s(s+0.5)(1 - e^{-2s})} = \frac{5(1 - e^{-s})}{s(s+0.5)(1 + e^{-s})}. \tag{5}$$

It is possible to obtain the complete solution from (5) in infinite series form by a method similar to that used for the circuit of Fig. 9.24. However, we shall leave this as an exercise for those interested.

A third method for obtaining a complete solution is to write the expression for v as a Fourier series, then write the expression for i_{ss} as a second Fourier series. Since the free response must be of the form $Ae^{-Rt/L}$ the coefficient A could then be determined by applying the initial condition of $i(0+) = 0$ to the complete solution.

9.8 CIRCUIT FICTION AND THE FREQUENCY SPECTRUM OF PULSES

In Chapter 2 we observed that the use of an inductance or capacitance as a lumped circuit element was permissible if the frequency were low enough so that the principal dimension of the circuit was not an appreciable fraction of a wavelength. In these last few sections we have found that a single rectangular pulse contains *all* frequencies, the magnitude of which fortunately decreases with frequency. What criterion shall we use to determine whether a particular circuit adequately represents a physical system under pulse conditions? The ultimate test is experimental; if the test results agree with those obtained from the circuit model, then the circuit model was adequate. However, we do have some guidelines. If a rectangular pulse occurs in a time increment of Δt seconds, then the frequency spectrum has its first zero at $1/\Delta t$ cycles per second, its second zero at $1/2 \Delta t$ cycles per second, etc. As one gains experience in a particular design area one develops the judgement to evaluate the adequacy of a particular circuit.

An impulse function is an impossible function as far as a circuit model is concerned. Since an impulse function has a frequency spectrum which is constant for all frequencies, no circuit with lumped elements can be satisfactory. However, we recall that an impulse function cannot exist physically. In the physical world an impulse function is a pulse of short duration. Whether or not a circuit model is adequate for a specific impulse requires experimental evidence.

9.9 DEPENDENT SOURCES

Our sources have thus far been considered as independent sources, meaning that they are not functions of other quantities. It was mentioned that the voltages of mutual inductance may be considered dependent sources since a voltage in one part of a system is a function of the current in another part of the system. However, mutual inductance is reciprocal by nature, $M_{12} = M_{21}$, and the word dependent source is generally restricted to those cases in which the action is not reciprocal.

Transistors and tubes offer common examples of dependent sources. These devices are inherently nonlinear; when they are restricted to operating in "linear" portions of the operating characteristics we are permitted to develop a linear equivalent circuit. Figure 9.25 shows a transistor in a connection known as the grounded-base connection. The direct voltages V_{ee} and V_{cc} are used to ensure that operation will occur in a linear region; we say that they supply the proper bias voltages. The voltage, v, represents a signal which is to be amplified.

If equations are written for the system of Fig. 9.25 in "linear"

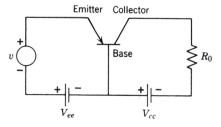

FIG. 9.25 Connections for grounded-base transistor.

regions, an equivalent circuit may be developed in which the direct voltages and currents are omitted. We shall not develop this circuit; such a development is appropriate in an electronics course. The equivalent circuit will be that shown in Fig. 9.26. Ki_1 represents a dependent voltage source. R_1, R_2, R_3, K, and R_0 are constants.

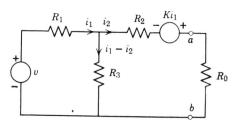

FIG. 9.26 Equivalent circuit for grounded-base transistor with voltage source, v, and load resistance R_0.

The voltage equations are

$$v = R_1 i_1 + R_3(i_1 - i_2) \tag{9.92}$$

$$R_3(i_1 - i_2) = (R_2 + R_0)i_2 - Ki_1 \tag{9.93}$$

These equations may be written in systematic form,

$$(R_1 + R_3)i_1 - R_3 i_2 = v \tag{9.94}$$

$$-(R_3 + K)i_1 + (R_0 + R_2 + R_3)i_2 = 0. \tag{9.95}$$

Equations 9.94 and 9.95 show clearly that for this circuit the mutual impedances are not equal; that is, $Z_{12} \neq Z_{21}$. Thus we note that this network is not reciprocal[3] and should expect that other networks with dependent sources are not reciprocal.

---

[3] See Section 4.44.

Equations 9.94 and 9.95 may be solved for i_2.

$$i_2 = \frac{v}{\dfrac{(R_1 + R_3)(R_0 + R_2 + R_3)}{R_3 + K} - R_3}$$

$$= \frac{\dfrac{v(R_3 + K)}{R_1 + R_3}}{R_0 + R_2 + R_3 - \dfrac{R_3(R_3 + K)}{R_1 + R_3}}. \qquad (9.96)$$

Equation 9.96 indicates an equivalent circuit as shown in Fig. 9.27.

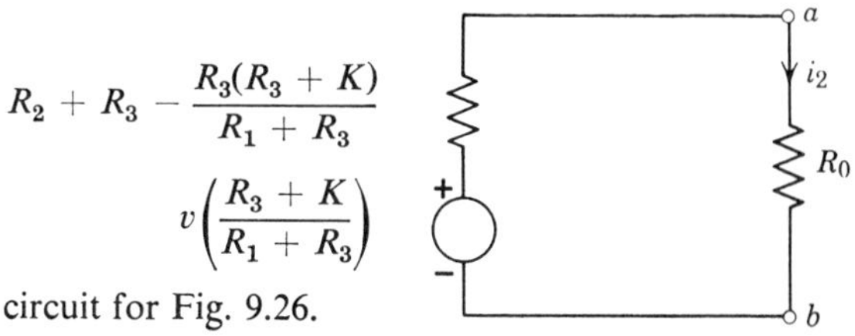

$$R_2 + R_3 - \frac{R_3(R_3 + K)}{R_1 + R_3}$$

$$v\left(\frac{R_3 + K}{R_1 + R_3}\right)$$

FIG. 9.27 Equivalent circuit for Fig. 9.26.

The student should note that Thévenin's impedance cannot be found as equal to the driving-point impedance with all internal sources, including dependent sources, made equal to zero. However, it is equal to the driving-point impedance with all internal independent sources made zero.

For circuits with dependent sources the power output may be larger than the power input. This does not mean that conservation of energy is being violated; the additional power is physically being supplied by d-c sources which provide proper bias and which have been ignored since we were interested in the response to some external driving function.

Figure 9.28 shows a "small signal" equivalent circuit for a vacuum tube circuit in which R_5 represents a feedback resistor connected between plate and grid of the tube. This circuit is capable of being a "multiplying" circuit in an analogue computer. v_g represents the time varying grid-cathode voltage. The resistance R_p in parallel with the dependent current source, $g_m v_g$, represents the plate-cathode circuit under time-varying conditions. g_m is called the mutual transconductance of the tube.

Kirchhoff's current law as applied to the nodes of this circuit results in

$$\left(\frac{1}{R_3} + \frac{1}{R_4} + \frac{1}{R_5}\right)v_g - \frac{1}{R_5}v_0 = \frac{v_1}{R_3}, \qquad (9.97)$$

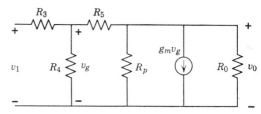

FIG. 9.28 "Multiplying" circuit using a vacuum tube.

and

$$\left(g_m - \frac{1}{R_5}\right)v_g + \left(\frac{1}{R_0} + \frac{1}{R_p} + \frac{1}{R_5}\right)v_0 = 0. \tag{9.98}$$

Solution of Equations 9.97 and 9.98 for v_0/v_1 results in

$$\frac{v_0}{v_1} = \frac{-1/R_3}{\dfrac{(1/R_3 + 1/R_4 + 1/R_5)(1/R_0 + 1/R_p + 1/R_5)}{g_m - 1/R_p} + \dfrac{1}{R_5}}. \tag{9.99}$$

If the circuit is constructed so that the first term in the denominator is small in comparison with the second term,

$$\frac{v_0}{v_1} \cong -\frac{R_5}{R_3}, \tag{9.100}$$

and the output voltage is approximately a constant, $-\dfrac{R_5}{R_3}$, times the input voltage.

9.10 SUMMARY

In this chapter the responses to single pulses of various shapes were obtained by both the classical and the Laplace transform methods. Impulses were introduced as a natural result of idealizing our circuits and as convenient approximations to natural phenomena. The need to include impulses at $t = 0$ caused a revision of the lower limit of the Laplace transform, so that $F(s) = \int_{0-}^{\infty} f(t)e^{-st}\,dt$. This meant no revision in the transforms used so far except that $\mathscr{L}[df(t)/dt] = sF(s) - f(0-)$. A list of transforms introduced in this chapter is included in Table 9.1 and in Appendix C.

The Fourier integral was presented as a special case of the bilateral transform with $\sigma = 0$. The Fourier transform is convenient for single pulses since the magnitude and phase of $F(j\omega_s)$ may be expressed as a function of the real variable, ω_s. If $f(t)$ has even symmetry at $t = 0$, the Fourier transform will be real; if $f(t)$ has odd symmetry at $t = 0$, the Fourier transform will be imaginary.

Table 9.1
Additional Transforms

$f(t)$	$F(s)$
$\dfrac{df(t)}{dt}$	$sF(s) - f(0-)$
$u(t)$	$\dfrac{1}{s}$
$\delta(t)$	1
$f(t-a)u(t-a)$	$e^{-as}F(s)$
Periodic function with period T	$\dfrac{\int_0^T f(t)e^{-st}\,dt}{1 - e^{-Ts}}$

FURTHER READING

For additional reading on the material of this chapter the student is referred to R. E. Scott's *Linear Circuits*, Chapters 21 and 22, Addison-Wesley, Reading, Mass., 1960, and to M. E. Van Valkenburg's *Network Analysis*, Chapter 8, Prentice-Hall, New York, 1955.

PROBLEMS

9.1 Write expressions for the pulses shown in Fig. P-9.1 in terms of step functions and shifted time functions.

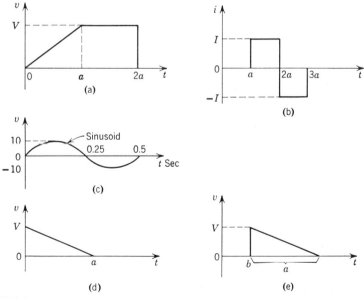

FIG. P-9.1

Answer: $(d)\,v = Vu(t) - (V/a)tu(t) + (V/a)(t-a)u(t-a)$.

9.2 For the circuit of Example 9.1 solve for v_{cd}, starting with current equation about node c. Use both the classical and Laplace transform methods.

9.3 For the circuit of Example 9.2, use $V = 10$ volts, $R = 1$ ohm, $L = 1$ henry, and $a = 2$ seconds. Evaluate each term in the response, i, of Equation 8 and sketch each term. Then sketch the net i for $0 < t < 5$ seconds, indicating how the various components add to give this resultant. Indicate clearly the values for i at $t = 1, 2,$ and 5 seconds.

9.4 Write the expression for the sinusoidal pulse of Fig. P-9.4 in terms of shifted time functions and then obtain its Laplace transform.

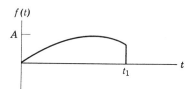

FIG. P-9.4

Partial answer:

$$F(s) = \frac{A\omega}{s^2 + \omega^2} - \frac{A\omega \cos \omega t_1 e^{-t_1 s}}{s^2 + \omega^2} - \frac{As \sin \omega t_1 e^{-t_1 s}}{s^2 + \omega^2}.$$

9.5 The voltage pulse shown in Fig. P-9.5a is applied to the circuit of Fig. P-9.5b. $i(0+) = 0$.

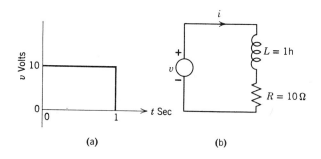

FIG. P-9.5   (a)   (b)

a Determine i as a function of time by either the classical or Laplace transform method.

b Sketch i vs. t, indicating numerical values at $t = 0.1, 0.5,$ and 1.1 seconds. Observe that steady state is essentially reached in 1 second.

c Now assume that the pulse of Fig. P-9.5a had a duration of 0.1 second instead of 1 second. From the sketch of part *b*, sketch i vs. t. Indicate numerical values for i at $t = 0.1$ and 0.2 seconds.

Partial answer: (a) $i = [1 - e^{-10t}]u(t) - [1 - e^{-10(t-1)}]u(t - 1)$ amp.

9.6 Same as problem 9.5, parts a and b, except use the pulse shown in Fig. P-9.6.

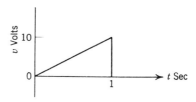

FIG. P-9.6

9.7 The voltage pulse shown in Fig. P-9.7a is applied to the circuit of Fig. P-9.7b, $v_{cd}(0+) = 0$.

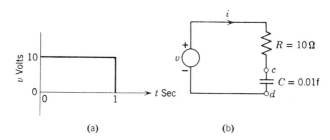

FIG. P-9.7   (a)   (b)

a Determine i as a function of time by either classical or Laplace transform method.
b Sketch i vs. t indicating numerical values at $t = 0.1$, 1, and 1.1 seconds.
c Now assume that the pulse of Fig. 9.7a had a duration of 0.1 second instead of 1 second. From your sketch of part b and the circuit equations, sketch i vs. t, indicating numerical values for i at $t = 0$, 0.1, and 0.2 seconds.

9.8 Same as problem 9.7, parts a and b, except use $v_{cd}(0+) = -10$ volts.

9.9 a For the circuit of Fig. P-9.9 determine the expression for i as a function of time.
b Sketch i vs. t, indicating numerical values of i at key points. Also show the current through L as a function of time. Observe that our ideal voltage source becomes an ideal short circuit if its magnitude is zero.

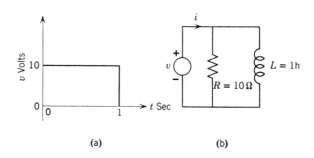

FIG. P-9.9   (a)   (b)

Pulses, Impulses, Dependent Sources   455

9.10 For the circuit of Fig. P-9.10, $v_{ab}(0+) = 0$, $i = 10[u(t) - u(t-1)]$ amp. Determine v_{ab} by both classical and Laplace transform methods. Then sketch v_{ab} vs. t, showing numerical values at key points.

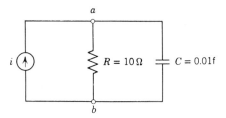

FIG. P-9.10

9.11 For the circuit shown in Fig. P-9.11, $i(0+) = 0$. Determine i and then sketch i vs. t for:

a $R = 10\ \Omega,\ L = 0$.
b $R = 0,\ L = 1$ henry.
c $R = 10\ \Omega,\ L = 1$ henry.

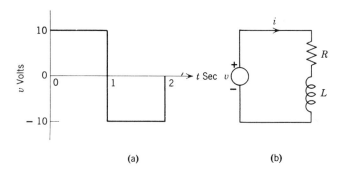

FIG. P-9.11   (a)   (b)

9.12 The circuit shown in Fig. P-9.12 has a single saw tooth of voltage applied. Find the current i by either the classical or Laplace transform method. $v_{cd}(0+) = 0$.

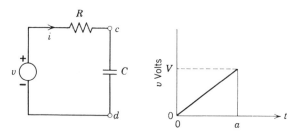

FIG. P-9.12

9.13 A single triangular pulse of current is impressed on a parallel RC circuit (Fig. P-9.13). $v_{cd}(0+) = 0$. Determine v_{cd}.

456

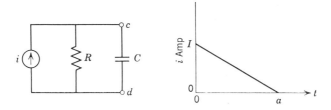

FIG. P-9.13

9.14 The voltage impressed on a series RL circuit is $v = A\delta(t) - B\delta(t-a)$. $i(0-) = 0$. Show that the current response is

$$i = \frac{A}{L} e^{-Rt/L} u(t) - \frac{B}{L} e^{-R(t-a)/L} u(t-a).$$

9.15 In Fig. P-9.15, steady state exists with the switch closed. The switch is opened at $t = 0$.

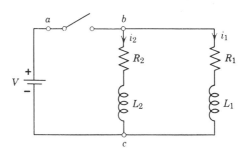

FIG. P-9.15

a By writing voltage equations which hold for all time and integrating these from $t = 0-$ to $t = 0+$, show that

$$\int_{0-}^{0+} v_{ab}\, dt = L_2 i_2(0-) - L_2 i_2(0+),$$

and that

$$i_1(0+) = \frac{L_1 i_1(0-) - L_2 i_2(0-)}{L_1 + L_2}.$$

b Using $V = 100$ volts, $L_1 = L_2 = 1$ henry, $R_1 = 10\ \Omega$, $R_2 = 20\ \Omega$, determine i_1 by the classical method.
c For the same data, determine i_1 by the Laplace transform method.
d Determine the voltage v_{bc}.
Partial answer: $i_1 = 2.5e^{-15t}u(t)$ amp.

9.16 In Fig. P-9.16, steady state exists with the switch closed. The switch is opened at $t = 0$. Determine the current i_1 as a function of time for $t > 0$. Also determine the magnitude and location of each voltage impulse. $V = 15$ volts, $R_1 = R_2 = R_3 = 10$ ohms, $L_1 = L_2 = 1$ henry.

Pulses, Impulses, Dependent Sources 457

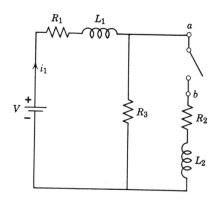

FIG. P-9.16

Partial answer:
$i_1 = [0.75 + 0.25e^{-20t}]u(t)$ amp.

9.17 In Fig. P-9.17, steady state exists with the switch closed. The switch is opened at $t = 0$. $V = 15$ volts, $R_1 = R_2 = R_3 = 10\ \Omega$, $L_1 = L_2 = L_3 = 1$ henry.

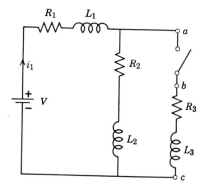

FIG. P-9.17

a Determine i_1.
b Determine the magnitude and location of each voltage impulse.
c Determine the expression for the voltage v_{ac} for $t > 0-$.

9.18 In Fig. P-9.18, steady state exists with the switch closed. The switch is opened at $t = 0$. $I = 10$ amp, $R = 2$ ohms, $L = 1$ henry. Determine v_{ab} for $t > 0-$.

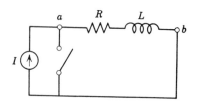

FIG. P-9.18

9.19 The voltage pulse shown in Fig. P-9.19a is applied to a series LR circuit shown in Fig. P-9.19b. $i(0-) = 0$. If the time interval, a, is small in comparison with the time constant, L/R, an approximate solution would be obtained by considering the integral $\int_0^a v\, dt = Va$ as a voltage impulse acting at $t = 0$.

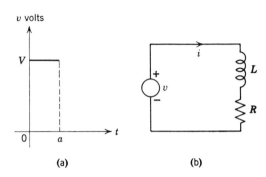

FIG. P-9.19   (a)   (b)

a Determine the expression for i at $t = a$ with the voltage pulse given.
b Determine the approximate expression for i at $t = a$ with voltage impulse $Va\,\delta(t)$.
Answer: $i = (Va/L)e^{-Ra/L}$.
c Determine the per-unit error in i at $t = a$ from $[i(\text{exact}) - i(\text{approx})]/[i(\text{exact})]$. Use the series expansion $e^\phi = 1 + \phi + \phi^2/2! + \phi^3/3!\ldots$, and assume $aR/L \ll 1$. Show that the error may be approximated by the expression

$$\left(\frac{aR/2L - a^2R^2/3L^2}{1 - aR/2L}\right).$$

Then calculate the error for $aR/L = 0.1$ and $aR/L = 0.5$. Note that the error would be less if the impulse were assumed to act at $t = a/2$.

9.20 In Fig. P-9.20 C_2 is uncharged; C_1 is charged to a voltage $v_{ab}(0-)$. The switch is closed at $t = 0$.

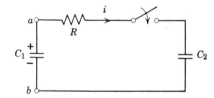

FIG. P-9.20

a Determine the solution for i and then show that $\int_0^\infty i\, dt$ is not a function of R. This means that the final voltage on the capacitors is not a function of R.
b From your solution for part *a*, show that $\int_0^\infty i^2R\, dt$, the energy dissipated

Pulses, Impulses, Dependent Sources   459

in R is the difference between the energy initially stored in the capacitance C_1 and the energies finally stored in the two capacitances.

Note that this energy difference exists if we assume $R = 0$. In case we are asked what happened to the energy we should realize that in real situations there will be some R or L; also we have ignored the energy which is radiated and this will increase as $L \to 0$, $R \to 0$.

9.21 A 1-μf capacitor charged to 100 volts is connected to a 2-μf capacitor charged to 200 volts, the connection being made $+$ to $-$, $-$ to $+$. Determine the final voltage. What is the magnitude of the current impulse in coulombs?

Partial answer: Final voltage is 100 volts.

9.22 In Fig. P-9.22, $v_{ab}(0-) = -50$ volts, $v_{bc}(0-) = 0$ volts. The switch is closed at $t = 0$. Determine the final voltage across each capacitor.

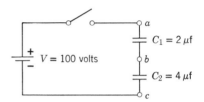

FIG. P-9.22

9.23 Three uncharged capacitors, 0.1 μf, 0.2 μf and 0.4 μf, are connected in series and charged to a total voltage of 500 volts. They are disconnected from the source. The 0.1-μf capacitor is connected to the 0.2-μf capacitor, $+$ terminal being connected to $+$ terminal and $-$ terminal to $-$ terminal. They are then disconnected and the 0.2-μf capacitor is connected to the 0.4-μf capacitor, $+$ terminal to $-$ terminal and $-$ terminal to $+$ terminal.
a What are the final voltages across each of the capacitors.
b How much energy was dissipated?

9.24 In Example 9.4 solve for the voltage v_{ab} by the Laplace transform method. Then sketch v_{ab} as a function of time.

9.25 Solve Example 9.5 by the Laplace transform method.

9.26 a If a d-c voltage source is impressed on a series RC circuit with the capacitor originally uncharged, a certain amount of energy is dissipated in the resistor R during the charging process. Prove that this energy is equal to the final energy stored in the electric field of the capacitor.
b It has been suggested that the efficiency would be improved by having a constant current source instead of a constant voltage source. Is this true? If so, should the current be small for a long period of time or large for a short period of time? Explain.

9.27 In Fig. P-9.27 steady state exists with the switch open. $v_{ab}(0-) = 0$. The switch is closed at $t = 0$. Determine the solution for i_1 for $t > 0-$.

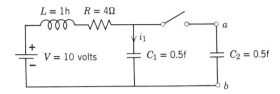

FIG. P-9.27

9.28 Steady state is reached in the circuit of Fig. P-9.28. The switch is closed at $t = 0$. Determine the voltages at $t = 0+$. $v_{ab}(0-) = 50$ volts, $v_{bc}(0-) = 50$ volts, and $v_{dc}(0-) = 0$.

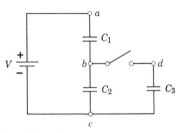

FIG. P-9.28 $V = 100$ volts; $C_1 = C_2 = C_3 = 1$ farad.

9.29 The switches in Fig. P-9.29 may be closed in either sequence. $v_{ae}(0-) = 50$ volts, $v_{be}(0-) = 0$ volts, $v_{cd}(0-) = 100$ volts, $v_{de}(0-) = -30$ volts, $C_1 = C_2 = C_3 = C_4 = 2$ farads.
a Show that the voltages on the capacitances after the switching is independent of the sequence with which the switches are closed.
b Determine the voltages on the capacitances with the switches closed.

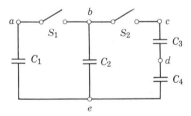

FIG. P-9.29

9.30 For Fig. P-9.30, $v_{ab}(0-) = 0$. Determine i for each of the following driving functions. If it is convenient, use a pole-zero diagram to evaluate coefficients in the partial fraction expansion.
a $v = \delta(t)$ volts.
b $v = u(t)$ volts.
c $v = e^{-t}u(t)$ volts.
d $v = \cos 2t u(t)$ volts.

FIG. P-9.30

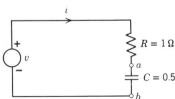

Pulses, Impulses, Dependent Sources   461

9.31 At $t = 0-$ the current through the inductance is zero (Fig. P-9.31). For each of the following driving functions, determine i. If it is convenient use a pole-zero diagram to evaluate coefficients in the partial fraction expansion.
a $v = 10e^{-t}u(t)$ volts.
b $v = 10e^{-5t}u(t)$ volts.
c $v = 10 \cos 5tu(t)$ volts.
Observe that there is no forced response for part b.

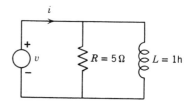

FIG. P-9.31

9.32 For Fig. P-9.32, initial conditions are zero. Determine i for:
a $v = 10e^{-2t}u(t)$ volts.
b $v = 10 \cos 2tu(t)$ volts.
For part (b) the forced response is zero. Why is this so?

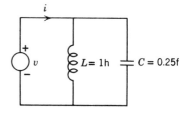

FIG. P-9.32

9.33 For Fig. P-9.33, $v_{bc}(0-) = 0$. Determine v_{ac} for:
a $i = 10 \, \delta(t)$ amperes.
b $i = 10e^{-2t}u(t)$ amperes.
c $i = 10 \cos 2tu(t)$ amperes.

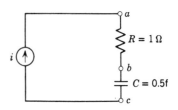

FIG. P-9.33

462

9.34 For Fig. P-9.34, initial conditions are zero. Determine i for:
a $v = 10u(t)$ volts.
b $v = 10 \cos 2tu(t)$ volts.
c $v = 10e^{-t} \cos 3tu(t)$ volts.

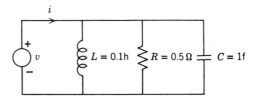

FIG. P-9.34

9.35 One black box has a resistance R. Another black box has the combination shown with $R = \sqrt{L/C}$. It is claimed that one cannot tell any difference between these circuits. Is this true? See if you can devise a test which will distinguish between them. Prove that your test will work.

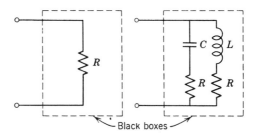

FIG. P-9.35 — Black boxes

9.36 A unit impulse driving function, $\delta(t)$, applied to a network that is initially at rest (no initial condition) produces the following response.

$r(t) = e^{-t} \sin tu(t)$.

Determine the response of this network to each of the following driving functions.
a $u(t)$.
b $e^{-t}u(t)$.

9.37 The driving function, $e^{-t}u(t)$, applied to a network with zero initial conditions, produces the response,

$r(t) = e^{-t} \sin tu(t)$.

Determine the response of this network to each of the following driving functions.
a $u(t)$.
b $\delta(t)$.

9.38 Figure P-9.38 shows a pole-zero diagram of a network function. The scale factor is one. For zero initial conditions, find the complete response of this network to the driving function, $(1 + \sin t)u(t)$.

Pulses, Impulses, Dependent Sources 463

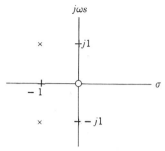

FIG. P-9.38

9.39 Show that the Fourier transforms of the pulses of Figs. P-9.39a and P-9.39b have the same magnitude characteristics but differ in phase by $\omega_s a/2$.

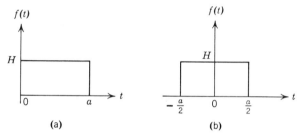

FIG. P-9.39 (a) (b)

9.40 Show that the Fourier transform for $i = Ie^{-at}u(t)$ is $I(j\omega_s) = I/(a + j\omega_s)$.

9.41 Show that the Fourier transform of the train of three rectangular pulses of equal duration is

$$F(j\omega_s) = Ha\left[\frac{\sin(\omega_s a/2)}{\omega_s a/2}\right](1 + 2\cos\omega_s T).$$

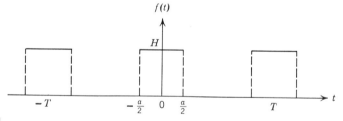

FIG. P-9.41

9.42 Determine the Fourier transform of the sinusoidal pulse consisting of three half cycles with an amplitude of unity and an angular velocity of unity. Sketch magnitude vs. ω_s.

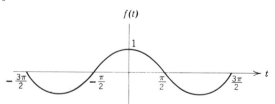

FIG. P-9.42

9.43 Determine the Laplace transform of the periodic function.

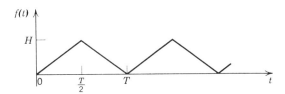

FIG. P-9.43

9.44 Determine the Laplace transform of the periodic function which is a half-wave rectified sine wave.

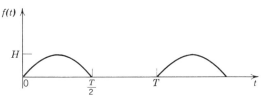

FIG. P-9.44

9.45 For the circuit of Fig. 9.26, determine Thévenin's equivalent circuit for that portion of the system to the left of terminals a, b. Determine Thévenin's impedance as the ratio of the open-circuited voltage to the short-circuited current. Check your answer with the circuit of Fig. 9.27.

9.46 For the circuit of Fig. 9.26, assume $v = 1.0 \cos \omega t$ volts, $R_1 = 200$ ohms, $R_2 = 20{,}000$ ohms, $R_3 = 300$ ohms, $K = 30{,}000$ ohms, $R_0 = 15{,}000$ ohms. Calculate the amplitudes of i_1 and i_2 and then calculate the average power input (avg vi_1) and the average power output (avg $v_{ab}i_2$).

9.47 In Equation 9.99, assume $g_m = 10^{-2}$ mhos, and that the resistances are each 10^6 ohms. What is the error in using Equation 9.100?

9.48 The circuit shown in Fig. P-9.48 is the equivalent circuit for a grounded-collector transistor.
  a Determine the solution for the current i_1.
  b Determine Thévenin's equivalent circuit for the portion of the network to the left of terminals a, b.

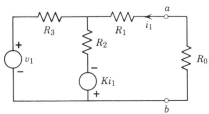

FIG. P-9.48

Pulses, Impulses, Dependent Sources   465

10 *Mutual Inductance and Transformers*

The subject of mutual inductance was introduced in Chapter 2 and has cropped up occasionally in other chapters. It embodies an important principle, that the magnetic field of one circuit may link a second circuit and that therefore transfer of energy may occur without electrical connection or without the phenomena of electromagnetic radiation. Unfortunately, the solution of circuits which involve mutual inductances tend to be complex enough so that one gets lost in the algebra. Furthermore, the reduction techniques developed in Chapter 4 do not seem to apply to magnetically coupled circuits. However, for the important case of two coupled coils known as a "transformer," some useful equivalent circuits can be obtained. The purpose of this chapter is primarily to study the equivalent circuits of transformers and to consolidate our knowledge of mutual inductance as a circuit element.

10.1 REVIEW OF MUTUAL INDUCTANCE

Consider two coils magnetically coupled as shown in Fig. 10.1. R_1 and R_2 represent the resistances of the windings. The presence of the core permits us to see the sense of the coupling. Kirchhoff's voltage equations are

$$v_1 = R_1 i_1 + \frac{d\lambda_1}{dt} = R_1 i_1 + \frac{d}{dt}(\lambda_{11} - \lambda_{12}), \tag{10.1}$$

and

$$0 = (R_2 + R_0)i_2 + \frac{d\lambda_2}{dt} = (R_2 + R_0)i_2 + \frac{d}{dt}(\lambda_{22} - \lambda_{21}). \tag{10.2}$$

λ_{11} is the flux linkage of circuit 1 caused by current i_1; λ_{12} is the flux linkages of circuit 1 caused by current i_2. The choice of current directions

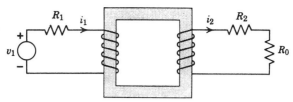

FIG. 10.1 Two circuits magnetically coupled.

is such in this case that the flux set up by current i_2 opposes the flux set up by current i_1; thus λ_{12} must be subtracted from λ_{11}. Similar definitions apply to λ_{22}, λ_{21} and the sign of λ_{21}. The following definitions are made.

$$L_1 = \frac{\lambda_{11}}{i_1}, \quad M_{12} = \frac{\lambda_{12}}{i_2}, \quad L_2 = \frac{\lambda_{22}}{i_2} \quad \text{and} \quad M_{21} = \frac{\lambda_{21}}{i_1}.$$

Then the voltage equations become

$$v_1 = R_1 i_1 + L_1 \frac{di_1}{dt} - M_{12} \frac{di_2}{dt}, \tag{10.3}$$

and

$$0 = (R_2 + R_0)i_2 + L_2 \frac{di_2}{dt} - M_{21} \frac{di_1}{dt}. \tag{10.4}$$

The flux linkages of the two circuits have been thus defined as

$$\lambda_1 = L_1 i_1 - M_{12} i_2, \tag{10.5}$$

and

$$\lambda_2 = L_2 i_2 - M_{21} i_1. \tag{10.6}$$

Appendix A proves from energy considerations that $M_{12} = M_{21} = M$ and that $M < \sqrt{L_1 L_2}$. Using

$$M = k\sqrt{L_1 L_2}, \tag{10.7}$$

the coefficient of coupling, k, can never be physically as large as 1.

The use of dot marking permits the omission of the core in Fig. 10.1. We could dot mark either both upper portions of the coils or both lower portions. The former choice is shown in Fig. 10.2.

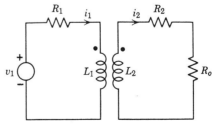

FIG. 10.2 Dot marking of coils in Fig. 10.1.

The convention is that, if the direction of the currents through the two coils is in the same sense with regard to the dot markings, the flux linkages are additive and the sign of the mutually induced voltage is the same as the self-induced voltage term.

The Laplace transform circuit for Fig. 10.2 can readily be sketched as shown in Fig. 10.3.

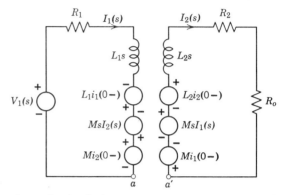

FIG. 10.3   Laplace transform circuit of Fig. 10.2.

Consider the transform circuit from the standpoint of the loop current method or the node method. In the loop current method there are two unknowns. In the node voltage method we may consider the two circuits electrically touching at some point; for example, the points a and a' may be connected together without affecting our calculations. Such circuit reduction does not help much, however, since the dependent voltages, $MsI_2(s)$ and $MsI_1(s)$, are functions of currents and elimination of these currents is not easy even for this simple illustration. Thus the use of the node voltage method is not attractive for the circuits of either Fig. 10.2 or Fig. 10.3. We shall postpone the use of the node voltage method until equivalent circuits are obtained which make this method attractive again.

10.2 ONE EQUIVALENT CIRCUIT FOR A TWO — WINDING TRANSFORMER

A transformer may be defined as a stationary device which couples magnetically two or more circuits which may or may not be electrically insulated from each other. We shall concentrate on the transformer consisting of two windings insulated from each other.

A typical transformer connection is shown in Fig. 10.2 in which a voltage source is applied to one winding and a resistive load placed on the second winding. Our first equivalent circuit is one which is commonly used for air-core transformers. For these transformers the coefficient of coupling is lower than for iron-cored transformers. For the development of the first equivalent circuit we make the sketch of Fig. 10.4 which shows the mutually induced voltages as dependent sources.

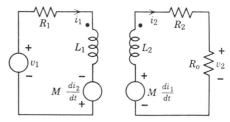

FIG. 10.4. Two circuits magnetically coupled (a transformer).

The voltage equations for these circuits are

$$v_1 = R_1 i_1 + L_1 \frac{di_1}{dt} - M \frac{di_2}{dt} \qquad (10.8)$$

$$0 = (R_2 + R_0) i_2 + L_2 \frac{di_2}{dt} - M \frac{di_1}{dt}. \qquad (10.9)$$

In order for these circuits to be joined together to form an electrically connected system it is essential that common voltages be found or developed. There are no common terms in the equations as written. However, if the term $M(di_1/dt)$ were to be added **and** subtracted from Equation 10.8, and if the term $M(di_2/dt)$ were to be added and subtracted from Equation 10.9, then there would be a common voltage, $M(d/dt)(i_1 - i_2)$. The equations may be written

$$v_1 = R_1 i_1 + L_1 \frac{di_1}{dt} - M \frac{di_1}{dt} + M \frac{di_1}{dt} - M \frac{di_2}{dt}, \qquad (10.10)$$

and

$$0 = (R_2 + R_0) i_2 + L_2 \frac{di_2}{dt} - M \frac{di_2}{dt} + M \frac{di_2}{dt} - M \frac{di_1}{dt}. \qquad (10.11)$$

The circuit represented by Equations 10.10 and 10.11 is shown in Fig. 10.5.

The terminals a and a' may be connected together first. Then the terminals b and b' may be connected together since there is no difference in potential between them. From bb' to aa' is a voltage $M(d/dt)(i_1 - i_2)$ and between these points is the current, $i_1 - i_2$, thus the section from bb' to aa' may be replaced by the element M. Also the voltage $M(di_1/dt)$, with current i_1 through it, may be replaced by element $-M$. With similar action for $M(di_2/dt)$, the circuit may be shown as in Fig. 10.6.

If we wish to solve the circuit of Fig. 10.6 in the Laplace transform domain we must use the initial conditions as shown in Fig. 10.7.

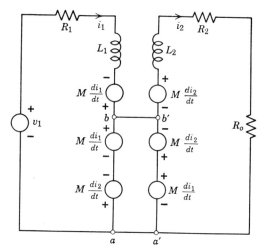

FIG. 10.5 The making of one electrical circuit.

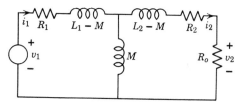

FIG. 10.6 An electrical network equivalent to Fig. 10.4.

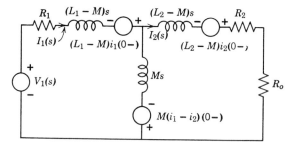

FIG. 10.7 Laplace transform circuit for Fig. 10.6.

This is a useful network since there are no dependent sources. The node voltage method requires only one unknown voltage. However, the elements, $L_1 - M$ and $L_2 - M$, will be negative if $\sqrt{L_1/L_2} < k$ or $\sqrt{L_2/L_1} < k$ respectively.

If the dot mark on coil L_2 of Fig. 10.4 was actually at the other end of the coil, and if one wished to maintain the same positions of input and output transformer connections, a development similar to that presented results in one of the equivalent circuits of Fig. 10.8.

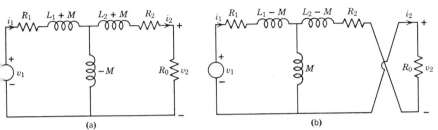

FIG. 10.8 Equivalent circuits for Fig. 10.4 assuming dot marking on L_2 was changed. (a) Circuit with negative value for M. (b) Circuit with crossover.

Example 10.1

GIVEN: The circuit of Fig. E-1.1 under steady-state sinusoidal conditions. The resistances of the windings are assumed to be negligible.

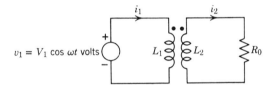

FIG. E-1.1

TO FIND: The solution for the phasor current $\hat{I}_2$ in literal form, and then its evaluation for $V_1 = 10.0$ volts, $\omega = 10$ radians/sec, $R_0 = 10$ ohms, $L_1 = 5$ henries, $L_2 = 1.2$ henries, $M = 1$ henry. Compare this solution with that obtained if the coupling coefficient were unity.

SOLUTION: The phasor circuit corresponding to the equivalent circuit of Fig. 10.6 may be sketched immediately as shown in Fig. E-1.2.

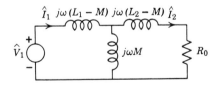

FIG. E-1.2

Kirchhoff's voltage equations for this circuit are

$$j\omega L_1 \hat{I}_1 - j\omega M \hat{I}_2 = \hat{V}_1, \quad (1)$$
$$-j\omega M \hat{I}_1 + (R_0 + j\omega L_2)\hat{I}_2 = 0. \quad (2)$$

The expression for $\hat{I}_2$ is

$$\hat{I}_2 = \frac{\begin{vmatrix} j\omega L_1 & \hat{V}_1 \\ -j\omega M & 0 \end{vmatrix}}{\begin{vmatrix} j\omega L_1 & -j\omega M \\ -j\omega M & R_0 + j\omega L_2 \end{vmatrix}}$$

$$= \frac{j\omega M \hat{V}_1}{j\omega L_1(R_0 + j\omega L_2) + \omega^2 M^2}$$

$$= \frac{\hat{V}_1(M/L_1)}{R_0 + j\omega L_2 + \dfrac{\omega^2 M^2}{j\omega L_1}} = \frac{\hat{V}_1(M/L_1)}{R_0 + j\omega\left[L_2 - \dfrac{M^2}{L_1}\right]}$$

Then

$$\hat{I}_2 = \frac{\hat{V}_1(M/L_1)}{R_0 + j\omega\left[\dfrac{L_1 L_2 - M^2}{L_1}\right]} \tag{3}$$

Equation 3 is written in the form shown to indicate that $\hat{I}_2$ is a constant times $\hat{V}_1$ divided by an impedance consisting of R_0 plus a quantity determined by the remainder of the circuit.

For the constants given, $L_1 - M = 5 - 1 = 4$ henries and $L_2 - M = 1.2 - 1 = 0.2$ henries so that the inductances in the equivalent circuit are positive for this case. $\hat{I}_2$ may be calculated by substituting numbers into Equation 3.

$$\hat{I}_2 = \frac{\hat{V}_1(M/L_1)}{R_0 + j\omega\left[\dfrac{L_1 L_2 - M^2}{L_1}\right]} = \frac{10(\frac{1}{5})}{10 + j\omega\left(\dfrac{6-1}{5}\right)}$$

$$= \frac{2e^{-j45°}}{10\sqrt{2}} = 0.141 e^{-j45°} \text{ amp} \tag{4}$$

For a coupling coefficient of unity, $M^2 = L_1 L_2$ and Equation 3 becomes

$$\hat{I}_2 = \frac{\hat{V}_1(M/L_1)}{R_0} = \frac{\hat{V}_1}{R_0}\sqrt{L_2/L_1}. \tag{5}$$

Substitution of numbers gives

$$\hat{I}_2 = \tfrac{10}{10}\sqrt{1.2/5} = 0.49 \text{ amp.} \tag{6}$$

Comparison of Equations 4 and 6 shows that the effect of having a coupling less than one is to reduce the magnitude as well as to cause a change in phase of the current $\hat{I}_2$.

Example 10.2

GIVEN: The circuit of Fig. E-2.1 under steady-state conditions with the switch closed. The switch is opened at $t = 0$.

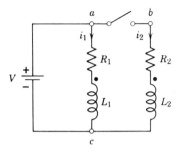

FIG. E-2.1

TO FIND: The location and magnitude of any voltage impulses. Assume $0 < k < 1$.

SOLUTION: We shall assume the switch is capable of breaking the circuit so i_2 becomes zero at $t = 0+$.

Kirchhoff's voltage equations may be written:

$$\left. \begin{array}{l} V = R_1 i_1 + L_1 \dfrac{di_1}{dt} + M \dfrac{di_2}{dt} \\ \\ V = v_{ab} + R_2 i_2 + L_2 \dfrac{di_2}{dt} + M \dfrac{di_1}{dt} \end{array} \right\} \quad (1)$$

Let us now integrate Equation 1 from $t = 0-$ to $t = 0+$ so as to determine impulses if any.

$$\left. \begin{array}{l} \int_{0-}^{0+} V\,dt = \int_{0-}^{0+} R_1 i_1\,dt + \int_{i_1(0-)}^{i_1(0+)} L_1\,di_1 + \int_{i_2(0-)}^{i_2(0+)} M\,di_2 \\ \\ \quad = \int_{0-}^{0+} v_{ab}\,dt + \int_{0-}^{0+} R_2 i_2\,dt + \int_{i_2(0-)}^{i_2(0+)} L_2\,di_2 + \int_{i_1(0-)}^{i_1(0+)} M\,di_1 \end{array} \right\} \quad (2)$$

i_2 is finite and i_1 must be finite since the magnitude of any voltage impulses is finite. Since $i_2(0+) = 0$, Equation 2 gives

$$\left. \begin{array}{l} 0 = L_1[i_1(0+) - i_1(0-)] - M i_2(0-) \\ \\ 0 = \int_{0-}^{0+} v_{ab}\,dt - L_2 i_2(0-) + M[i_1(0+) - i_1(0-)] \end{array} \right\} \quad (3)$$

From Equation 3

$$i_1(0+) = i_1(0-) + \frac{M}{L_1} i_2(0-), \quad (4)$$

and

$$\int_{0-}^{0+} v_{ab} \, dt = L_2 i_2(0-) + M i_1(0-) - M i_1(0+),$$

$$= L_2 i_2(0-) - \frac{M^2}{L_1} i_2(0-). \tag{5}$$

From Equation 4 we observe that i_1 will not have a discontinuity if $M = 0$ ($k = 0$), but otherwise i_1 is discontinuous at $t = 0$. From Equation 5 there will be impulsive voltages across the switch and across L_2. The impulse is a maximum for $k = 0$ and will become 0 for $k = 1$.

Inspection of Fig. E-2.1 shows there can be no impulsive voltage from a to c since this voltage is fixed at V. The voltage across coil 1, $L_1(di_1/dt) + M(di_2/dt)$, must be finite, whereas the voltage across coil 2, $L_2(di_2/dt) + M(di_1/dt)$, must be impulsive, and is equal but opposite to the impulsive voltage across the switch.

The equivalent circuit, as shown in Fig. E-2.2, would not have helped. The implication would be voltage impulses across $L_1 - M$ and M, but no net voltage impulse across the combination. (In a physical situation, with impulse voltages across L_2, we would expect high voltages across some of the turns of L_1.)

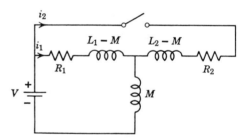

FIG. E-2.2

0.3 A SECOND EQUIVALENT CIRCUIT FOR A TWO – WINDING TRANSFORMER

The equivalent circuit of Section 10.2 is adequate for coils which are not coupled closely; a second equivalent circuit is better suited to closely coupled coils such as exist in iron-cored transformers. We start with our original circuit of Fig. 10.2 which is resketched in Fig. 10.9.

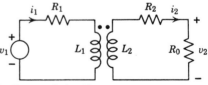

FIG. 10.9 Circuit with two magnetically coupled coils.

Kirchhoff's voltage equations are

$$v_1 = R_1 i_1 + L_1 \frac{di_1}{dt} - M \frac{di_2}{dt} \tag{10.12}$$

$$0 = (R_2 + R_0)i_2 + L_2 \frac{di_2}{dt} - M \frac{di_1}{dt} \tag{10.13}$$

Let us change scale for current i_2, changing i_2 to i_2/a in which a is a constant. Then we shall change scale for the voltages concerned so that we will have exactly the same power as before. First, multiply Equation 10.13 by i_2 to obtain a power equation

$$0 = (R_2 + R_0)i_2^2 + L_2 i_2 \frac{di_2}{dt} - M i_2 \frac{di_1}{dt}. \tag{10.14}$$

Then multiply and divide Equation 10.14 by a^2 with the restrictions that a be real, finite, and nonzero. Adjust terms so that the current, i_2/a, always appears.

$$0 = a^2(R_2 + R_0)\left(\frac{i_2}{a}\right)^2 + a^2 L_2 \left(\frac{i_2}{a}\right)\frac{d(i_2/a)}{dt} - aM\left(\frac{i_2}{a}\right)\frac{di_1}{dt}. \tag{10.15}$$

Factor out the current, i_2/a, to obtain a voltage equation,

$$0 = a^2(R_2 + R_0)\left(\frac{i_2}{a}\right) + a^2 L_2 \frac{d(i_2/a)}{dt} - aM \frac{di_1}{dt}. \tag{10.16}$$

Equation 10.12 is modified to be

$$v_1 = R_1 i_1 + L_1 \frac{di_1}{dt} - aM \frac{d(i_2/a)}{dt}. \tag{10.17}$$

Equations 10.16 and 10.17 are similar to our original equations except i_2 has become i_2/a, M has become aM, $R_2 + R_0$ has become $a^2(R_2 + R_0)$ and L_2 has become $a^2 L_2$. By analogy with the development of the equivalent circuit of Fig. 10.6, we can sketch the equivalent circuit for Equations 10.16 and 10.17 as in Fig. 10.10.

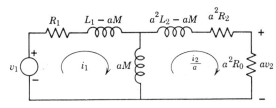

FIG. 10.10 A circuit equivalent to that of Fig. 10.9 with respect to power.

In case one wonders about what happened to i_2, it is possible to claim we have invented, mathematically, an "ideal" transformer. This is shown in Fig. 10.11, and is defined as a two-port passive network in which the instantaneous power input is equal to the power output, and in which the ratio of input voltage to output voltage is equal to a real constant.

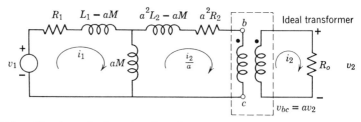

FIG. 10.11. Circuit equivalent showing "ideal transformer."

The value generally chosen for a is that value which equates the terms, $L_1 - aM$ and $a^2L_2 - aM$, of Fig. 10.11. This means $a^2 = L_1/L_2$, or

$$a = \sqrt{L_1/L_2}. \tag{10.18}$$

If we use this relationship and $M = k\sqrt{L_1L_2}$, Fig. 10.10 may be redrawn as shown in Fig. 10.12.

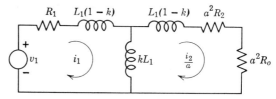

FIG. 10.12 Equivalent circuit in terms of k.

Our transformer would become an "ideal transformer" if (1) $R_1 = 0$, $R_2 = 0$ (this is possible if operating under conditions of superconductivity); (2) $k = 1$ (this is really not possible since it means the two windings must be in the same space); (3) $L_1 = \infty$ and $L_2 = \infty$ (this is not possible either).

It is possible with iron-cored transformers to obtain values of k nearly equal to 1. For these the ratio L_1/L_2 is very nearly $(N_1/N_2)^2$, in which N_1 and N_2 are the number of turns of L_1 and L_2 respectively. This can be shown by noting that L_1 is

$$L_1 = \frac{\lambda_{11}}{i_1} = \frac{N_1\phi_{11}}{i_1} = \frac{N_1N_1\phi_{11}}{N_1i_1} = N_1^2\left(\frac{\phi_{11}}{N_1i_1}\right)$$

in which ϕ_{11} is an average flux linkage per turn set up by the ampere turns N_1i_1, and

$$L_2 = \frac{\lambda_{22}}{i_2} = \frac{N_2\phi_{22}}{i_2} = \frac{N_2N_2\phi_{22}}{N_2i_2} = N_2^2\left(\frac{\phi_{22}}{N_2i_2}\right),$$

and since the ratios $\phi_{11}/N_1 i_1$ and $\phi_{22}/N_2 i_2$ would be equal for perfect coupling,

$$\frac{L_1}{L_2} \simeq \left(\frac{N_1}{N_2}\right)^2, \tag{10.19}$$

or

$$\sqrt{\frac{L_1}{L_2}} = a \simeq \frac{N_1}{N_2}^1 \tag{10.20}$$

For a transformer with $k \simeq 1$ the equivalent circuit may be sketched as shown in Fig. 10.13.

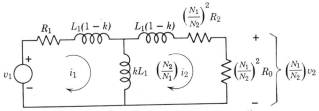

FIG. 10.13 Equivalent circuit for transformer with good coupling.

A second circuit could have been developed in which v_2 and i_2 were retained instead of v_1 and i_1. "By inspection", this circuit would be that shown in Fig. 10.14.

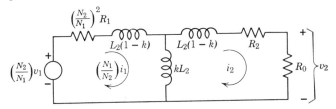

FIG. 10.14 Second equivalent circuit for transformer with good coupling.

The use of iron in transformers has several effects. The principal effect is that the self-inductances L_1 and L_2 are increased, thus decreasing the magnetizing current, that current flowing in one winding with no current in the other winding. However, because of the nonlinear magnetic properties of the iron the magnetizing current is nonsinusoidal with impressed sinusoidal voltages. This means that it becomes impractical to define or measure L_1, L_2, M and thus k. However, the inductance represented as $L_1(1 - k)$ or $L_2(1 - k)$ in Figs. 10.11 and 10.12 is that portion of the self-inductance represented by "leakage" flux which has a path which is not completely in iron. This inductance

---

[1] There are conditions other than good coupling for which $L_1/L_2 \simeq N^2$.

is therefore linear. The inductance represented as kL_1 or kL_2 is definitely nonlinear.

There are many applications in which we are justified in treating the iron-cored transformer as an ideal transformer. As we let $R_1 = R_2 = 0$, $k = 1$ and $L_1 = L_2 = \infty$, we observe from Figs. 10.13 and 10.14 that such idealization implies

$$\frac{v_1}{v_2} = \frac{N_1}{N_2}, \tag{10.21}$$

and

$$\frac{i_1}{i_2} = \frac{N_2}{N_1}. \tag{10.22}$$

From a practical standpoint, iron-cored transformers with sinusoidal sources under steady state are usually treated by the equivalent circuit shown in Fig. 10.15 which should be compared with Fig. 10.13.

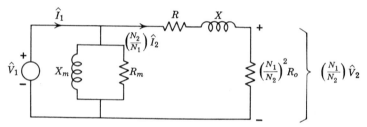

FIG. 10.15 Equivalent circuit for iron-cored transformer.

The resistance R is essentially $R_1 + (N_1/N_2)^2 R_2$, the reactance X is essentially $2\omega L_1(1 - k)$. The reactance X_m corresponds approximately to $\omega k L_1$, R_m is a value such that $V_1^2/2R_m$ correctly gives the power loss in the iron. The current taken by the parallel branch of X_m and R_m is certainly nonsinusoidal; this fact is not indicated by the equivalent circuit.

A second circuit could be sketched which is similar to Fig. 10.14. This is shown in Fig. 10.16. We observe that if we idealize the iron-cored transformer under these steady-state conditions,

$$\frac{\hat{V}_1}{\hat{V}_2} = \frac{N_1}{N_2}, \tag{10.23}$$

and

$$\frac{\hat{I}_1}{\hat{I}_2} = \frac{N_2}{N_1}. \tag{10.24}$$

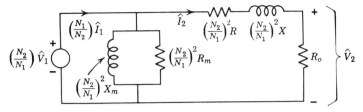

FIG. 10.16 Second equivalent circuit for iron-cored transformer.

Example 10.3

GIVEN: The circuit of Fig. E-3.1. v is sinusoidal.

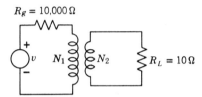

FIG. E-3.1

TO FIND: The turns ratio, N_1/N_2, in order to maximize the power to R_L. Assume the transformer is ideal.

SOLUTION: Refer to Fig. 10.13 for the equivalent circuit of a transformer. First assume $R_1 = R_2 = 0$, and $k = 1$. Then Fig. E-3.1 may be shown as Fig. E-3.2. Then assume $L_1 = \infty$ and $a^2 \cong (N_1/N_2)^2$. This results in Fig. E-3.3, for which maximum power in R_L requires

$$\left(\frac{N_1}{N_2}\right)^2 = \frac{R_g}{R_L} = \frac{10,000}{10},$$

or $N_1/N_2 = 31.6$.

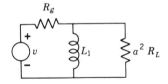

FIG. E-3.2

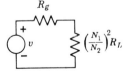

FIG. E-3.3

Mutual Inductance and Transformers

Example 10.4

GIVEN: The circuit of Fig. E-4.1 which has the same constants as that of Example 10.1 with the exception that the coupling coefficient is now 0.99. Thus $V = 10.0$ volts, $\omega = 10$ radians/sec, $R_0 = 10$ ohms, $L_1 = 5$ henries, $L_2 = 1.2$ henries, $M = 0.99\sqrt{6}$ henries.

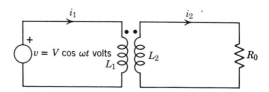

FIG. E-4.1

TO FIND: The equivalent circuit corresponding to that of Fig. 10.12 under steady-state conditions and the solution for $\hat{I}_2$.

SOLUTION: The equivalent circuit may be sketched as shown in Fig. E-4.2. Substitution of numbers into the circuit of Fig. E-4.2 results in the circuit of Fig. E-4.3. Observe that $a = \sqrt{L_1/L_2} = \sqrt{5/1.2} = 2.04$.

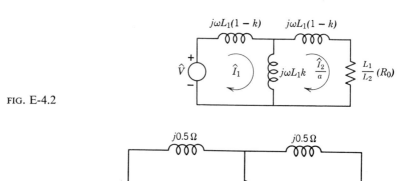

FIG. E-4.2

FIG. E-4.3

An exact solution for $\hat{I}_2$ may be obtained from this circuit or from Equation 3 of Example 10.1. However, since the impedance of $j0.5$ ohms is about 1% of the other impedances, let us assume these to be negligible. Then

$$\frac{\hat{I}_2}{2.04} \cong \frac{\hat{V}}{41.7} = \frac{10}{41.7},$$

or

$\hat{I}_2 \cong 0.49$ amp.

This agrees with the solution obtained in Example 10.1 for perfect coupling. We would expect that the correct $\hat{I}_2$ to our problem would be about 1% less in magnitude and would have a negative phase angle of less than 1 degree.

Example 10.5

GIVEN: The circuit of Fig. E-5.1 which shows three single-phase transformers designed to transmit 1000 kva between the two balanced three-phase lines.

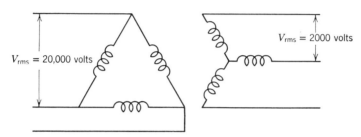

FIG. E-5.1

TO FIND: The turns ratio of the transformer and the rms current rating of each winding.

SOLUTION: The low voltage winding of each transformer has a voltage of $2000/\sqrt{3} = 1160$ volts rms. Therefore the ratio of the number of turns on the high voltage winding to those on the low voltage winding are

$$\frac{N_1}{N_2} \simeq \frac{V_1}{V_2} = \frac{20,000}{1160} = 17.3. \tag{1}$$

Since 1000/3 kva is transmitted by each transformer, and since the volt-ampere rating is $V_{rms}I_{rms}$, the current rating of the high voltage winding is,

$$I_{rms} = \frac{\frac{1,000,000}{3}}{V_{rms}} = \frac{1,000,000}{3(20,000)} = 16.7 \text{ amp.}$$

Similarly the current rating of the low voltage winding is

$$I_{rms} = \frac{1,000,000}{3V_{rms}} = \frac{1,000,000}{3(1160)} = 287 \text{ amp.}$$

Example 10.6

GIVEN: The following tests were made at 60 cycles per second on a power distribution transformer rated 15,000 volt amperes, 2400/240 volts, 6.25/62.5 amperes.

(*1*) With the high voltage terminals open circuited and rated voltage applied to the low voltage terminals the power input was 80 watts at 240 volts rms and 1.5 amp rms. The current is not sinusoidal.

(*2*) With the low voltage terminals shorted and a source connected to the high voltage terminals, the power input was 200 watts at 60 volts rms and 6.25 amp rms.

TO FIND: The constants of the equivalent circuits as seen from both the high voltage and low voltage sides and to discuss the operating characteristics of this device.

SOLUTION: Test *1* was made from the low voltage side at rated voltage. The open circuit voltage of the high voltage side would have been 2400 volts rms. If the test had been made from the high voltage side at rated voltage the power input would have still been 80 watts but the current would have been 0.15 amp rms.

With reference to Fig. 10.15 and considering $\hat{V}_1$ and $\hat{I}_1$ as the voltage and current on the low voltage side, test *1* was made with high voltage side open circuited and so we can readily calculate R_m and X_m.

$$P = \frac{V_{rms}^2}{R_m}, \quad R_m = \frac{V_{rms}^2}{P} = \frac{240^2}{80} = 720 \text{ ohms}.$$

The circuit may be shown in Fig. E-6.1.

$$I_{a_{rms}} = \frac{V_{rms}}{R_m} = \frac{240}{720} = 0.33 \text{ amperes}.$$

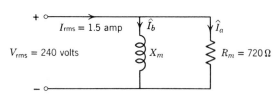

FIG. E-6.1

Since $\hat{I}_a$ and $\hat{I}_b$ differ in phase by 90°,

$$I_{b_{rms}}^2 + I_{a_{rms}}^2 = 1.5^2$$

$$I_{b_{rms}} = \sqrt{1.5^2 - 0.33^2} = 1.46 \text{ amp}.$$

$$X_m = \frac{V_{rms}}{I_{b_{rms}}} = \frac{240}{1.46} = 162 \text{ ohms}.$$

These calculations are of questionable value since the current is not sinusoidal.

Test 2 was made from the high voltage side with the low voltage side shorted. Reference to Fig. 10.16 indicates that the equivalent circuit is as shown in Fig. E-6.2. Note that R and X are values referred to the low voltage side. The circuit is a simple series circuit. The magnitude of the impedance as seen from the high side is

$$Z_{high} = \frac{V_{2rms}}{I_{2rms}} = \frac{60}{6.25} = 9.6 \text{ ohms.}$$

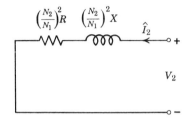

FIG. E-6.2

The resistance and reactance as seen from the high side are

$$R_{high} = \frac{P}{I_{2rms}^2} = \frac{200}{(6.25)^2} = 5.07 \text{ ohms}$$

and

$$X_{high} = \sqrt{Z_{high}^2 - R_{high}^2} = \sqrt{9.6^2 - 5.07^2} = 8.17 \text{ ohms}$$

Then

$$\left(\frac{N_2}{N_1}\right)^2 R = R_{high}, \quad R = \frac{5.07}{(10)^2} = 0.0507 \text{ ohms}$$

and

$X = 0.0817$ ohms.

The equivalent circuit for the low side of the transformer may be sketched as in Fig. E-6.3.

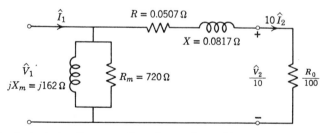

FIG. E-6.3 Equivalent circuit as seen from low voltage side.

If this transformer is operating at rated current and voltage, since rated current is 62.5 amp and the magnetizing current is 1.5 amp, we are justified in ignoring the parallel branches R_m and X_m as far as calculations of voltages and currents are concerned. The transformer becomes simply a series impedance. However, for calculations of efficiency we could not ignore the power dissipated in R_m, the iron power loss.

A second equivalent circuit may be sketched for the high side. This corresponds to Fig. 10.16 and becomes that of Fig. E-6.4.

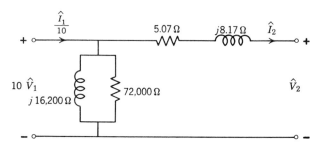

FIG. E-6.4 Equivalent circuit as seen from high voltage side.

10.4 SUMMARY

Two-winding transformers may be represented by an infinite number of circuits which are equivalent on the basis of power. However, only two are commonly used, one for coils which are loosely coupled and one for coils which are closely coupled. Iron-cored transformers may sometimes be approximated by ideal transformers which are two-port circuits for which the instantaneous net power into the circuit is always zero and for which the ratio of the voltages at the two ports is a constant.

FURTHER READING

Nearly all introductory circuit texts present the equivalent circuits for the two-winding transformer. For an excellent detailed treatment of this subject, see M.I.T. Electrical Engineering Staff, *Magnetic Circuits and Transformers*, Wiley, New York, 1943.

PROBLEMS

10.1 Two self-inductances, L_1 and L_2, which are mutually coupled, are connected in series.
  a Show that the equivalent inductance is $L_0 = L_1 + L_2 \pm 2M$.
  b Assume $L_1 = 4L_2$ and $k = 0.5$. Then determine the two possible values of L_0 as a function of L_2.
  c Same as part *b* except use $k = 1.0$.

d Assume $L_1 = L_2$ and $k = 1.0$. Then determine the two possible values of L_0 as a function of L_2.

10.2 a For the two coils connected in parallel as shown in Fig. P-10.2a, assume resistances are negligible. Prove that this parallel combination may be replaced by an equivalent inductance equal to

$$\frac{L_1 L_2 - M^2}{L_1 + L_2 - 2M}.$$

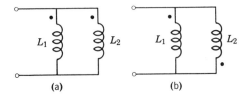

FIG. P-10.2   (a)   (b)

b Same as part *a* except use Fig. P-10.2b and show that the equivalent inductance is

$$\frac{L_1 L_2 - M^2}{L_1 + L_2 + 2M}.$$

10.3 a Two self-inductances, $L_1 = 4$ henries and $L_2 = 1$ henry, have a coefficient of coupling, k, of 0.5. Determine all values of inductance that may be obtained by connecting these inductances in series or in parallel. Use the results of problems 10.1 and 10.2.
b Same as part *a* except use $k = 0.99$.

Answer: *b*, 1.04 h, 8.96 h, 0.0089 h, and 0.076 h.

10.4 a For the circuit of Fig. P-10.4 show that the phasor driving-point impedance, $\hat{V}_1/\hat{I}_1$, is

$$\frac{\hat{V}_1}{\hat{I}} = j\omega L_1 + \frac{\omega^2 M^2}{R_0 + j\omega L_2}.$$

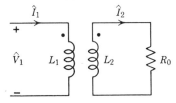

FIG. P-10.4

b Assume that $\hat{V}_1$ is held constant and that R_0 is either ∞ or zero. Show that for these two values of R_0 there are two values of $\hat{I}_1$ which differ in magnitude by the factor $(1 - k^2)$. Observe that for $k = 0.5$, the magnitude of $\hat{I}_1$ with $R_0 = \infty$ is 0.75 of its magnitude with $R_0 = 0$.

Mutual Inductance and Transformers 485

10.5 For the circuit shown in Fig. P-10.4, assume $L_1 = L_2 = 1$ henry, $R_0 = 10$ ohms, $\omega = 10$ radians, $\hat{V}_1 = 100 + j0$ volts.
 a With $k = 0.5$, sketch the equivalent circuit, labeling each impedance value. Calculate $\hat{I}_1$ and $\hat{I}_2$.
 b Repeat part a for $k = 1$.
 c Compare the average powers in R_0 for parts a and b.
 Partial answer: a, $\hat{I}_2 = 4e^{-j36.9°}$ amp.

10.6 For the circuit shown in Fig. P-10.6, $R_1 = 1\ \Omega$, $L_1 = L_2 = 1$ henry, $k = 0.5$, $i_1(0-) = 0$, $v = 10 \cos 4tu(t)$. Determine v_{ab}.

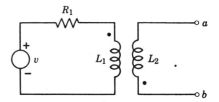

FIG. P-10.6

10.7 Same as problem 10.6, except assume $R_1 = 0$.

10.8 In Fig. P-10.8 steady state has been reached with the switch closed. The switch is opened at $t = 0$.

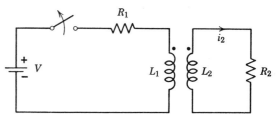

FIG. P-10.8 $R_1 = 1\ \Omega$, $R_2 = 10\ \Omega$, $L_1 = L_2 = 1$ henry, $V = 10$ volts.

 a Assume the magnetic coefficient of coupling to be 0.8. Determine the solution for the current i_2. Also determine the magnitude and location of any voltage impulses.
 b Assume the magnetic coefficient of coupling to be 1.0. Show that there are now no voltage impulses. Explain with the aid of an equivalent circuit.
 Partial answer: a, $i_2 = -8e^{-10t}u(t)$ amp.

10.9 For the circuit of Example 10.2, Fig. E-2.1, determine i_1 as a function of time after the switch is opened. Assume $V = 100$ volts, $R_1 = R_2 = 10$ ohms, $L_1 = 1$ henry, $L_2 = 10$ henries, $k = 0.8$.

10.10 a For the circuit of Fig. 10.10, page 475, assume $a = L_1/M$ and show that the equivalent circuit is that of Fig. P-10.10.

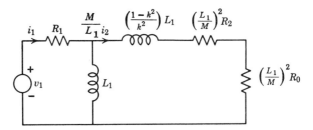

FIG. P-10.10

b For the circuit of Fig. 10.11, assume $a = M/L_2$ and sketch the resulting circuit.

10.11 For the circuit of Fig. P-10.11 assume that the transformer is ideal. Determine the turns ratio, N_1/N_2, which will permit maximum power transfer to the 10-ohm resistance.

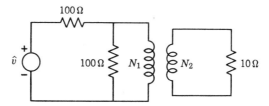

FIG. P-10.11

10.12 A 1000-cps voltage source has an internal impedance (resistive) of 500 ohms. It is to supply a resistive load of 10 ohms.

a If an ideal transformer were used to insure maximum power transfer to the load, what should be the turns ratio of this transformer?

b Assume the transformer to be ideal except that the self inductances are finite ($R_1 = R_2 = 0$, $k = 1$). The self-inductance on the high voltage side is 0.1 henry. Show that the turns ratio will not be the same as calculated for part *a*. Calculate the proper turns ratio for this case.

Partial answer: *b*, turns ratio is 6.3.

10.13 For the transformer of Fig. P-10.13: $L_1 = 4$ henries. $L_2 = 1$ henry, $R_1 = 1$ Ω. $R_2 = 0.25$ Ω, $k = 0.99$. $R_0 = 25$ Ω, $\omega = 100$ radians/sec. $\hat{V}_1 = 100 + j0$ volts. $N_1/N_2 = 2$.

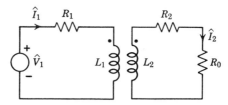

FIG. P-10.13

a Sketch the phasor equivalent circuit corresponding to Fig. 10.13. Insert numerical values.

Mutual Inductance and Transformers 487

b Calculate $\hat{I}_1$ and $\hat{I}_2$. Would appreciable error have been introduced by neglecting the current through $j\omega k L_1$? Would appreciable error have been introduced by assuming $R_1 = R_2 = 0$ and that $k = 1.0$?

c Determine efficiency of power transfer from the formula:

$$\text{Efficiency} = \frac{I_2^2 R_0}{I_2^2 R_0 + I_2^2 R_2 + I_1^2 R_1} = 1 - \frac{I_2^2 R_2 + I_1^2 R_1}{I_2^2 (R_0 + R_2) + I_1^2 R_1}$$

10.14 The transformer of Fig. P-10.14 is ideal with a turns ratio $N_1/N_2 = 1/10$. Determine $\hat{I}_1$ and $\hat{I}_2$ by any method. $\hat{V}_1 = 10 e^{j0°}$ volts.

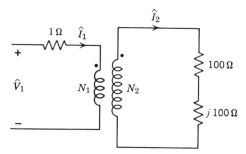

FIG. P-10.14

10.15 Three identical single-phase transformers are connected together to form a three-phase transformer as shown in Fig. P-10.15. The load is a balanced 3ϕ load of 100 kva, 1000 volts (rms) line to line. The line-to-line voltage of the power system is 10,000 volts (rms). Assume transformers are ideal.

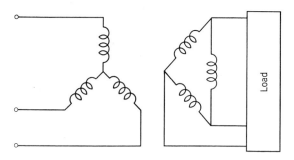

FIG. P-10.15

a Determine the turns ratio of each transformer.
b Determine the rms values of the currents in each winding of the transformer and in the lines to the load.

Partial answer: *a*, 5.77.

10.16 An autotransformer is sometimes used, particularly if the voltage change desired is small. An autotransformer is a single winding with a tap.

a For the autotransformer shown in Fig. P-10.16, assume it is ideal, and calculate the amplitudes of the currents in the windings if $V_{ac} = 110$ volts, $V_{bc} = 100$ volts.

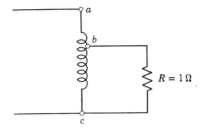

FIG. P-10.16

b What advantages would such an autotransformer have over an ordinary transformer in which two windings are insulated from each other?
c What possible disadvantages would an autotransformer have?

Partial answer: *a*, current in coil *ab* has amplitude of 90.9 amp.

10.17 For the transformer shown in Fig. P-10.17, determine poles and zeros of the transfer function

$$T(s) = \frac{V_{cd}(s)}{V_{ab}(s)}$$

for each of the conditions given (use same *s* plane). Then sketch magnitude and phase of $\hat{T}$ vs. ω for each of the conditions given.

a $R_1 = R_2 = 0.1$ ohm, $L_1 = L_2 = 1$ henry, $k = 0.99$.
b Same as part *a* except $R_1 = R_2 = 0$.
c Same as part *b* except $k = 1.0$.
d Same as part *c* except $L_1 = L_2 = \infty$.

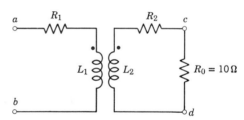

FIG. P-10.17

10.18 In supplying power to a growing industrial region it is suggested that two single-phase transformers be used in an open-delta (see Fig. P-10.18) during the early stages of development, the third transformer to be added later when needed. Show that the suggestion is feasible, that 3ϕ power can be supplied in this manner. Are there any disadvantages to this proposal?

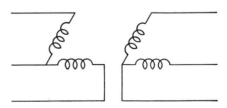

FIG. P-10.18

10.19 Some of the literature shows an equivalent circuit, Fig. P-10.19, which differs somewhat from that shown in Fig. 10.15. In both cases the data were obtained from open-circuit tests and short-circuit tests such as those of Example 10.6. Try to justify one of these circuits in preference to the other. (An excellent discussion of this subject is in *Magnetic Circuits and Transformers*, M. I. T. Electrical Engineering Staff, Wiley, New York, 1943.)

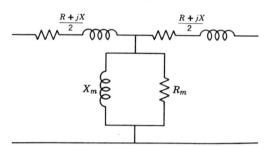

FIG. P-10.19

10.20 For Example 10.6, using Fig. E-6.4, assume $\hat{V}_{2\,rms} = 2400 + j0$ volts and that $\hat{I}_{2\,rms}$ has a magnitude of 6.25 amp and lags $\hat{V}_2$ by 30°

a Calculate $V_{1\,rms}$ and $I_{1\,rms}$. In calculating $V_{1\,rms}$ try to obtain an accurate value since the difference between this value and 240 represents voltage change from no-load to full-load conditions.

b Calculate the efficiency, noting that the power dissipated in the transformer in this case is the sum of the power loss in the conductors (200 watts) plus the power loss in the iron (80 watts). Note the high efficiency, an important factor in making a-c transmision economically feasible. To obtain high accuracy use efficiency $= 1 - \dfrac{\text{power losses}}{\text{power input}}$.

11 Analogues (Duals)

If two systems obey equations of the same form the systems are said to be analogous. If two analogous systems are in the same discipline, such as electric circuits, the systems are said to be duals. For electrical systems such duality involves an interchange between voltage and current, between inductance and capacitance, between resistance and conductance, between series and parallel elements, between mesh and nodal methods.

The study of analogues serves to conserve time and effort. A solution obtained for one system applies to analogous systems. A concept in one system is more readily learned if we are familiar with analogous concepts in other systems. Knowledge of response functions in one system is helpful in design and development work in analogous systems.

11.1 ELECTRICAL DUALS

The voltage-current relationship for circuit elements are shown in Fig. 11.1 equations of the same form being placed in the same row.

$v = L \dfrac{di_1}{dt}$	$i = C' \dfrac{dv_1}{dt}$
$v = Ri_1$	$i = Gv_1$
$v = \dfrac{1}{C} \int_0^t i_1 \, dt + v(0-)$	$i = \dfrac{1}{L'} \int_0^t v_1 \, dt + i(0-)$
$\sum v = 0$	$\sum i = 0$

FIG. 11.1 Voltage-current relationships for electric circuits.

The equations in the left column of Fig. 11.1 become the equations in the right column if the following quantities are interchanged; v and i, v_1 and i_1, L and C', R and G, C and L'. Each of these quantities is said to be the dual of the other. If this interchange is applied to complete circuits, the mesh equations for one circuit become the node equations for its dual circuit and vice versa. Therefore, one way to develop the dual of a given electrical circuit is to write the mesh or node equations for this circuit, write new equations with dual terms, and then sketch the dual electrical system from these new equations. If each pair of dual quantities have the same numerical values, the two circuits are said to be "exact" duals.

Not all circuits have realizable duals. There is no dual for mutual inductance when expressed as a dependent source voltage; however, if the mutual inductance terms can be represented by passive inductive elements in an equivalent circuit (see Chapter 10), the duals of such circuits can be constructed. Furthermore, we shall find that no duals exist for nonplanar networks.

The following examples illustrate how duals may be determined. We emphasize exact duals since these are of most practical value.

Example 11.1

GIVEN: The series L, R, C circuit of Fig. E-1.1.

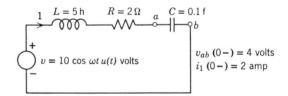

FIG. E-1.1

TO FIND: The exact dual of this circuit.

SOLUTION: The voltage equation for this circuit is

$$v = L\frac{di_1}{dt} + Ri_1 + \frac{1}{C}\int_0^t i_1\,dt + v_{ab}(0-). \tag{1}$$

The dual of Equation 1 may be written

$$i = C'\frac{dv_1}{dt} + Gv_1 + \frac{1}{L'}\int_0^t v_1\,dt + i_{ab}(0-). \tag{2}$$

We construct a circuit from Equation 2 by realizing that each term is a current and that the same voltage, v_1, is across C', G, and L' thus inferring that these elements are in parallel. Furthermore, for the dual to be an exact dual, C' must numerically equal L, etc. The circuit is shown in Fig. E-1.2.

492

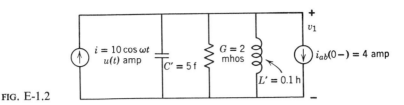

FIG. E-1.2

The original circuit had an initial current, $i_1(0-)$, of 2 amperes through the inductance, L; this should become an initial voltage, $v_1(0-)$, of 2 volts across the capacitance C'. In case we question the polarity of this voltage it may be desirable to develop the dual Laplace transform circuit for which initial conditions are automatically included. The Laplace transform of Equation 1 is

$$V(s) = \left(Ls + R + \frac{1}{Cs}\right)I_1(s) - Li_1(0-) + \frac{v_{ab}(0-)}{s}. \tag{3}$$

The dual of Equation 3 is

$$I(s) = \left(C's + G + \frac{1}{L's}\right)V_1(s) - C'v_1(0-) + \frac{i_{ab}(0-)}{s}. \tag{4}$$

For Equation 4 the circuit is as shown in Fig. E-1.3.

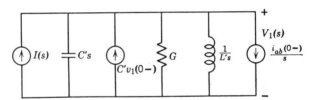

FIG. E-1.3

The direction of the current $C'v(0-)$ is fixed and thus the polarity of the initial voltage on C' is fixed, $v_1(0-)$ being 2 volts.

The exact dual is so constructed that the current response to the voltage driving function $10 \cos \omega t\, u(t)$ volts, in Fig. E-1.1 is the same as the voltage response to the current driving function, $10 \cos \omega t\, u(t)$ amp, in Fig. E-1.2. The inductance of 5 henries becomes a capacitance of 5 farads, the resistance of 2 ohms becomes a conductance of 2 mhos (or a resistance of 0.5 ohms) and the capacitance of 0.1 farads becomes an inductance of 0.1 henries. The two circuits have the same number of elements and "two elements in parallel" have as their dual "two elements in series."

Example 11.2

GIVEN: The circuit of Fig. E-2.1.
TO FIND: The exact dual of this circuit.

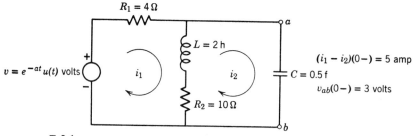

FIG. E-2.1

SOLUTION: The Laplace transform circuit corresponding to Fig. E-2.1 may be sketched immediately as in Fig. E-2.2.

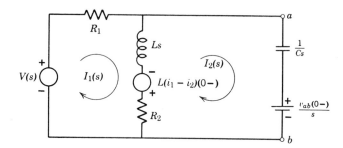

FIG. E-2.2

The voltage equations around the meshes of Fig. E-2.2 are

$$(Ls + R_1 + R_2)I_1(s) - (Ls + R_2)I_2(s) = V(s) + L(i_1 - i_2)(0-) \quad (1)$$

$$-(Ls + R_2)I_1(s) + \left(Ls + R_2 + \frac{1}{Cs}\right)I_2(s)$$
$$= -\frac{v_{ab}(0-)}{s} - L(i_1 - i_2)(0-). \quad (2)$$

The duals of these equations may be written

$$(C's + G_1 + G_2)V_1(s) - (C's + G_2)V_2(s) = I(s) + C'(v_1 - v_2)(0-) \quad (3)$$

$$-(C's + G_2)V_1(s) + \left(C's + G_2 + \frac{1}{L's}\right)V_2(s)$$
$$= -\frac{i_{ab}(0-)}{s} - C'(v_1 - v_2)(0-). \quad (4)$$

Note that the duals of mesh equations are node equations. The circuit may be constructed from these last equations and is shown in Fig. E-2.3.

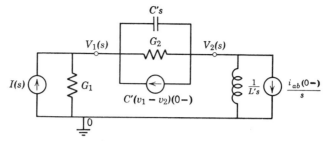

FIG. E-2.3

We are used to using double subscripts for voltages, the voltage v_{ab} meaning the difference in voltage between terminals a and b or $v_{ab} = v_a - v_b$. The dual of v_{ab} is i_{ab} which must be interpreted as a mesh current. We have been using only single subscripts with currents since this is adequate to describe a branch or mesh current. The dual of a current, such as $I_1(s)$, will be written $V_1(s)$ but must be interpreted as the voltage difference between node 1 and some reference node, such as 0 in Fig. E-2.3.

All immittance terms in Fig. E-2.3 are admittances. If we desire to sketch the circuit in the time domain, and insert the proper values for the quantities, we obtain the circuit of Fig. E-2.4, which is the dual of Fig. E-2.1.

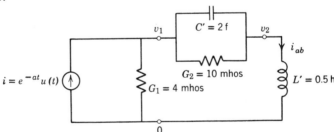

FIG. E-2.4 $i_{ab}(0-) = 3$ amp; $(v_1 - v_2)(0-) = 5$ volts.

Although we started with mesh equations, we could also have started with node equations. For this example there is only one node equation for the transform circuit of Fig. E-2.2.

$$V_{ab}(s)\left(\frac{1}{R_1} + \frac{1}{Ls + R_2} + Cs\right)$$
$$= \frac{V(s)}{R_1} - \frac{L(i_1 - i_2)(0-)}{Ls + R_2} + Cv_{ab}(0-). \quad (5)$$

The dual equation is

$$I_{ab}(s)\left[\frac{1}{G_1} + \frac{1}{C's + G_2} + L's\right] = \frac{I(s)}{G_1}$$
$$- \frac{C'(v_1 - v_2)(0-)}{C's + G_2} + L'i_{ab}(0-). \quad (6)$$

Analogues (Duals) 495

The circuit for this equation is that of Fig. E-2.5, which is really the same as Fig. E-2.3 with all current sources changed to equivalent voltage sources, and with the mesh current $I_{ab}(s)$ through the impedances, $L's$, $1/G_1$, and $1/(C's + G_2)$.

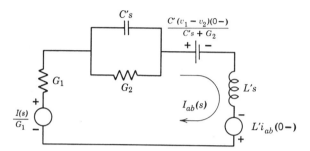

FIG. E-2.5

It is possible to sketch the dual circuit without writing equations. From our experience with the examples, we can see that a planar network with n mesh currents will have a dual network with n node voltages. A reference node is implied and so it is possible to associate a node with each mesh and in addition have the reference node associated with the area outside the planar network. (If we think of the network as sketched on a spherical surface we have as many meshes as nodes.) An element which has a current through it of $i_m - i_n$ will become a dual element which has a voltage across it of $v_m - v_n$. We shall need a convention which relates positive values of voltage and current and shall adopt the convention that a mesh current, i_n, which is positive in a clockwise sense will have a dual node voltage, v_n, which is positive with respect to the reference. In accord with this, a voltage difference in the mesh equation which is $+$ to $-$ in the clockwise sense will have, as its dual, a current which is positive leaving the node. We demonstrate with the transform circuit of Fig. E-2.2 which is resketched in Fig. 11.2. The dual of this circuit is shown in dashed lines on the same sketch. For the mesh which has $I_1(s)$ around its contour, a node is placed inside this mesh and marked $V_1(s)$, implying that for positive values of $I_1(s)$, $V_1(s)$ will be positive with respect to the reference node. The same is done for the mesh with current $I_2(s)$. The reference node is represented by a line surrounding the entire circuit. Lines are then drawn through each element connecting the nodes of adjacent areas, and in this line the dual element is placed. For example, a line is drawn from the node marked $V_1(s)$ through the source $V(s)$ to the reference node; in this line a current source is placed. Since the polarity of $V(s)$ is $-$ to $+$ in a clockwise sense, the positive direction of $I(s)$ is shown into the node marked $V_1(s)$. The resulting circuit is identical to that of Fig. E-2.3.

It is also possible to sketch a dual for a circuit by first labeling nodes

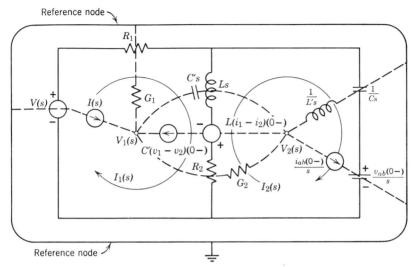

FIG. 11.2 Illustration of the sketching of a dual circuit.

and then obtaining a mesh dual. To do this it is desirable to start with the circuit as sketched from the node voltage equations, a circuit which has only current sources. As an illustration we might start with Equation 5 of Example 11.2, sketch the circuit corresponding to this equation and then develop the dual circuit which would be that of Fig. E-2.5.

The reason that nonplanar networks do not have duals may be deduced as follows. A branch, which has the difference of two mesh currents through it has a dual which is a branch having the difference of two node voltages across it. In a nonplanar network one cannot label independent currents without making one branch current the algebraic sum of three loop currents. This would require a dual branch which has its two terminals connected to three separate nodes. This is impossible and thus the dual of a nonplanar network does not exist.

11.2 ELECTROMECHANICAL ANALOGUES

In mechanics as in electromagnetism, the fundamental laws are field laws. In mechanics as in electromagnetism there are many applications in which lumped elements may be used to obtain approximate results. It is the purpose of this section to demonstrate that certain mechanical problems can be solved rapidly by sketching a mechanical circuit, sketching an analogous electrical circuit and then using electrical circuit theory. We shall limit ourselves to systems in which the driving and response functions have only one direction in space, the emphasis being on simple translational motion.

It is assumed that the student has some background in mechanics from his courses in physics; therefore the introductory remarks and definitions are minimal. However, since we wish to define mechanical circuit elements which may be connected together to form mechanical circuits, the viewpoint taken in this section differs from that of most physics texts. We shall try to justify the use of each idealized element as well as to state its principal limitations.

In mechanics the driving and response functions are forces, torques, and velocities. These are vector quantities but since they will have components in only one direction for our problems, we shall treat the quantities as scalars. For translation the driving and response functions are force, f, with the units of newtons, and velocity, u, with the units of meters per second. For rotation the driving and response functions are torque, τ, with the units of newton-meters, and angular velocity, ω_r, with the units of reciprocal seconds. The subscript r is used with ω_r to indicate rotational angular velocity.

If we choose force or torque as the analogue of voltage, velocity is the analogue of current. However, if we choose force or torque as the analogue of current, velocity is the analogue of voltage. Should we emphasize one analogue? Is there one analogue which makes the sketching of a mechanical circuit easier because of our background in electrical circuits? The answer is yes to both questions. It is the one that makes force or torque the analogue of current, velocity the analogue of voltage. The reasons for preferring this analogue will become apparent as we develop our mechanical circuit elements; the following argument may make the choice plausible. For an electrical circuit element, current is considered continuous **through** the element whereas voltage is considered as a voltage difference **across** the element. Similarly, for a mechanical circuit element, force or torque must be considered as continuous **through** the element whereas velocity is considered as a velocity difference **across** the element.

Our first mechanical circuit elements are ideal sources. The force or torque source is shown symbolically as a circuit element in Fig. 11.3a.

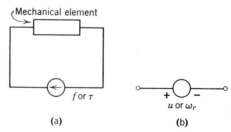

FIG. 11.3 Symbolic representation of mechanical sources. (a) Ideal force or torque source applied to mechanical element. (b) Ideal velocity source.

It is sketched in the same manner as a current source and is shown connected to a mechanical circuit element (as yet undefined) simply to indicate that the force or torque must always be continuous. The velocity source is sketched as shown in Fig. 11.3b and resembles the voltage source of our electrical circuits.

.2.1 Passive Elements

The three passive, lumped, and idealized mechanical circuit elements which are useful to define are: a mass or inertial element, a compliance element, and a viscous damper element.

For translational motion our idealized mass element, M, has the units of kilograms and is defined[1] by the relationship (known as Newton's second law) between the net force and the acceleration:

$$f = M \frac{du}{dt}. \tag{11.5}$$

In Equation 11.5 it is assumed that the mass M is not a function of velocity and that f and u are positive in the same sense. We are also assuming that the velocity u is a velocity difference with respect to some reference (usually earth). This inertial reference will be indicated by the symbol, ▓▓▓▓, which corresponds to "ground" in an electrical circuit. Symbolically, we show the circuit element, M, as indicated in Fig. 11.4a, the L-shaped side being the grounded side.

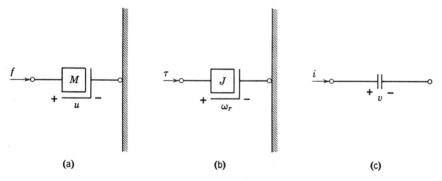

FIG. 11.4 Symbolic representation of mechanical inertial elements and electrical capacitance. (a) Mass M: $f = M(du/dt)$; (b) Moment of inertia, J: $\tau = J(d\omega_r/dt)$; (c) Capacitance C: $i = C(dv/dt)$.

---

[1] It is possible to consider mass as a postulate and consider Equation 11.5 as the defining expression for either force or acceleration.

Analogues (Duals) 499

For rotational motion, a relationship similar to Equation 11.5 is

$$\tau = J \frac{d\omega_r}{dt}, \tag{11.6}$$

in which J is the moment of inertia with the units of kilograms-meter^2 or newton-meter-seconds^2. In Fig. 11.4b is shown the symbolic representation of the circuit element, J; in Fig. 11.4c is shown the symbolic representation of the electrical circuit element, C.

It is important to show each inertial element as having one terminal connected to the zero velocity reference or to ground. It is important also to realize that our inertial elements are lumped or point elements and not distributed quantities.

In order to demonstrate the usefulness of the concepts mentioned so far, consider the body of mass M which is free to slide on the stationary horizontal plane shown in Fig. 11.5a. The driving function is the applied force, f_a, and the response function is the velocity u. We use the subscript a to indicate the applied or driving function. If the driving function were a velocity source, we would place the subscript a on the velocity, u_a, and omit it from the response function, f. Although it is not necessary to choose the response function as positive in the same sense as the driving function, it is convenient to do so. In Fig. 11.5b is shown the mechanical circuit which is sketched by combining the ideal force source with the ideal mass. It is assumed that our mass is perfectly rigid and that there is no friction of any kind. The polarity markings for u are such that if u is plus, the ungrounded side being positive with respect to the grounded side, u is in the direction indicated as positive

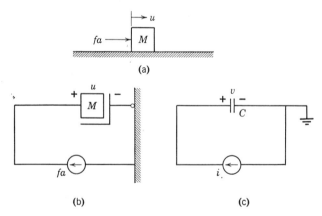

FIG. 11.5 Mechanical circuit diagram, and electrical analogue, for force applied to mass. (*a*) Mechanical system. (*b*) Mechanical circuit. (*c*) Analogous electrical circuit.

in Fig. 11.5a. The analogous electrical circuit is sketched in Fig. 11.5c and can clearly be drawn "by inspection" from Fig. 11.5b.

There is a rotational system with the same form of circuit. Consider the body with moment of inertia J which is free to rotate without friction as shown in Fig. 11.6a. The mechanical circuit diagram is sketched in exactly the same manner as for Fig. 11.5, the translational quantities being replaced by the rotational quantities.

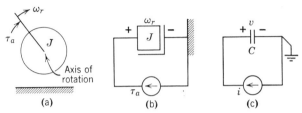

FIG. 11.6 Mechanical circuit diagram, and electrical analogue, for torque applied to moment of inertia. (a) Mechanical system. (b) Mechanical circuit. (c) Analogous electrical circuit.

Example 11.3

GIVEN: The mass M on the horizontal plane of Fig. E-3.1.

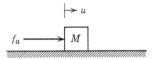

FIG. E-3.1

TO FIND: The velocity,[2] u, if $f_a = 2e^{-t}u(t)$ newtons, $u(0-) = 1$ meter per sec, and $M = 4$ kilograms. All friction may be neglected.

SOLUTION: The mechanical and electrical circuit diagrams are sketched as shown in Fig. E-3.2a and Fig. E-3.2b. If we consider the

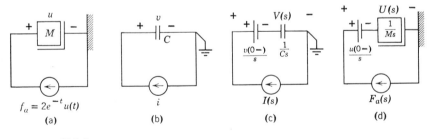

FIG. E-3.2

---

[2] There may be some possibility of confusing velocity, u, and the unit step function, $u(t)$. We shall avoid using functional notation for velocity, and thus $u(t)$ will mean unit step function.

electrical circuit as the exact analogue of the mechanical circuit, the solution for v in volts should be numerically the same as the solution for u in meters per second. We shift to the transform domain for the electrical circuit in Fig. E-3.2c and then to the mechanical circuit of Fig. E-3.2d. Since we have more confidence in solving the electrical circuit we use Fig. E-3.2c.

$$V(s) = \frac{v(0-)}{s} + \frac{I(s)}{Cs}. \tag{1}$$

Since

$$I(s) = \mathscr{L}[2e^{-t}u(t)] = \frac{2}{s+1}, \quad C = 4, \text{ and } v(0-) = 1,$$

$$V(s) = \frac{1}{s} + \frac{2}{4s(s+1)} = \frac{s+1.5}{s(s+1)} = \frac{A}{s} + \frac{B}{s+1}.$$

Since

$$A = \frac{s+1.5}{s+1}\bigg|_{s=0} = 1.5, \text{ and } B = \frac{s+1.5}{s}\bigg|_{s=-1} = -\frac{1}{2},$$

$$v = [1.5 - 0.5e^{-t}]u(t) \text{ volts}. \tag{2}$$

Therefore the solution for the velocity may be written

$$u = [1.5 - 0.5e^{-t}]u(t) \text{ meters per second}. \tag{3}$$

Another mechanical circuit element is the compliance element which is labeled C_m for translational motion and C_r for rotational motion. It represents an idealized spring which is assumed to be massless and to follow Hooke's law. This law states that force and displacement are proportional for translational motion and that torque and angular displacement are proportional for rotational motion. For the sketch of Fig. 11.7a, this relation is

$$f = \frac{1}{C_m} \int_0^t (u_1 - u_2) \, dt + f(0-), \tag{11.7}$$

while for the sketch of Fig. 11.7b, the relation is

$$\tau = \frac{1}{C_r} \int_0^t (\omega_{r_1} - \omega_{r_2}) \, dt + \tau(0-). \tag{11.8}$$

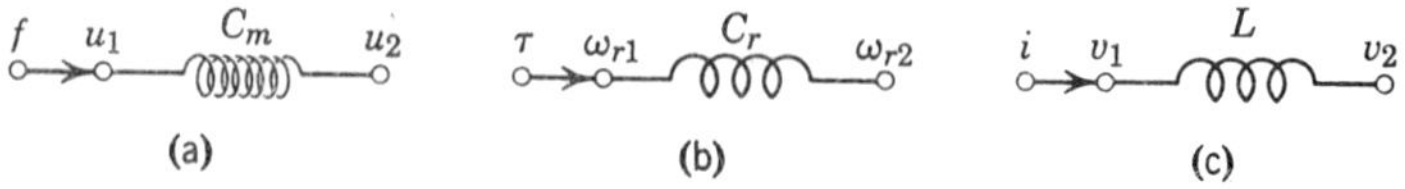

FIG. 11.7 Symbolic representation of compliance element and its electrical analogue. (a) Translational compliance. (b) Rotational compliance. (c) Inductance.

For the mechanical system indicated in Fig. 11.8a, an applied force is acting on a helical spring which is connected to a mass which is free to slide without friction. If we neglect all friction and neglect the mass of the spring we can sketch the mechanical circuit of Fig. 11.8b. The compliance and mass elements are in series because they have the same force; this is because the force, f_a, is continuous through the compliance, C_m.

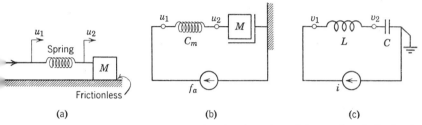

FIG. 11.8 A mechanical system, its circuit and analogous electrical circuit. (a) Mechanical system. (b) Mechanical circuit. (c) Analogous electrical circuit.

A mechanical system which may seem to be very similar is shown in Fig. 11.9a. The applied force is acting on the mass which pushes against the spring. We again idealize the system by neglecting the mass of the spring and all friction. In constructing the mechanical circuit we need to realize that all parts of the mass have the same velocity and that this

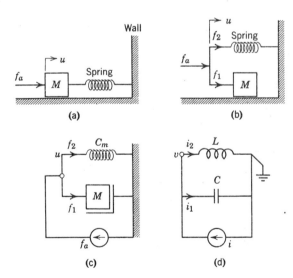

FIG. 11.9 Another mechanical system and its circuit. (a) Mechanical system. (b) Resketch of system. (c) Mechanical circuit. (d) Analogous electrical circuit.

Analogues (Duals) 503

velocity is the same as the left side of the spring. Thus a better sketch of the mechanical system is shown in Fig. 11.9b, the mass M being turned so that the force, f_a, acts at a node which has a single velocity. The force, f_a, may be considered as dividing into two portions, one part acting on the mass M, the other part acting on the spring. A force equation in accord with Newton's second law may be written at the node.

$$f_a = M \frac{du}{dt} + \frac{1}{C_m} \int_0^t u \, dt + f_2(0-). \tag{11.9}$$

Observe that this equation correctly expresses the force relationship for both Fig. 11.9a and Fig. 11.9b. We may find it easier to solve the electrical circuit of Fig. 11.9d for which Kirchhoff's current law gives

$$i = C \frac{dv}{dt} + \frac{1}{L} \int_0^t v \, dt + i_2(0-). \tag{11.10}$$

Although the spring and mass seemed to be in series in both Fig. 11.8a and Fig. 11.9a we have found that in the latter case the mass and spring are really in parallel. We conclude that two mechanical elements are in series if they have the same force and they are in parallel if their terminals have the same velocities.

The two-terminal mass element is a convenient invention, since it permits the rapid sketching of the mechanical circuit and its electrical analogue. We should recognize the "fictional" character of this element, however. As an illustration, if we are asked for the force on the wall of Fig. 11.9a, let us observe that this is the force exerted by the spring only and thus corresponds to the response i_2 of Fig. 11.9d. The mechanical circuit of Fig. 11.9c may lead one to think that the total applied force, f_a, acts on the wall. It should not surprise us that a useful circuit may not be in accord with physical reality. This was also true for the element of inductance in electrical circuits, which was invented for the purpose of having the voltage difference between points independent of the path. The two-terminal mass is as useful a circuit "fiction" as the element inductance.

The third passive mechanical element is the viscous damper which is labeled[3] G_m for translational motion and G_r for rotational motion. The previous two elements are developed to account for the energy stored in the motion of bodies or in the deflection of elastic bodies; this element accounts for the energy dissipated as heat. For the sketches of Figs. 11.10a and 11.10b, the defining equations are

$$f = G_m(u_1 - u_2), \tag{11.11}$$

---

[3] Other letters used in the literature to represent the viscous damper are B, D, and R_m.

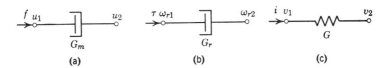

FIG. 11.10 Symbolic representation of viscous damper element and its electrical analogue. (*a*) Translational damper. (*b*) Rotational damper. (*c*) Electrical conductance.

and

$$\tau = G_r(\omega_{r_1} - \omega_{r_2}). \tag{11.12}$$

R_m and R_r are defined as the reciprocals of G_m and G_r respectively.

The term viscous is used because the ratio between the retarding force and relative velocities is approximately constant in many cases involving the relative motion of solids and fluids. Such a constant ratio is not usually found in the sliding friction of two solid bodies. The word damper is used because the effect of this element is to dampen the oscillations that occur otherwise. The dashpot symbol is used because the dashpot (or shock absorber) is a practical device which, for a limited range of velocities and displacements, has characteristics that are approximately those of the viscous damper element.

Figure 11.11*a* shows a velocity driving function, u_a, moving a mass M which is connected to the inertial reference by the parallel combination of a spring and dashpot. The masses of the spring and dashpot are neglected and the dashpot is assumed to be a viscous damper. Figure 11.11*b* shows the mechanical circuit, and Fig. 11.11*c* shows the analogous electrical circuit.

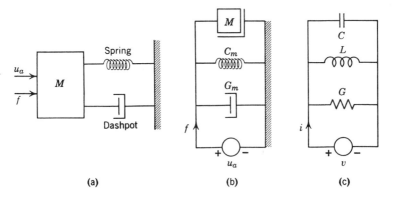

FIG. 11.11 Mechanical system with velocity driving function. (*a*) Mechanical system. (*b*) Mechanical circuit. (*c*) Analogous electrical circuit.

The ideal velocity source is one for which the velocity is not a function of the resulting force. A practical example which approximates a velocity source is the velocity resulting from cam action.

11.2.2 Laws of Mechanics and Sketching of Mechanical Circuits

For electrical circuits we used Kirchhoff's current and voltage laws to write the necessary equations. In mechanics the analogue to Kirchhoff's current law is a modified form of Newton's second law which is known as D'Alembert's principle and which is written

$$\sum f = 0, \qquad (11.13)$$

or

$$\sum \tau = 0. \qquad (11.14)$$

These equations are interpreted to mean that the sum of the forces (or torques) leaving a node in a mechanical circuit must be zero.

The analogue to Kirchhoff's voltage law may be expressed by saying that the sum of the velocity differences around a closed path in a mechanical circuit is zero. Mathematically we write

$$\sum u = 0, \qquad (11.15)$$

and

$$\sum \omega_r = 0. \qquad (11.16)$$

For the mechanical analogue that we have been stressing, with force analogous to current and velocity analogous to voltage, it seems natural to define transform mechanical admittance, $Y_m(s)$, as

$$Y_m(s) = \frac{F(s)}{U(s)}. \qquad (11.17)$$

Although we shall use this definition for mechanical admittance, many writers prefer to define mechanical admittance as the reciprocal of this ratio.

A summary of all the relations obtained is shown in Fig. 11.12.

The implications of Fig. 11.12 are that for any one circuit it should be possible to develop five analogous circuits. Of these six circuits there are three sets of duals. Not all of these circuits may be physically realizable.

From a sketch of a mechanical system, we are able to sketch the mechanical circuit and then sketch one analogous electrical circuit. By sketching the dual of the electrical circuit we can sketch another mechanical circuit which should be the dual of the first mechanical

$v = L\dfrac{di}{dt}$	$v = Ri$	$v = \dfrac{1}{C}\int_0^t i\,dt + v(0-)$	$\sum v = 0$
$i = C\dfrac{dv}{dt}$	$i = Gv$	$i = \dfrac{1}{L}\int_0^t v\,dt + i(0-)$	$\sum i = 0$
$f = M\dfrac{du}{dt}$	$f = G_m(u_1 - u_2)$	$f = \dfrac{1}{C_m}\int_0^t (u_1 - u_2)\,dt + f(0-)$	$\sum f = 0$
$u = C_m\dfrac{df}{dt}$	$u_1 - u_2 = R_m f$	$u = \dfrac{1}{M}\int_0^t f\,dt + u(0-)$	$\sum u = 0$
$\tau = J\dfrac{d\omega_r}{dt}$	$\tau = G_r(\omega_{r_1} - \omega_{r_2})$	$\tau = \dfrac{1}{C_r}\int_0^t (\omega_{r_1} - \omega_{r_2})\,dt + \tau(0-)$	$\sum \tau = 0$
$\omega_{r_1} - \omega_{r_2} = C_r\dfrac{d\tau}{dt}$	$\omega_{r_1} - \omega_{r_2} = R_r \tau$	$\omega_r = \dfrac{1}{J}\int_0^t \tau\,dt + \omega_r(0-)$	$\sum \omega_r = 0$

duals (between rows 1–2, 3–4, 5–6)

FIG. 11.12 Analogous relations in electric circuits, mechanical translation, and mechanical rotation.

circuit. We demonstrate with the system of Fig. 11:11 which is repeated in parts *a*, *b*, and *c* of Fig. 11.13. Fig. 11.13*d* is sketched as the dual of Fig. 11.13*c*. Then the mechanical circuit is shown in Fig. 11.13*e* and, from this, is sketched the mechanical system of Fig. 11.13*f*. The analogous rotational circuits and systems are also readily sketched since these have a strong correlation with the translational circuits and systems.

The advantages in using the force-current analogue are now apparent. This analogue permits a rapid entry to mechanical circuits from mechanical systems and then permits the sketching of analogous circuits.

We can always check the accuracy of an analogue. For example, the force equation for Fig. 11.13*b* is

$$f = M\frac{du_a}{dt} + G_m u_a + \frac{1}{C_m}\int_0^t u_a\,dt + f_1(0-). \tag{11.18}$$

If we replace all quantities by their duals, we obtain

$$u = C'_m\frac{df_a}{dt} + R_m f_a + \frac{1}{M'}\int_0^t f_a\,dt + u_1(0-). \tag{11.19}$$

Equation 11.19 correctly gives the velocity equation for the circuit of Fig. 11.13*e*.

Not all mechanical duals can be physically realized by the tools we have developed. This is partly because all mass elements must have one

Analogues (Duals)   507

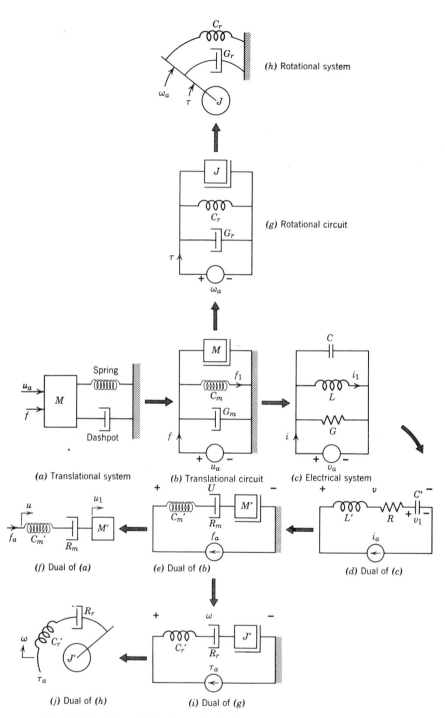

FIG. 11.13 A translational mechanical system and its analogues.

terminal connected to the inertial reference. Such limitations can be overcome by floating-lever systems which are beyond the scope of this text.

Example 11.4

GIVEN: The mechanical system of Fig. E-4.1.

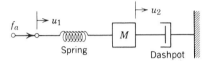

FIG. E-4.1

TO FIND: The dual of this system.

SOLUTION: The mechanical circuit is sketched in Fig. E-4.2 and the electrical analogue for this circuit is sketched in Fig. E-4.3. The dual of the electrical circuit is sketched (Fig. E-4.4) and then the dual of the mechanical circuit (Fig. E-4.5). Finally the dual mechanical system is shown in Fig. E-4.6.

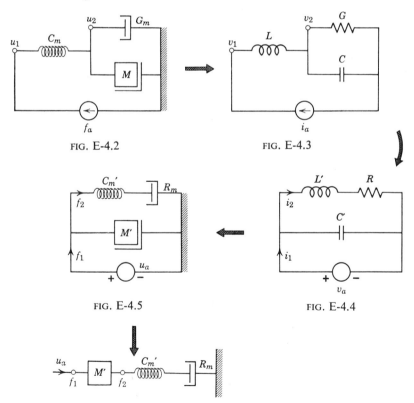

FIG. E-4.6 Dual mechanical system.

Analogues (Duals) 509

The force equation for Fig. E-4.2. is

$$f_a = \frac{1}{C_m}\int_0^t (u_1 - u_2)\,dt + f_c(0-) = M\frac{du_2}{dt} + G_m u_2, \qquad (1)$$

in which $f_c(0-)$ is the force in the spring at $t = 0-$.
The dual of this equation is

$$u_a = \frac{1}{M'}\int_0^t (f_1 - f_2)\,dt + u_M(0-) = C'_m\frac{df_2}{dt} + R_m f_2. \qquad (2)$$

Equation 2 is in accord with the mechanical system of Fig. E-4.6.

Example 11.5

GIVEN: The mechanical system of Fig. E-5.1.

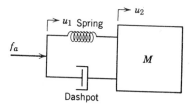

FIG. E-5.1

TO FIND: The transform circuit showing initial conditions.

SOLUTION: The mechanical circuit and its analogous electrical circuit are shown in Figs. E-5.2 and E-5.3, respectively.

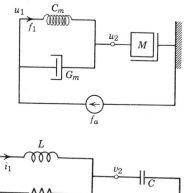

FIG. E-5.2

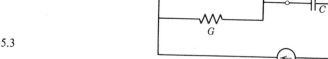

FIG. E-5.3

It is possible to sketch the transform circuits immediately (Figs. E-5.4 and E-5.5). Since the driving function is a current, we may prefer to make each passive element an admittance and each initial condition a current source. The electrical circuit acts as an aid.

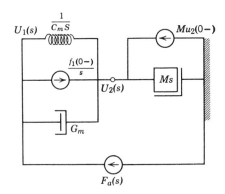

FIG. E-5.4

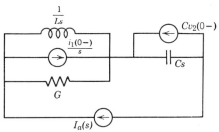

FIG. E-5.5

As a check, we can write the force equation for Fig. E-5.1.

$$f_a = G_m(u_1 - u_2) + \frac{1}{C_m}\int_0^t (u_1 - u_2)\, dt + f_1(0-) = M\frac{du_2}{dt}. \quad (1)$$

The transform of Equation 1 is

$$F_a(s) = G_m[U_1(s) - U_2(s)] + \frac{U_1(s) - U_2(s)}{C_m s}$$

$$+ \frac{f_1(0-)}{s} = MsU_2(s) - Mu_2(0-). \quad (2)$$

We observe that Equation 2 is in accord with the transform circuit of Fig. E-5.4.

Example 11.6

GIVEN: The mechanical circuit of the previous example, which is shown again in Fig. E-6.1.

TO FIND: The transfer function $\hat{F}_1/\hat{F}_a$ under sinusoidal steady-state conditions.

← Analogues (Duals) 511

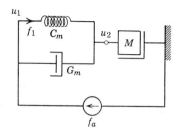

FIG. E-6.1

SOLUTION: Steady-state sinusoidal conditions means that steady state exists with $f_a = \text{Re}(\hat{F}_a e^{j\omega t})$. Also $f_1 = \text{Re}(\hat{F}_1 e^{j\omega t})$. A phasor circuit diagram may be sketched as in Fig. E-6.2 with each element expressed as an admittance.

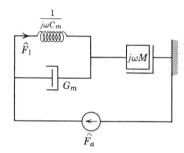

FIG. E-6.2

Since the admittances $1/j\omega C_m$ and G_m are in parallel, we may consider force division as analogous to current division and write

$$\frac{\hat{F}_1}{\hat{F}_a} = \frac{1/j\omega C_m}{1/j\omega C_m + G_m} = \frac{1}{1 + j\omega C_m G_m}. \tag{1}$$

11.2.3 Gravitational Force

The force of gravity must be considered whenever it has a component in the direction of summation of forces. For most systems on earth we are justified in considering the force of gravity as a constant. We shall try to develop some circuit concepts involving this force.

Consider the mass M falling in a viscous medium as indicated in Fig. 11.14a. The force of gravity is Mg in which g is the acceleration of gravity in meters per second². This force must equal the force of acceleration, $M(du/dt)$, plus the viscous damper force $G_m u$. Thus

$$f = Mg = M\frac{du}{dt} + G_m u. \tag{11.20}$$

The circuit for Equation 11.20 is shown in Fig. 11.14b. The analogous electrical circuit is shown in Fig. 11.14c. The steady-state voltage for this circuit is

$$v_{ss} = \frac{Cg}{G}, \tag{11.21}$$

and thus the steady-state velocity for the mass is

$$u_{ss} = \frac{Mg}{G_m}. \tag{11.22}$$

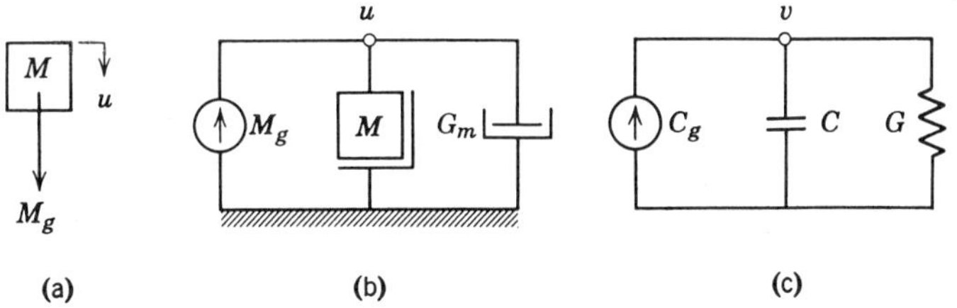

FIG. 11.14 Circuits representing falling mass. (a) Mass falling in viscous medium. (b) Mechanical circuit. (c) Analogous electrical circuit.

As a second illustration, consider the physical situation of Fig. 11.15a in which the mass M is initially supported at such a position that the force in the spring is zero. The displacement x is measured downward from this position. At $t = 0$ the support for the mass is removed and we may consider the applied force as $Mgu(t)$. The mechanical circuit and the analogous electrical circuit both indicate that the steady-state velocity will be zero and that the steady-state force on the floor is entirely in the spring. The steady-state deflection, x_{ss}, can be determined since for the spring,

$$f_1 = \frac{1}{C_m}\int_0^t u\,dt + f_1(0-). \tag{11.23}$$

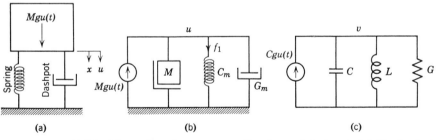

FIG. 11.15 Mass released on support of spring and dashpot. (a) Mechanical System. (b) Mechanical circuit. (c) Analogous electrical circuit.

In this case $f_1(0-) = 0$, and the steady-state force is

$$f_{1_{ss}} = \frac{1}{C_m}(x_{ss}) = Mg.$$

Analogues (Duals) 513

Thus

$$x_{ss} = C_m M g \qquad (11.24)$$

It may be observed that if we assumed C_m to be zero, this would correspond to L being zero in the electrical circuit and would mean that the mass M was resting on a rigid plane.

For a third illustration, consider the mechanical system of Fig. 11.16a for which the mechanical circuit is sketched in Fig. 11.16b. It is important to observe that each mass must have one terminal connected to the inertial reference.

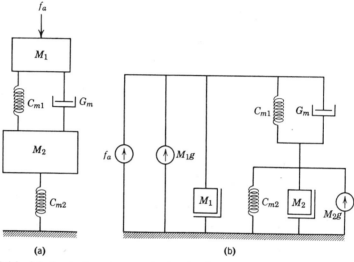

FIG. 11.16 Mechanical system and circuit showing gravitational forces. (a) Mechanical system. (b) Circuit for (a).

Since our systems are linear we may use superposition and consider the effect of the gravitational forces as separate from that of other sources. In many cases the major effect of gravity is to establish spring forces and deflections of constant values, about which occur the variations of interest to us.

11.3 ELECTRONIC ANALOGUE COMPUTER

We have been using mathematical models for electrical and mechanical systems. The student may wonder whether the use of these models is worthwhile; would it not be easier to put the parts of the system together and perform some tests? Some experience in the laboratory will be helpful in pointing out that there are difficulties in measurement and that the time required to assemble parts and perform tests is not small. An engineer gradually develops the judgment to determine whether an analytic or experimental method is better.

How about systems which obey differential equations of higher order than the second order and how about systems in which some nonlinearity cannot be ignored? Is it then always better to use experimental methods than analytical methods? The electronic analogue computer[4] is a tool to assist the analyst in this area, allowing him to extend the analysis to equations which are of high order and which have some measure of nonlinearity. For special purposes, mechanical or electrical analogue computers may be preferred, but for general purposes the electronic analogue computer is best.

Our study of the electronic analogue computer will be limited to how it works and how it may be used to solve differential equations.

Inspection of the mesh equations for circuit analyses reveals such terms as Ri, $L(di/dt)$, $1/C \int_0^t i\, dt$, and some voltage-driving functions, the sum of all voltages being zero. In order for a computer to solve this equation it must be able to multiply, add, integrate, and differentiate.

The heart of the electronic analogue computer is the "operational" amplifier, which is outlined in Fig. 11.17. It is designed to have a high negative ratio, A, between output and input voltages,

$$A = \frac{v_0}{v_g}, \qquad (11.25)$$

and also to have a small input current.

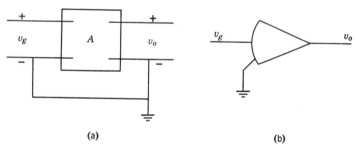

(a) (b)

FIG. 11.17 Operational amplifier. (a) Connection diagram. (b) Standard diagram.

For our purposes we shall idealize the amplifier and assume that the input current is zero. We shall also assume that $v_g = 0$; the use of the amplifier is such that v_0 is finite and our assumption then is that $A = -\infty$ or that A is so large that v_g is negligibly small.

---

[4] The digital computer is another tool which the engineer may use for this work, and with which he should become familiar.

A "scale changer" may consist of two resistors and an amplifier as shown in Fig. 11.18. Since $v_g = 0$, and since the current input to the amplifier is assumed zero

$$i = \frac{v_1}{R_1} = -\frac{v_0}{R_0},$$

or

$$v_0 = -\left(\frac{R_0}{R_1}\right)v_1. \tag{11.26}$$

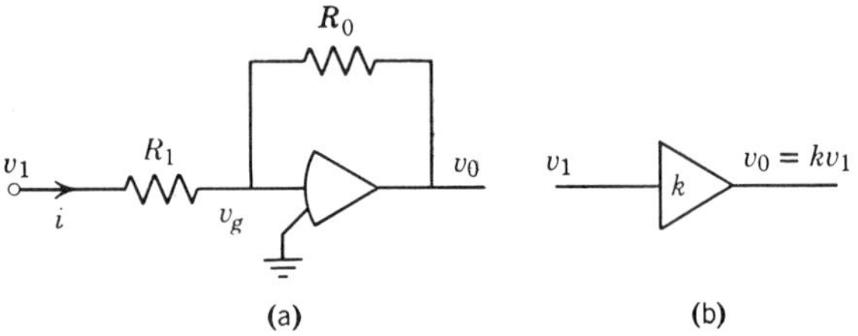

FIG. 11.18 Scale changer unit. (*a*) Amplifier and resistor connections. (*b*) Diagram for computer use.

Note that, for the computer, driving and response functions are **both** voltages. Note also that there is an automatic sign reversal and that if $R_1 = R_0$ the unit multiplies by -1.

An "adder" may be built as shown in Fig. 11.19 for adding three voltages. The current i_0 is equal to the sum of the input currents, so we may write

$$i_0 = i_1 + i_2 + i_3 = -\frac{v_0}{R_0} = \frac{v_1}{R_1} + \frac{v_2}{R_2} + \frac{v_3}{R_3},$$

or

$$v_0 = -\left(\frac{R_0 v_1}{R_1} + \frac{R_0 v_2}{R_2} + \frac{R_0 v_3}{R_3}\right). \tag{11.27}$$

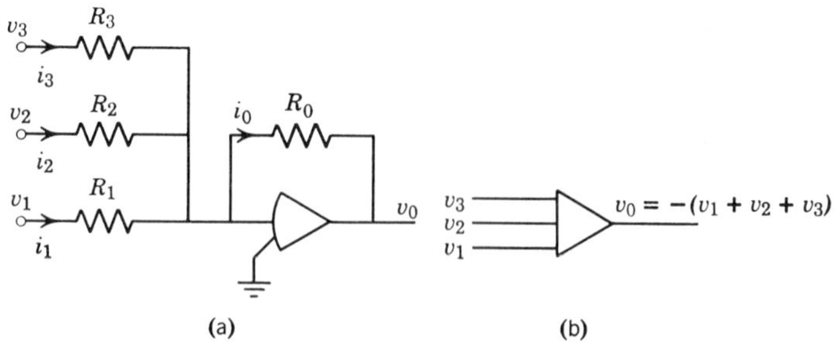

FIG. 11.19 An adder unit. (*a*) Amplifier and resistor connections. (*b*) Diagram for computer use (scale factor = 1).

If $R_1 = R_2 = R_3 = R_0$ the unit is a simple adder; if these resistances are different the unit is a scale changer–adder.

An "integrator" unit may be built as shown in Fig. 11.20a.

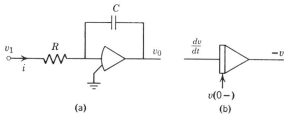

FIG. 11.20 An integrator unit. (a) Amplifier and circuit elements. (b) Diagram for computer use (scale factor = 1).

The expression for the output voltage is

$$v_0 = -\frac{1}{C}\int_0^t i\,dt + v_0(0-) = -\frac{1}{C}\int_0^t \frac{v_1}{R}\,dt + v_0(0-)$$

$$= -\frac{1}{RC}\int_0^t v_1\,dt + v_0(0-). \tag{11.28}$$

If $RC = 1$ ($R = 10^6$ ohms, $C = 10^{-6} f$ are convenient), the output voltage is the negative integral of the input voltage. The initial condition $v_0(0-)$ would need to be introduced as the voltage to which the capacitor C was charged prior to starting the problem on the computer. Suitable switches are necessary to connect the charged capacitors to the integrating units at the same time other driving functions are connected, and at which time the computing process starts.

Figure 11.20b shows an integrator diagram for computer use. The sign of the initial condition is reversed by the integrator unit in this case; this is the convention we shall use, although not all commerical integrating units operate in this manner.

A potentiometer is used to multiply a voltage by a real constant less than one. It is shown pictorially in Fig. 11.21.

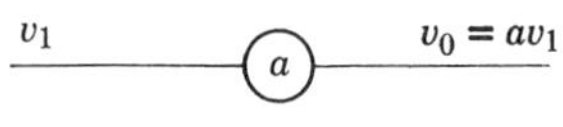

FIG. 11.21 Potentiometer unit, $0 < a < 1$.

Differentiating units are normally not used since these accentuate "noise" or unwanted signals and because they are more easily subject to output voltages in excess of the amplifier rating.

Since it is possible to develop electronic computing units in which the output may be related to the input in some nonlinear manner, the computer is capable of handling some nonlinear systems.

The various units are connected together to satisfy the differential equations. The technique is demonstrated in the following examples.

Although adders and integrators can also multiply, we shall assume that their multiplication factor is one.

We shall also assume that the time scaling and magnitude scaling which is necessary to match the physical problem to the computer capabilities has resulted in coefficients of all terms being less than one.

Example 11.7

GIVEN: The circuit shown in Fig. E-7.1.

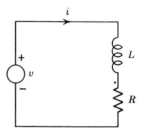

FIG. E-7.1

TO FIND: The block diagram simulating the computer for this circuit.
SOLUTION: The voltage equation is normally written

$$v = L\frac{di}{dt} + Ri, \tag{1}$$

but now it is best to rearrange, placing the highest order of the derivative on the left side of the equation. Then

$$L\frac{di}{dt} = v - Ri. \tag{2}$$

The right side of the equation is satisfied by multiplying i by R, then subtracting Ri from v. We assume that we have available the variable i; we shall operate on it with a potentiometer (R) and add the output to v. This is shown in Fig. E-7.2.

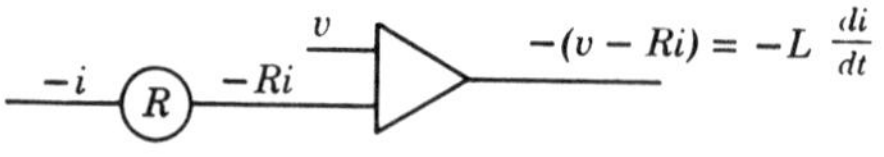

FIG. E-7.2

The output of the adder must be $-(v - Ri) = -L(di/dt)$. What remains to be done is to operate on $-L(di/dt)$ so as to obtain i. Multiplying by $1/L$ results in $-di/dt$ and integrating results in $+i$. Another multiplier is needed to change the sign to $-i$ and then the connection can be made as shown in Fig. E-7.3.

To start the computer, $i(0-)$ and v must be applied simultaneously.

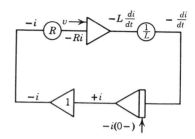

FIG. E-7.3

Example 11.8

GIVEN: The circuit shown in Fig. E-8.1.

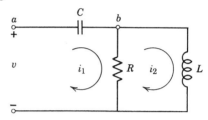

FIG. E-8.1

TO FIND: The block diagram simulating the computer for this circuit.
SOLUTION: The equations for the circuit are

$$v = \frac{1}{C}\int_0^t i_1\, dt + v_{ab}(0-) + R(i_1 - i_2) \tag{1}$$

$$L\frac{di_2}{dt} = R(i_1 - i_2) \tag{2}$$

One approach is to rewrite Equation 1 as

$$R(i_1 - i_2) = v - \frac{1}{C}\int_0^t i_1\, dt - v_{ab}(0-) \tag{3}$$

and then use Equations 2 and 3 to develop the diagram. The diagram is shown in Fig. E-8.2.

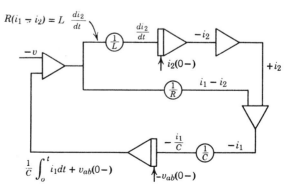

FIG. E-8.2

Analogues (Duals) 519

11.4 ELECTROMECHANICAL DEVICES

We are all familiar with various devices which convert electrical energy into mechanical energy or vice versa. Some examples are turbogenerators, loudspeakers, and electric motors.

In Fig. 11.22 is shown a sketch of the armature circuit of a d-c machine which is driving a fan. The shunt-field circuit is not shown; it is assumed that this circuit has been closed a long time and that there is no mutual coupling between the shunt-field circuit and the armature circuit. Under this condition, $v_1 = K\omega_r$, (K is a constant), and a voltage equation may be written

$$v = Vu(t) = L\frac{di}{dt} + Ri + K\omega_r. \tag{11.29}$$

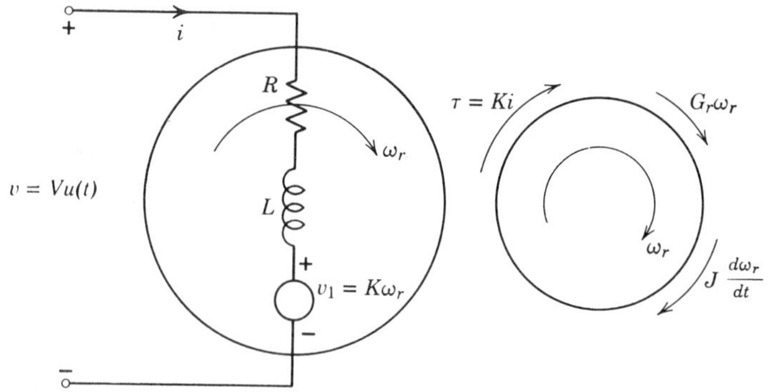

FIG. 11.22 Sketch of d-c armature for motor driving fan load.

The power equation is obtained by multiplying Equation 11.29 by the current i.

$$vi = Li\frac{di}{dt} + Ri^2 + K\omega_r i. \tag{11.30}$$

The last term represents power being supplied to the mechanical system, which is also equal to $\tau\omega_r$.

Then

$$\tau\omega_r = K\omega_r i$$

or

$$\tau = Ki, \tag{11.31}$$

and we find that the constant K relating voltage and angular velocity in Equation 11.29 is the same K which relates torque and current in Equation 11.31.

J represents the polar moment of inertia of the armature and load, G_r represents the viscous damping of the armature as well as the fan load.

A torque equation may be written

$$Ki = J\frac{d\omega_r}{dt} + G_r\omega_r \qquad (11.32)$$

Let us assume $i(0-) = 0$ and $\omega_r(0-) = 0$ and take the transforms of Equations 11.29 and 11.32,

$$\frac{V}{s} = (Ls + R)I(s) + K\Omega_r(s) \qquad (11.33)$$

$$KI(s) = (Js + G_r)\Omega_r(s) \qquad (11.34)$$

Elimination of $\Omega_r(s)$ results in the equation

$$\frac{V}{s} = (Ls + R)I(s) + \frac{I(s)}{Js/K^2 + G_r/K^2}. \qquad (11.35)$$

An electrical circuit may be sketched to represent Equation 11.35. This is done in Fig. 11.23. We note that J/K^2 represents a capacitance and G_r/K^2 represents a conductance.

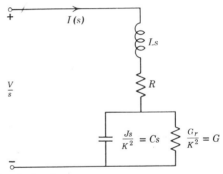

FIG. 11.23 Equivalent electric circuit for d-c armature.

11.5 SUMMARY

Some mechanical systems can be solved readily by using that electrical analogue which makes current the analogue for force and voltage the analogue for current. Then the electrical circuit has the same appearance as the mechanical circuit. Duality permits the rapid determination of other systems whose behavior is governed by equations of the same form.

Some of the elementary concepts of electronic analogue computers were introduced.

FURTHER READING

Additional material on the subjects of this chapter will be found in Lynch and Truxal's *Introductory System Analysis*, Chapters 3 and 7,

McGraw-Hill, New York, 1961. More advanced material on the subject of mechanical analogues will be found in Ley, Lutz, and Reyberg's *Linear Circuit Analysis*, Chapter 5, McGraw-Hill, New York, 1959, and in Gardner and Barnes's *Transients in Linear Systems*, Chapter 2, Wiley, New York, 1942.

PROBLEMS

11.1 *a* Determine the dual of the circuit of Fig. P-11.1. $i(0-) = 0$.
  b Show that $Z(s)$ of one circuit has the same form as $Y(s)$ of the dual.

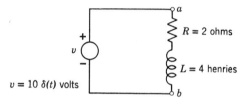

FIG. P-11.1

11.2 Determine the dual of the circuit of Fig. P-11.2 by starting with mesh equations.

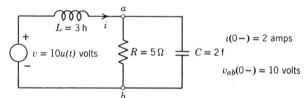

FIG. P-11.2

11.3 For the circuit of Fig. P-11.3, write transform current equations about the nodes *a* and *c*. Then write the transform equations for the dual circuit and from this construct the dual circuit showing numerical values for constants.

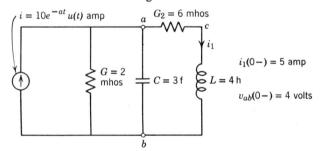

FIG. P-11.3

11.4 Determine the dual of the circuit of Fig. P-11.2 by sketching the transform circuit and then sketching the dual circuit. Show numerical values of quantities and positive sense of all sources and initial conditions.

11.5 Determine the dual of the circuit of Fig. P-11.3 by sketching the transform circuit and then sketching the dual circuit. Show numerical values of quantities and positive sense of all sources and initial conditions.

11.6 Showing your method clearly, sketch the dual of the circuit shown in Fig. P-11.6.

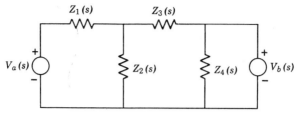

FIG. P-11.6

11.7 Showing your method clearly, sketch the dual of the circuit shown in Fig. P-11.7.

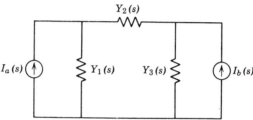

FIG. P-11.7

11.8 *a* For translational motion, the power into a mass M is $p = fu$ watts, with $f = M(du/dt)$. By integrating power with respect to time, show that the energy stored in a mass in translational motion is $\frac{1}{2}Mu^2$.

b Same as part *a* except start with $p = \tau\omega_r$ and show that the energy stored in a body rotating about an axis, which is at the inertial reference, is $\frac{1}{2}J\omega_r^2$.

11.9 The mass M of 0.5 kilograms is at rest on the frictionless plane as shown in Fig. P-11.9. Determine the expression for the velocity, u, and sketch u versus t for:

a $f_a = u(t) - u(t-1)$ newtons.

b $f_a = \delta(t)$ newtons.

Explain why the velocity obtained in part *a* for $t > 1$ is the same as in part *b* for $t > 0$.

FIG. P-11.9

11.10 Use Fig. P-11.9 and the data of problem 11.9 but assume that $f_a = \cos 2tu(t)$ newtons. Determine the expression for the velocity u. Sketch both f_a and u versus time for one period. Observe that these quantities differ in phase by 90°.

11.11 For the mechanical system shown in Fig. P-11.11, assume that the spring has a compliance, C_m, and that steady state exists. By using the analogous electrical circuit show that the velocity u_1 will have an amplitude of zero if $\omega C_m = 1/\omega M$. Also show that the amplitude of u_2 is inversely proportional to ω. $f_a = F \cos \omega t$.

Analogues (Duals) 523

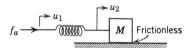

FIG. P-11.11

11.12 For the mechanical system shown in Fig. P-11.12, the mass is 0.25 kilograms, and the spring has a compliance of 1 meter per newton. Initial conditions are zero. Determine the expressions for u and f_2 for:

a $f_a = \delta(t)$ newtons.

b $f_a = u(t)$ newtons.

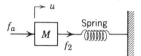

FIG. P-11.12

Partial answer: (a) $u = 4 \cos 2t u(t)$ meters per second; $f_2 = 2 \sin 2t u(t)$ newtons.

11.13 For each of the mechanical systems shown in Fig. P-11.13, sketch the mechanical circuit and then sketch an analogous electrical circuit labeling all quantities. Note: force source in circuit should enter node at which force is applied, regardless of translational direction of force.

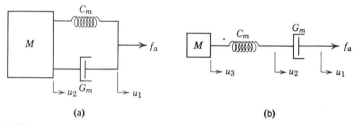

FIG. P-11.13

11.14 For each of the electrical circuits shown in Fig. P-11.14, sketch one analogous mechanical circuit and then sketch the mechanical system.

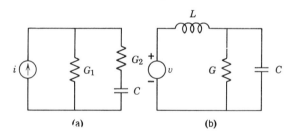

FIG. P-11.14

11.15 In each of the mechanical systems shown in Fig. P-11.15, $f_a = e^{-2t}u(t)$, the velocity of the mass is zero at $t = 0$, $M = 1$ kilogram, $G_m = 2$ newton seconds per meter. Determine the solution for u_1 in each case. Explain the differences in the solution. You may use the electrical analogue as an aid in the explanation.

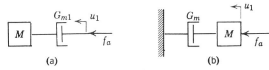

FIG. P-11.15

(a)   (b)

11.16 For each of the mechanical systems shown in problem 11.13 develop the dual mechanical system. Use any method that appeals to you, but be certain to show that the two systems satisfy equations of the same form.

11.17 Figure P-11.17 represents an idealized mechanical system consisting of a dashpot and spring fastened to a wall at one end and to a tow bar of zero mass at the other end. The system is initially at rest. $G_m = 1$ newton second per meter; $C_m = 1$ meter per newton.

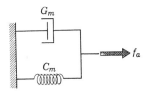

FIG. P-11.17

a Determine the expression for the velocity of the tow bar for:

  (1) $f_a = \cos tu(t)$ newtons.
  (2) $f_a = 10u(t)$ newtons.

b For $f_a = 10u(t)$, there will be a time, t_1, at which the force in the viscous damper will equal the force in the spring. Calculate this time in seconds.

c Also, for $f_a = 10u(t)$, determine how far the tow bar will have moved at $t = 1$ second.

11.18 Figure P-11.18 shows an idealized mechanical system that is initially at rest (zero initial conditions).

a Sketch the mechanical circuit.
b Sketch two analogous electrical circuits.
c Assume $M = 1$ kilogram, $C_m = \frac{1}{2}$ meter per newton, $G_m = 3$ newton seconds per meter, and $f_a = 10u(t)$ newtons. Find the complete solution for the velocity of the mass.

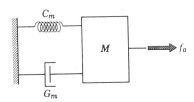

FIG. P-11.18

11.19 For the idealized mechanical system shown in Fig. P-11.19:

a Determine the mechanical circuit and an analogous electrical circuit.
b Determine the complete solution for the velocity of the mass if $G_m = 1$

Analogues (Duals) 525

newton seconds per meter, $M = 1$ kg, $G_{m2} = 3$ newton seconds per meter, $C_m = \frac{1}{2}$ meter per newton and $f_a = 1u(t)$ newtons. Initial conditions are zero.

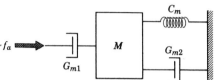

FIG. P-11.19

11.20 In Fig. P-11.20 the mass M is 2 kilograms. The deflection of the spring caused by the weight of this mass is 0.02 meters. Weight is Mg, g being the acceleration of gravity = 9.8 meters per sec².
 a Calculate C_m.
 b The mass is pulled down an additional distance of 0.02 meters beyond the location where it hangs freely on the spring, and is then released. Determine the velocity, u, as a function of time. (Note that one analogue for displacement is flux linkage.)

FIG. P-11.20

11.21 A machine produces a vibrating force which causes excessive vibration to be transmitted through the building when the machine is bolted directly to the floor. It is suggested that the machine be mounted on springs, possibly with some dashpot to provide viscous damping. Figure P-11.21 represents this situation. M represents the mass of the machine. The force may be thought to be sinusoidal, $f_a = F \cos \omega t$.
 a Develop one electrical analogue for this system.
 b Determine the ratio, $\hat{F}_0/\hat{F}_a$, in which $\hat{F}_0$ is the complex force on the floor.
 c Assume $G_m = 0$. Under this condition should $1/\sqrt{MC_m}$ be greater or less than ω in order that the ratio $|\hat{F}_0/\hat{F}_a| \ll 1$. Would you expect any operating problems?

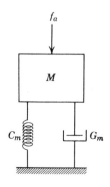

FIG. P-11.21

11.22 A delicate galvanometer will not operate properly because of the vibration of the bench on which the galvanometer rests. It is suggested that the galvanometer be mounted on springs to reduce the motion (Fig. P-11.22). The velocity of the bench may be considered to be sinusoidal, $u_1 = U_1 \cos \omega t$.

a Determine one electrical analogue for this mechanical system.

b Determine the complex ratio, $\hat{U}_2/\hat{U}_1$. How do you explain the fact that the magnitude of $\hat{U}_2$ may be larger than the magnitude of $\hat{U}_1$? What must be the relation between M, C_m, and ω in order for the magnitude of $\hat{U}_2$ to be one tenth the magnitude of $\hat{U}_1$?

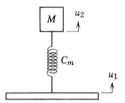

FIG. P-11.22

11.23 It is claimed that the system shown in Fig. P-11.23 will change an ideal voltage source to a constant current source if $\omega L = 1/\omega C$. This means that the current $\hat{I}$ would not be a function of R.

a Show that this claim is true.

b Develop the network's dual, a system which will convert a constant current source to a constant voltage output.

c Develop a mechanical analogue for converting a constant force source to a constant velocity output.

d Develop a mechanical analogue for converting a constant angular velocity source to a constant torque output.

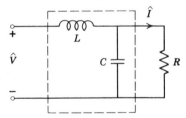

FIG. P-11.23

11.24 Impulses occur in idealized electrical systems such as shown in Fig. P-11.24. For each of these systems determine one analogous mechanical system. Discuss the implications of the switching operation for the mechanical system.

11.25 In Fig. P-11.25 the mass M_1 is at rest under the influence of gravity. The falling mass strikes M_1 at $t = 0$ with a velocity of $u_2(0-)$. The collision is inelastic; that is, M_1 and M_2 have the same velocities for $t > 0$. Determine an electrical circuit condition which is analogous to this mechanical situation. Determine the expression for the velocity at $t = 0+$.

Analogues (Duals) 527

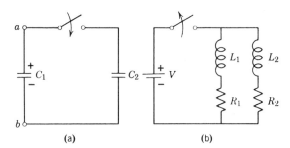

FIG. P-11.24 (a) (b)

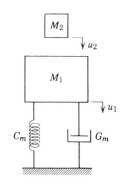

FIG. P-11.25

11.26 Determine the analogous electrical circuit for the mechanical system shown in Fig. P-11.26.

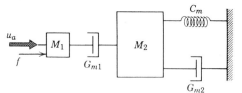

FIG. P-11.26

11.27 Develop the block diagram simulating the electronic analogue computer for the circuit shown in Fig. P-11.27.

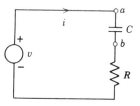

FIG. P-11.27

11.28 Develop the block diagram simulating the electronic analogue computer for the circuit shown in Fig. P-11.28.

11.29 Develop the block diagram simulating the electronic analogue computer for the circuit shown in Fig. P-11.29.

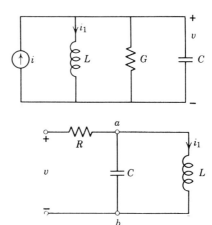

FIG. P-11.28

FIG. P-11.29

11.30 In the use of analogue computers it is often desirable to change the time scale. One reason for this is that the physical system may have more rapid variations than the recording system of the computer is capable of following. If in the equations of a system the variable time, t, be changed to the variable t', defined by, $t = kt'$, the nature of the response is not changed. Show that this type of change, as far as the parameters L, R, and C are concerned, may be thought of as changing L to a new value $L' = L/k$, and C to a new value $C' = C/k$.

11.31 For the motor drive of Fig. 11.23 (page 521) assume $V = 100$ volts, $R = 2$ ohms, L assumed negligibly small, $G_r = 0.01$ newton-meter-seconds, $K = 0.5$ newton meters per ampere. $J = 0.2$ kilograms-meters^2, $i(0-) = 0$, $\omega_r(0-) = 0$.
  a Determine i as a function of time.
  b Determine ω_r as a function of time.

11.32 Develop the block diagram simulating the electronic analogue computer for the motor drive of Section 11.4. This means the representation of Equations 11.29 and 11.32.

Appendix A
Proof that $M_{12} = M_{21}$ and that $k \leqslant 1$

For the circuit shown in Fig. App. A-1 the voltage equations are

$$v_1 = R_1 i_1 + L_1 \frac{di_1}{dt} + M_{12} \frac{di_2}{dt}, \tag{1}$$

and

$$v_2 = R_2 i_2 + L_2 \frac{di_2}{dt} + M_{21} \frac{di_1}{dt}. \tag{2}$$

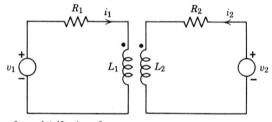

FIG. App. A-1 $i_1(0-) = 0$, and $i_2(0-) = 0$.

The power expressions are obtained by multiplying Equation 1 by i_1 and Equation 2 by i_2. The energy equations are obtained by integrating these. Then

$$\int_0^t v_1 i_1 \, dt = \int_0^t R_1 i_1 \, dt + \int_0^{I_1} L_1 i_1 \, di_1 + \int_0^{I_2} M_{12} i_1 \, di_2, \tag{3}$$

and

$$\int_0^t v_2 i_2 \, dt = \int_0^t R_2 i_2 \, dt + \int_0^{I_2} L_2 i_2 \, di_2 + \int_0^{I_1} M_{21} i_2 \, di_1. \tag{4}$$

In these equations, I_1 and I_2 are some fixed values at which we shall evaluate the energy stored in the magnetic fields. This energy will be constant regardless of the manner in which I_1 and I_2 are reached, since there is a unique solution for the fields, thus making $\int \frac{1}{2} \mu H^2 \, d\tau$

a constant. The total energy supplied to the two circuits is

$$\int_0^t v_1 i_1 \, dt + \int_0^t v_2 i_2 \, dt = \int_0^t R_1 i_1^2 \, dt + \int_0^t R_2 i_2^2 \, dt + \tfrac{1}{2} L_1 I_1^2$$
$$+ \tfrac{1}{2} L_2 I_2^2 + \int_0^{I_2} M_{12} i_1 \, di_2 + \int_0^{I_1} M_{21} i_2 \, di_1. \quad (5)$$

We recognize the first two terms of the right side of the equation as energy dissipated as heat and the remaining terms as energy stored in the magnetic field.

One possible way to reach values of I_1 and I_2 is to increase i_1 from 0 to I_1, holding $i_2 = 0$, and then increase i_2 from 0 to I_2, holding $i_1 = I_1$. If this is done, the energy stored in the field, w_m, is

$$w_m = \tfrac{1}{2} L_1 I_1^2 + \tfrac{1}{2} L_2 I_2^2 + M_{12} I_1 I_2, \quad (6)$$

the last integration being zero.

If now the process is changed to have i_2 increase from 0 to I_2, holding $i_1 = 0$, and then have i_1 increase from 0 to I_1, holding $i_2 = I_2$, the energy stored is

$$w_m = \tfrac{1}{2} L_1 I_1^2 + \tfrac{1}{2} L_2 I_2^2 + M_{21} I_1 I_2. \quad (7)$$

In order for Equations 6 and 7 to be equal,

$$M_{12} = M_{21} = M. \quad (8)$$

Now consider the energy equation,

$$w_m = \tfrac{1}{2} L_1 I_1^2 + \tfrac{1}{2} L_2 I_2^2 + M I_1 I_2. \quad (9)$$

Divide Equation 9 by $L_1 I_2^2 / 2$ to obtain

$$\left(\frac{I_1}{I_2}\right)^2 + \frac{2M}{L_1}\left(\frac{I_1}{I_2}\right) + \frac{L_2}{L_1} = \frac{2 w_m}{L_1 I_2^2}. \quad (10)$$

The left-hand side of this equation cannot be negative since the right-hand side is never negative. Consider I_1 / I_2 as a new variable x, which may have any real value from $-\infty$ to $+\infty$. Then

$$x^2 + \frac{2M}{L_1} x + \frac{L_2}{L_1} > 0 \quad (11)$$

or

$$(x - x_1)(x - x_2) > 0, \quad (12)$$

with

$$x_1 = -\frac{M}{L_1} + \sqrt{\left(\frac{M}{L_1}\right)^2 - \frac{L_2}{L_1}} = \frac{-M + \sqrt{M^2 - L_1 L_2}}{L_1}$$

and

$$x_2 = \frac{-M - \sqrt{M^2 - L_1 L_2}}{L_1}.$$

In order for $(x - x_1)(x - x_2)$ to be greater than zero as x varies in real value from $-\infty$ to $+\infty$, x_1 and x_2 must be complex conjugates.

This means

$M^2 < L_1 L_2$,

or

$M < \sqrt{L_1 L_2}$,

or

$k < 1$.

The limiting case occurs when the roots are real. Then

$M^2 = L_1 L_2$,

and

$k = 1$.

Appendix B
Proof of the Uniqueness Theorem for Second-Order Linear Differential Equation with Constant Coefficients

The equation may be written

$$\frac{d^2x}{dt^2} + a_1 \frac{dx}{dt} + a_0 x = f(t). \tag{1}$$

Assume that there are two solutions, x_1 and x_2, each of which satisfies the differential equation. Then

$$\frac{d^2x_1}{dt^2} + a_1 \frac{dx_1}{dt} + a_0 x_1 = f(t),$$

and

$$\frac{d^2x_2}{dt^2} + a_1 \frac{dx_2}{dt} + a_0 x_2 = f(t).$$

By subtraction we find:

$$\frac{d^2}{dt^2}(x_1 - x_2) + a_1 \frac{d}{dt}(x_1 - x_2) + a_0(x_1 - x_2) = 0.$$

This is a homogeneous equation for which $m = -\frac{a_1}{2} \pm \sqrt{\left(\frac{a_1}{2}\right)^2 - a_0}$.

If $m_1 \neq m_2$,

$$(x_1 - x_2) = A_1 e^{m_1 t} + A_2 e^{m_2 t}.$$

Application of the initial conditions,

$$x_1(0+) = x_2(0+), \text{ and } \frac{dx_1}{dt}(0+) = \frac{dx_2}{dt}(0+),$$

yields:

$$0 = A_1 + A_2$$
$$0 = m_1 A_1 + m_2 A_2,$$

the solution for which is

$A_1 = A_2 = 0$.

Then

$x_1 - x_2 = 0$, or $x_1 \equiv x_2$.

If

$m_1 = m_2 = m$,

$(x_1 - x_2) = e^{mt}(A_1 + A_2 t)$.

The application of initial conditions gives

$0 = A_1$,

and

$0 = mA_1 + A_2$,

for which again

$A_1 = A_2 = 0$,

and

$x_1 \equiv x_2$.

A similar proof applies to higher order equations of this type.

Appendix C
Table of Transforms

	$f(t)$	$F(s)$ for $s = \sigma + j\omega_s$
1.	i	$I(s)$
2.	v	$V(s)$
3.	$f_1(t) + f_2(t)$	$F_1(s) + F_2(s)$
4.	$\Sigma i = 0$	$\Sigma I(s) = 0$
5.	$\Sigma v = 0$	$\Sigma V(s) = 0$
6.	$Af(t)$	$AF(s)$
7.	A	$\dfrac{A}{s}$
8.	$\dfrac{df(t)}{dt}$	$sF(s) - f(0+)$
9.	$\displaystyle\int_0^t f(t)\,dt$	$\dfrac{F(s)}{s}$
10.	e^{-at}	$\dfrac{1}{s+a}$
11.	te^{-at}	$\dfrac{1}{(s+a)^2}$
12.	$\dfrac{t^{n-1}e^{-at}}{(n-1)!}$	$\dfrac{1}{(s+a)^n}$
13.	$\cos \omega t$	$\dfrac{s}{s^2 + \omega^2}$
14.	$\sin \omega t$	$\dfrac{\omega}{s^2 + \omega^2}$
15.	$\cos(\omega t + \phi)$	$\dfrac{s\cos\phi - \omega\sin\phi}{s^2 + \omega^2}$
16.	$\sin(\omega t + \phi)$	$\dfrac{s\sin\phi + \omega\cos\phi}{s^2 + \omega^2}$
17.	$e^{-at}\cos(\omega t)$	$\dfrac{(s+a)}{(s+a)^2 + \omega^2}$
18.	$e^{-at}\sin(\omega t)$	$\dfrac{\omega}{(s+a)^2 + \omega^2}$

If lower limit of Laplace transform is interpreted as $0-$, so $F(s) = \int_{0-}^{\infty} f(t)e^{-st}\,dt$ impulse functions at $t = 0$ are included. Transform No. 11 must be modified to become No. 19.

19.	$\dfrac{df(t)}{dt}$	$sF(s) - f(0-)$
20.	$u(t)$	$\dfrac{1}{s}$
21.	$\delta(t)$	1
22.	$(t-a)u(t-a)$	$e^{-as}F(s)$
23. Periodic function with period T		$\dfrac{\int_{0-}^{T} f(t)e^{-st}\,dt}{1 - e^{-Ts}}$

Appendix D

The Basis of Operation of Certain Electrical Instruments

We consider here only several instruments whose deflections are caused by the torque arising between two circuits which are magnetically coupled.

In Appendix A, for the circuit sketched, it is shown that the energy stored in the magnetic field is

$$w_m = \tfrac{1}{2}L i_1^2 + \tfrac{1}{2}L_2 i_2^2 + M i_1 i_2. \tag{1}$$

It can be shown, by methods of virtual displacement, that the magnitude of the torque, τ, tending to turn the coils in the angular sense of θ is

$$\tau = \frac{\partial w_m}{\partial \theta}.$$

Since L_1 and L_2 are not usually functions of θ,

$$\tau = i_1 i_2 \frac{dM}{d\theta},$$

the units of τ being newton meters, θ being in radians.

In a D'Arsonval instrument a coil carrying a current i_1 is placed in a magnetic field set up by a permanent magnet. We may consider the magnet as consisting of current loops (the net effect of electron spin and orbital motion), giving the effect of a current i_2. The design is made such that $dM/d\theta$ tends to be constant. The coil is prevented from free rotation by a spring so that with a given current i_1 a certain deflection will occur. Then $\tau \sim i_1$.

This instrument will measure direct current and will follow slow variations of current. As the frequency is increased, the instrument will eventually fail to follow the current variations because of the inertia of the moving parts and will take a position corresponding to the average torque or average current.

An electrodynamometer type of instrument has two coils, one a stationary coil and the other a movable one restrained by a spring as for the D'Arsonval instrument. Again the design is such as to make $dM/d\theta$ approximately constant.

This instrument may be used to measure rms values of current or voltage by placing the coils in series.

Then

$\tau \sim i^2$.

Because of the inertia of moving parts, the movable coil will take up a position of average torque. Thus, the deflection, θ, is

$\theta \sim \tau_{avg} \sim \text{avg}(i^2)$.

The rms value of current or voltage is defined as the square root of the average of the function squared. The square root factor can be placed on the instrument by properly marking the scale.

For power measurements the moving coil is placed in series with a large value of resistance and this combination connected across the voltage which is involved. The current involved passes through the stationary coil.

Then

$\tau \sim vi$,

and again because of the inertia of moving parts the movable coil takes a position of average torque. Thus

$\theta \sim \tau_{avg} \sim \text{avg}(vi)$.

and thus the scale can be calibrated in terms of average power.

Appendix E
Proof of Thévenin's Theorem

We wish to prove that a one terminal pair (one-port) network which is linear and which may have any number of independent and/or dependent transform sources (as long as the dependent sources are not functions of quantities outside the network) may be replaced by a transform voltage source in series with a transform impedance. The transform voltage source is the voltage across the terminal pair when these are open circuited and the transform impedance is the ratio of this transform voltage to the transform current which flows between these terminals when short-circuited.

We shall consider the independent sources as voltage sources, converting current sources to voltage sources if necessary. Dependent sources are written as functions of currents. We shall consider the network as having n unknown branch currents, and we shall treat $I_1(s)$ as the current in a branch (a, b), for which we will replace the rest of the network by Thévenin's equivalent circuit. See Fig. App. E-1.

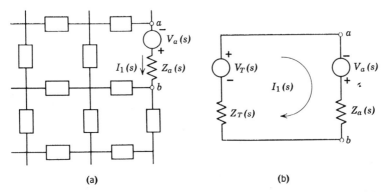

FIG. App. E-1 Sketch of a network with n unknown currents and Thévenin's equivalent circuit for one of the branches. (a) Network. (b) Thévenin's equivalent circuit.

The transform voltage equations for the network may be written

$$Z_{11}(s)I_1(s) + Z_{12}(s)I_2(s) + \ldots Z_{1n}(s)I_n(s) = V_1(s),$$
$$Z_{21}(s)I_1(s) + Z_{22}(s)I_2(s) + \ldots Z_{2n}(s)I_n(s) = V_2(s), \quad (1)$$

and

$$Z_{n1}(s)I_1(s) + Z_{n2}(s)I_2(s) + \ldots Z_{nn}(s)I_n(s) = V_n(s).$$

The solution for the transform current, $I_1(s)$, may be written

$$I_1(s) = \frac{V_1(s)\Delta_{11}(s) + V_2(s)\Delta_{21}(s) + V_3(s)\Delta_{31}(s) \ldots V_n(s)\Delta_{n1}(s)}{Z_{11}(s)\Delta_{11}(s) + Z_{21}(s)\Delta_{21}(s) + Z_{31}(s)\Delta_{31}(s) \ldots Z_{n1}(s)\Delta_{n1}(s)}. \quad (2)$$

If Equation 2 is divided by $\Delta_{11}(s)$,

$$I_1(s) = \frac{V_1(s) + \sum_{j=2}^{n} \dfrac{V_j(s)\Delta_{j1}(s)}{\Delta_{11}(s)}}{Z_{11}(s) + \sum_{j=2}^{n} \dfrac{Z_{j1}(s)\Delta_{j1}(s)}{\Delta_{11}(s)}}. \quad (3)$$

We let

$$Z_{11}(s) = Z_a(s) + Z'_a(s),$$

and

$$V_1(s) = V_a(s) + V'_a(s),$$

in order that we may separate $Z_a(s)$ and $V_a(s)$ from the other impedances and voltages, respectively, of mesh 1. We note that $Z_a(s)$ appears only in the first term of the denominator, $Z_{11}(s)$, and that $V_a(s)$ appears only in the first term of the numerator $V_1(s)$. $V_a(s)$ cannot be a dependent voltage source.

Equation 3 becomes

$$I_1(s) = \frac{V_a(s) + V'_a(s) + \sum_{j=2}^{n} \dfrac{V_j(s)\Delta_{j1}(s)}{\Delta_{11}(s)}}{Z_a(s) + Z'_a(s) + \sum_{j=2}^{n} \dfrac{Z_{j1}(s)\Delta_{j1}(s)}{\Delta_{11}(s)}}, \quad (4)$$

and the equivalent circuit for this equation is shown in Fig. App. E-2.

$$V'_a(s) + \sum_{j=2}^{n} \frac{V_j(s)\Delta_{j1}(s)}{\Delta_{11}(s)} = V_T(s)$$

$$Z'_a(s) + \sum_{j=2}^{n} \frac{Z_{n1}(s)\Delta_{j1}(s)}{\Delta_{11}(s)} = Z_T(s)$$

FIG. App. E-2

If the terminals a and b are shorted, Fig. App. E-2 becomes that of Fig. App. E-3.

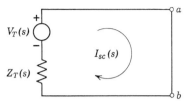

FIG. App. E-3

The expression for the short circuit current is

$$I_{sc}(s) = \frac{V_T(s)}{Z_T(s)}, \qquad (5)$$

and from Equation 5 we see that

$$Z_T(s) = \frac{V_T(s)}{I_{sc}(s)}. \qquad (6)$$

It is apparent that if a switch is placed in the branch a-b, of Fig. App. E-2, the circuit may be shown as in Fig. App. E-4, and the

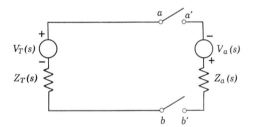

FIG. App. E-4

open-circuit voltage across terminals a and b is Thévenin's voltage $V_T(s) = V_{ab}(s)$ (with branch a-b open-circuited).

Appendix E 541

Index

Abscissa of convergence, 104
Active element, 142
Active network, 142
Admittance, transform, 142
  matrix, 174
  phasor, 234
Algebra, complex (phasor), 225
Ampere's circuital law, 8
Amplitude of sinusoid, 217
Analogue, 491
  electromechanical, 497
  electronic computer, 514
  for d-c motor, 520
  sketching, 508, 509
Angular velocity (ω), 10
Antiresonant frequency, 317
Autotransformer, 488
Auxiliary equation, 88

Bandwidth, 314
Block diagram, 518
Branch, 17
Branch, current method (loop), 20, 25, 28, 177–180
  example, 26–36, 180–181, 248
Bridge circuits, 161

Capacitance, 47
  initial conditions, 48, 64
Capacitor, 49
Characteristic equation, 88
Charge, 1
Circuit element, 14
Circuit fiction, 11, 39, 449, 504
Circuit theory, limitations of, 11, 37

Circular locus, 297
Classical method, homogeneous equations, 87–97
  nonhomogeneous equation, 97–102, 124–128
  pulses and impulses, 403–427
Coefficient of coupling, 56
Cofactor, 174
Complementary function, 98
Complex frequency, 103
Complex frequency plane, 303
Complex quantities, 218, 225
  graphical addition, 228
Complex time functions, 221, 261
Compliance element, 502
Conductance, 241
Conducting medium, definition, 10
  power dissipated, 10
Conductivity, 9, 24
Conjugate, 226
Conjugate method for average power, 252, 369
Conservation of charge, 422
Conservation of flux linkages, 416
Cramer's rule, 29
Critically damped solution, 89, 95
Current, definition, 8, 14
  direction of, 14, 15
Current density, 9, 23
Current ratio method, 150

D'Alembert's principle, 506
D'Arsonval instrument, 280, 346, 537
Delta-wye equivalence, 156–162
  example, 247

Dependent source, 449
Determinant, 30, 174, 179
Dielectric medium, definition, 10
Dirichlet conditions, 334
Dot marking, 57
Driving point immittance, transform, 142
  phasor, 239
Duals, 491
  sketching of dual circuits, 496

Effective values, 255
Electric circuit, definition of, 14
Electric field, 4
  energy of, 10
  flux, 4
  flux density, 4
  intensity, 4
Electric force, 2
Electrodynamometer instrument, 253, 257, 346, 377, 538
Electronic analogue computer, 512
Equivalent circuits, 144
  for transformers, 470, 477
  idealized independent sources, 196
  immittances in parallel, 148–152
  immittances in series, 144–148
  Norton's, 166
  sources, 162
  symmetrical networks, 187
  Thévenin's, 165
  wye-delta equivalence, 156–162
Euler's formula, 89, 219
Even function, 336
Exponential driving functions, 433

Faraday's law of magnetic induction, 8
Field, electromagnetic, 2
Flux linkage, 7, 38, 54
Force, 498
Forced response, 98
Fourier series, 333
  frequency spectrum of rectangular pulses, 348
  exponential form, 351
Fourier transform, 439
Free response, 98
Frequency, 10
Frequency characteristics, 301
Fundamental frequency, 333
Fundamentals, meaning of, 3

Gain function, **144**
Gate function, 401
Gauss's law, 7
Gibbs phenomenon, 334
Graph, 184
Gravitational force, 512

Half-wave symmetry, 335
Half-power points, 314
Harmonics, 333
  three-phase, 381
Heaviside's expansion theorem, 114
Homogeneous medium, 4
Homogeneous differential equations, 87

Imaginary part of, 220
Immittance, transform, 142
  phasor, 234
Impedance, transform, 142
  matrix, 179
  phasor, 234
Impulse, 410
  capacitive, 420
  inductive, 411
Independent equations, current, 21, 172–175
  voltage, 20, 177–180, 183
Independent unknown currents, 20, 177, 183
Independent unknown voltages, 20, 172
Inductance, 36
  initial conditions, 40, 64
  mutual, *see* Mutual inductance
Inductor, 41
Initial conditions, examples, 66–70
  for capacitance, 48, 64
  for inductance, 40, 64
Isotropic medium, 4

j, defined, 89

Kirchhoff's current law, 9, 18
Kirchhoff's voltage law, 19

Ladder network, 189
Laplace transformation, 102
  application to circuits, 109–129, 403–427
  definition, 102
  of impulse function, 413
  of periodic function, 445

544

Laplace transformation, of shifted time function, 404
   table of transforms, 108, 453, 535–536
Lenz's law, 57
Link current method, 183
Loci of phasor network functions, 286–301
Loop current method (branch), 20, 25, 28, 177–180
   example, 26–36, 180–181, 248

Magnetic force, 2
Magnetic field, 4
   energy of, 10, 38
   field intensity, 4
   flux, 4
   flux density, 4
Mass, 1, 499
Matrix, 30, 174, 179
Maxwell's equations, 4, 5–9
Mechanical circuit elements, 498–506
Mesh current method, 181
Moment of inertia, 499
Mutual inductance, 54, 466, 530
   dot marking, 57
   examples, 59–63, 471–474, 479–484

Network function, transform, 141
   phasor, 234
Network reduction, *see* Equivalent circuits
Newton's law, 3, 499
Node, 16
Node voltage method, 20, 25, 171–175
   example, 26–36, 175–176, 237, 249
Nonhomogeneous differential equations, 97
Nonplanar network, 182, 497
Normalizing, 193

Odd function, 336
Ohm's law, 23
   point form of, 9
Open circuit, 17
Operator, 63, 97
Orthogonality of trigonometric functions, 335
Overdamped solution, 89, 96

Parallel connection, 17
Partial fraction expansion, 113–116

Particular integral, 98
Passive element, 142
Passive network, 142
Periodic function, 332
Permittivity, 4, 48
Phase angle of driving point impedance, 240
Phase angle of sinusoid, 43, 217
Phasor, 234
Phasor diagram, 235
   double subscript voltage notation, 230, 364
   examples, 235–238, 242–244, 365–368, 372–377
Phasor network functions, 234
   frequency characteristics of, 301–311
Phasor-time relationships, 230
   for circuit, 239
Phasor-transform relationship, 260
Phase sequence, 362
Phase sequence indicator, 375
Poles and zeros, of network functions, 301–311
   of reactive networks, 318
   of response functions, 428–438
Power, 10, 15
   average, for Fourier series, 345
   average, for sinusoidal steady state, 250
   three-phase, 368, 377
Power factor, 252
Power factor angle, 252
Power, maximum average value, 239
Pulse, 400

Q, reactive volt amperes, 369
Quadratic equation, roots of, 88
Quality factor, 314

Ramp function, 407
Reactance, 240
Reactive networks, immittance of, 317
Reactive volt amperes (vars), 370
Realizability in delta-wye conversion, 247
Real part of, 219
Reciprocity theorem, 191
Reference phasor, 236
Resistance, definition of, 23
Resistor, 25
Resonance, 312

Right-hand rule, 7, 8, 54
Root-mean square values, 254
  of Fourier series, 345
  of sinusoids, 257

s Plane, 303
Scaling, 193
Semiperiodic function, 445
Series connection, 16
Shifted time function, 401
Short circuit, 16
Sinusoidal functions of time, 216
  addition of, 217
Sketching of circuits, 186
Source response, 98
Sources, dependent, 449
  independent current, 22, 196, 197
  independent voltage, 21, 196, 197
  mechanical, 498
Spring, *see* Compliance element
Steady-state response, 98
Step function, 400
Superposition, 86
  example, 152–154, 404–410
Susceptance, 241
Switch, 17
Symmetrical networks, 187
Symmetry of waves, 335

Thévenin's theorem, 165
  examples, 166–171, 245–247
  proof, 539
Three-phase machine, 361
Time constant, 91, 97
Time-phasor relationships, 230
  for circuit, 239
Time-transform relationships, 102
  for circuit, 239
  for elements, 112–113
Topology, 183

Torque, 500
Transfer function, transform, 143
  phasor, 234
Transformers, equivalent circuits for, 470, 477
  ideal, 475
Transform-phasor relationship, 260
Transform-time relationships, 102
  for circuit, 239
  for elements, 112–113
Transient response, 98
Tree, 184
Two-wattmeter method, 378

Underdamped solution, 90, 93
Undetermined coefficients, method of, 99
Units, 4
Unit impulse function, 412
Unit step function, 400
Uniqueness of solution, 86, 533

Vector, 4, 234
Velocity, angular, 10
  rotational angular, 498
  translational, 498
Viscous, damper element, 504
Voltage difference, 15
Voltage drop convention, 16
Voltage ratio method, 146
Volt-ampere relationships for elements, time domain, 71
  time-transform domain, 112–113

Wattmeter, 253
Wye-delta equivalence, 156–162
  example, 247

Zeros, *see* Poles and zeros

546

ALPHABETICAL LIST OF SYMBOLS

(This list does not include symbols primarily used in field theory of Chapters 1 and 2 nor symbols used with mechanical circuits in Chapter 11.)

Symbol	Designation and Remarks
B	Susceptance, Im $\hat{Y}$
C	Capacitance
f	Frequency in cycles per second
$f(t)$	Function of time
$F(s)$	Laplace transform of $f(t)$, $F(s) = \mathscr{L}f(t)$
$F_{\mathrm{II}}(s)$	Bilateral Laplace transform
$F(j\omega_s)$	Fourier transform of $f(t)$
G	Conductance, Re $\hat{Y}$
$H(s)$	Transform network function
$\hat{H}, H(j\omega)$	Phasor network function, $\hat{H} = H(j\omega)$ $= H(s)|_{s=j\omega}$
i	Current, real function of time
I	Current, constant value
$\hat{\imath}$	Current, complex function of time
$I(s)$	Laplace transform of i, $I(s) = \mathscr{L}(i)$
$\hat{I}(s)$	Laplace transform of $\hat{\imath}$, $\hat{I}(s) = \mathscr{L}(\hat{\imath})$
$\hat{I}$	Phasor current, $i_{ss} = \mathrm{Re}\,(\hat{I}e^{j\omega t})$, $\hat{I} = Ie^{j\phi}$
$\hat{I}_{\mathrm{rms}}$	Rms value of phasor current, $i_{ss} = \mathrm{Re}\,(\sqrt{2}\hat{I}_{\mathrm{rms}}e^{j\omega t})$
i_f	Free component of current response
i_{ss}	Steady-state component or forced component of current response
j	Imaginary unit, $\sqrt{-1}$
k	Coefficient of magnetic coupling, $k = M/\sqrt{L_1 L_2}$
L	Self-inductance

M	Mutual inductance
p, P	Instantaneous power, average power
q	Charge
Q	Vars, $Q = \frac{1}{2} \text{Im}(\hat{V}\hat{I}^*)$
Q	Quality factor
R	Resistance
s	Complex frequency variable of Laplace transform $s = \sigma + j\omega_s$
t	Time
t_0	Time constant
T	Period of periodic function
$T(s)$	Transform transfer function
$\hat{T}, T(j\omega)$	Phasor network function, $\hat{T} = T(j\omega) = T(s)\|_{s=j\omega}$
$u(t)$	Unit step function
v	Voltage, real function of time
V	Voltage, constant
$\hat{v}$	Voltage, complex function of time
$V(s)$	Laplace transform of v
$\hat{V}(s)$	Laplace transform of $\hat{v}$
$\hat{V}$	Phasor voltage, $v_{ss} = \text{Re}(\hat{V}e^{j\omega t})$, $\hat{V} = Ve^{j\phi}$
$\hat{V}_{\text{rms}}$	Rms value of phasor voltage, $v_{ss} = \text{Re}(\sqrt{2}\hat{V}_{\text{rms}}e^{j\omega t})$
v_f	Free component of voltage response
v_{ss}	Steady-state component or forced component of voltage response
X	Reactance, $X = \text{Im}\,\hat{Z}$
$Y(s)$	Transform admittance
$\hat{Y}, Y(j\omega)$	Phasor admittance, $\hat{Y} = Y(j\omega) = Y(s)\|_{s=j\omega}$
$Z(s)$	Transform impedance
$\hat{Z}, Z(j\omega)$	Phasor impedance, $\hat{Z} = Z(j\omega) = Z(s)\|_{s=j\omega}$
$\delta(t)$	Unit impulse function
θ	Angle of driving-point phasor impedance
σ	$\text{Re}(s)$
ω	Angular frequency in radians per second, $\omega = 2\pi f$
ω_s	$\text{Im}(s)$, also variable of the Fourier transform